Technological Superpower China

To Krystyna

Technological Superpower China

Jon Sigurdson

Professor of Research Policy and Director, East Asia Science and Technology and Culture Programme, European Institute of Japanese Studies, Stockholm School of Economics and Associate Faculty Professor, Stockholm School of Entrepreneurship, Sweden

in collaboration with
Jiang Jiang, Dr Xinxin Kong, Dr Yongzhong Wang and Dr Yuli Tang

Edward Elgar
Cheltenham, UK • Northampton, MA, USA

© Jon Sigurdson, 2005

All rights reserved. No part of this publication may be reproduced, stored in a retrieval system or transmitted in any form or by any means, electronic, mechanical or photocopying, recording, or otherwise without the prior permission of the publisher.

Published by
Edward Elgar Publishing Limited
Glensanda House
Montpellier Parade
Cheltenham
Glos GL50 1UA
UK

Edward Elgar Publishing, Inc.
136 West Street
Suite 202
Northampton
Massachusetts 01060
USA

A catalogue record for this book
is available from the British Library

Library of Congress Cataloguing in Publication Data

Sigurdson, Jon.
    Technological superpower China/by Jon Sigurdson;
  in collaboration with Jiang Jiang et al.
      p. cm.
  Includes bibliographical references and index.
    1. Technology—China. 2. Technology and state—China.
I. Jiang, Jiang. II. Title.
T27.C5S58 2005
338.951'06—dc22
                                      2005047803

ISBN 1 84542 376 3

Typeset by Manton Typesetters, Louth, Lincolnshire, UK
Printed and bound in Great Britain by MPG Books Ltd, Bodmin, Cornwall

Contents

List of figures vii
List of tables and boxes viii
Preface ix
List of abbreviations xiii

1. China becoming a technological superpower: a narrow window of opportunity 1
2. National reform programmes and human resources development 29
3. Technology access through FDI and technology transfer 69
4. Research and technological mastery in the corporate sector 100
5. The information and communication technologies: example of institute reform 126
6. Rising technological capability 157
7. Space and defence technologies 191
8. Regional innovation systems in China 215
9. Shanghai: from development to knowledge city 248
10. China regaining its position as a source of learning 279

Appendix: The 2020 Plan on Science and Technology 306

Bibliography 313
Index 327

Figures

2.1	National innovation system – key institutions in China	30
2.2	National science and technology programmes	39
2.3	Output value of enterprises in national hi-new technology development zones	48
2.4	Exports of enterprises in national hi-new technology development zones	48
7.1	China's defence industry within the political system	194
8.1	Innovation system structure in China, 2004	229
8.2	NIS evolution in China	245
10.1	Estimate of China's accumulated graduates up to 2020	290
10.2	Salary comparison between China and Finland and Germany	293

Tables and boxes

TABLES

2.1	Distribution of 863 Programmes by field in 2002	43
2.2	Regional distribution of projects of 863 Programmes in 2002	44
2.3	The distribution of projects of the 973 Programme by fields from 1998 to 2002	50
2.4	Number of student enrolments by level and type of school	55
2.5	Enrolled students by field of study in regular institutions of higher learning	56
2.6	Graduates by field of study in regular institutions of higher learning	56
2.7	Employment in large- and middle-size enterprises	57
2.8	Employment of S&T personnel in R&D institutions	58
2.9	S&T personnel of higher learning institutions	58
2.10	Number of postgraduates and students studying abroad	66
5.1	Employment in China's semiconductor industry (June 2002)	133
5.2	Total semiconductor production in China including exports 2003	134
8.1	Urbanization level in China, 1949–98	232
8.2	Enrolment in institutions of higher education and specialized secondary schools by region (2002)	234
8.3	Three types of patent applications examined and granted by region (2002)	235
8.4	Statistics for China's five city regions (2003)	238
9.1	Fudan University enrolment	252
10.1	Institutions of higher learning in Ningbo City	288

BOXES

3.1	R&D work of Novozymes in China	89
6.1	The semiconductor industry	183
7.1	The European Union Country Strategy Paper	205

Preface

I arrived in Beijing on 15 October 1964, the day before China successfully tested its first atomic device.[1] The Chinese capital in those days reminded the visitor of a vastly overgrown village town without any high-rise buildings and traffic completely dominated by bicycles. Reporting on science and technology progress from the Swedish Embassy, I could hardly imagine China rising to the status of a technological superpower within 60 years. Much less did my thinking move in that direction when as a member of a Swedish government delegation in 1979 I was visiting run-down factories and over-staffed and poorly managed research institutes. When publishing *Technology & Science in the People's Republic of China* (Pergamon) in 1980 I sensed that China was involved in a massive task of reorganizing and developing her technological and scientific institutions. However, because I was focusing on developments in Japan, it took me another 20 years to realize that changes in China were changing the technological landscape in the country in a major way, and affecting the whole world.

The starting point of this book was an invitation by Professor Wang Tongsan at the Institute of Quantitative and Technological Economics of the Chinese Academy of Social Sciences (CASS) to give a lecture to students at the CASS Graduate School. My seminar was well attended, although with an extremely quiet audience listening to my naïve views on how China would become a technological superpower. At the end there were no questions until I started to put questions to the students. Suddenly, everyone wanted to make comments and ask questions, and we continued our discussions in a smaller seminar room. Several students wanted to participate in my project to understand China's technological emergence.

Three of the students from that session, at the time PhD students, have assisted me in a major way with preparing this book. They are Dr Xinxin Kong, who is now working for the Ministry of Science and Technology, Dr Yongzhong Wang, who is now a postdoctoral researcher at the Chinese Academy of Social Sciences, and Ms Jiang Jiang, who is now completing her PhD studies. We were joined later on by Dr Yuli Tang of the Ministry of Science and Technology. Their contributions to the project 'The Emergence of New Knowledge Systems in China and their Global Interactions' have been incorporated into Chapters 2–5.

I would never have written this book without having been cultural attaché in Beijing for three years during the mid-1960s, and I landed in this position after having been sent to Hong Kong by my Chinese language teacher. The stay in Hong Kong resulted in my recruitment to the Swedish Embassy in Beijing, as the ambassador regarded my engineering background and Chinese language capability, although rudimentary, as exceptional qualifications. Shortly afterwards my interest in China's science and technology was reinforced by Professor Sven Brohult, the Director of the Swedish Academy of Engineering Sciences (IVA). He posed captivating questions on how far the Chinese had reached in developing their own integrated circuits in the 1960s. In many areas, China's scientists and engineers were at the time closer to frontier developments than they were in the early 1980s. China's dramatic changes, not just during the past decades, have prompted me to turn my kaleidoscope around many times. This book is my perspective, after having found an exciting and very colourful image of China's future as an emerging knowledge nation.

My perspectives on China and the role of science and technology in national development were strongly swayed by Stevan Dedijer, who in 1966 established the Research Policy Institute (RPI) at the University of Lund; at the time, this became a beacon for studies on technological change in developing countries. When moving on to social intelligence and the character of the intelligent corporation he offered me the post of RPI director, which I filled for more than ten years.

Earlier draft versions of this book were made available to a limited audience; as a result, the book has been more-or-less completely amended and revised to suit the overall objective of highlighting China's move forward to become a technological superpower. The chapters have been arranged in a sequence that should enable a reader to study and understand changes.

The introductory chapter and the chapter on regional innovation systems (Chapter 8) have appeared as working papers at the European Institute of Japanese Studies of the Stockholm School of Economics. A final chapter on China as a knowledge economy (Chapter 10) has appeared as a working paper at the Institute of East and Southeast Asian Studies at Lund University. A further chapter on China's technological capability (Chapter 6) was presented at a conference organized by the Asian Development Bank Institute in Beijing in early December 2004. These four chapters were drafted while Visiting Senior Research Fellow at the East Asian Institute of the National University of Singapore. I want to express my thanks to Professor John Wong and Professor Wang Gungwu for having been able to spend a very productive period of research in preparing drafts for this book.

Finally, I want to extend my thanks to the European Institute of Japanese Studies at the Stockholm School of Economics, where the former director,

Professor Jean-Pierre Lehmann, invited me to join his team as visiting professor on Japan's science and technology. This project was generously supported by the Astra (now AstraZeneca) pharmaceutical company. My excellent research conditions at EIJS continued when Professor Magnus Blomström became EIJS director in 1997, and gradually my research interests shifted towards China and its appearance in the global technological landscape. During this period I was coordinator for a programme to develop scholarly contacts between China and Sweden, supported by the Swedish Foundation for International Cooperation in Research and Higher Education (STINT). Furthermore I have on several occasions received generous support from the Swedish Agency for Innovation Systems (VINNOVA) for projects and workshops directly related to the contents of this book.

My studies on recent developments in China have increasingly been guided by a sense of the growing role of Chinese regional ambitions and initiatives, which I have tried to illustrate in various chapters of the book. In this endeavour I was fortunate to meet the Vice-Mayor of Ningbo, Mr Wu Hemin, and the Vice-Mayor of Shanghai, Mr Zhou Yupeng, who each arranged for me to make week-long study visits in their cities in the late stage of preparing my manuscript.

In understanding regional development and the role of science and technology in regional innovation systems (RIS) I have benefited from the support of numerous people. First I would like to mention Mr Chen Zexing of the Beijing City Foreign Economic Relations and Trade Commission, who spent a sabbatical year at the Stockholm School of Economics under the STINT sponsorship programme. Subsequently, he has facilitated my meetings with staff of Beijing companies and research organizations on many occasions. Similarly I have on a couple of occasions been provided excellent opportunities to meet with staff of high-tech companies in Shenzhen, with great support from Mr Richard Zhong of the Foreign Affairs Office in the Shenzhen Municipal Government.

When I returned from my survey of the science and technology system in Shanghai in early January 2005, I was immediately offered a quiet writing corner in the premises of the Stockholm School of Entrepreneurship with a single focus on finalizing the manuscript for this book. I am very grateful to the director, Professor Per Olof Berg, and his staff for their compassionate thoughtfulness.

Early ideas and various manuscript drafts on China as an emerging technological superpower would never have materialized into this book without unfailing support from my wife Krystyna. Undiscouraged by reading early and immature drafts, she constantly suggested new ideas and enhancements, until the completion of the manuscript in early 2005. She also provided insightful support when participating in several of my study tours in China during 2004, a year that also took us on extended tours to Japan, Australia and Singapore.

NOTE

1. On 16 October 1964 China joined the nuclear club by conducting its first atomic test, which was the beginning of a series of increasingly sophisticated tests. The first one was an implosion fission device with U-235, for which uranium had been enriched by the electromagnetic process after partial enrichment in a gaseous diffusion facility at Lanchou.

Abbreviations

863	High-Tech Research Development Programme
973	National Key Basic Research Programme
ACAC	Avic-1 Commercial Aircraft Co.
AMD	Advanced micro devices
ASEAN	Association of Southeast Asian Nations
ASIC	application-specific integrated circuits
ATI	Advanced Technology Industries
AVIC I	China Aviation Industry Corporation I (CAICI)
AVIC II	China Aviation Industry Corporation II (CAICII)
AWACS	airborne early warning and control systems
BCG	Boston Consulting Group
BDA	Beijing Economic and Technical Development Area
BGI	Beijing Genomics Institute
BGU	Beijing Geely University
BHR	BoHai Rim region
BI	business intelligence
BJSMC	Beijing Semiconductor Manufacturing Corporation
BOE	Beijing Oriental Enterprise
BUPT	Beijing University of Post and Telecommunications
CAE	Chinese Academy of Engineering
CAICI	China Aviation Industry Corporation I
CAICII	China Aviation Industry Corporation II
CALT	China Academy of Launch Technology
CAMEC	China Aerospace Machinery and Electronics Corporation
CAS	Chinese Academy of Sciences
CASC	China Aerospace Corporation
CASIC	China Aerospace Science and Industry Corporation
CASS	Chinese Academy of Social Sciences
CAST	China Academy of Space Technology
CASTC	China Aerospace Science and Technology Corporation
CATIC	China National Aero-Technology Import and Export Corporation
CBD	Shanghai's central business district
CBERS	China–Brazil Earth Resources Satellite

CCID	China Centre of Information Industry Development
CDB	China Development Bank
CEA	Chinese Economic Area
CENC	China–Europe Global Navigation Satellite System Technical Training and Cooperation Center
CGWIC	China Great Wall Industry Corporation
CIMC	China International Marine Container Group
CIMT	China Institute of Mining and Technology
CISRI	Central Iron and Steel Research Institute
CME	China Ministry of Education
CMI	Civilian–military integration
CMM	Capability maturity model
CN	Core network
CNSA	China National Space Administration
COSTIND	State Commission of Science, Technology, and Industry for National Defence
COSTINF	Commission on Science and Technology and Industry for Defence
CPIT	China Putian Institute of Technology
CPMEIC	China Precision Machinery Import and Export Corporation
CPU	central processor units
CRT	cathode ray tubes
CUMT	China University of Mining and Technology
DAB	digital audio broadcasting
DSP	Digital signal processing
EADS	European Aeronautic Defence and Space Company
Embraer	Empresa Brasileira de Aeronáutica
ERA	European Research Area
EU	European Union
EUREKA	a pan-European network for market-oriented, industrial R&D
EVD	Enhanced Versatile Disc
FDI	Foreign direct investment
FUSP	Fudan University Science Park
GaAs	Gallium-Arsenide
GAD	General Armaments Department
GIS	global innovation system
GERD	Gross Expenditure on R&D
GLF	Great Leap Forward
GM	General Motors
GPN	global production networks
GRINM	General Research Institute for Nonferrous Metals
GriTek	GRINM Semiconductor Materials Co. Ltd

GS	graphic synthesizer
GSM	Groupe Spécial Mobile (GSM) to study and develop a pan-European public land mobile system
GSMC	Grace Semiconductor Manufacturing International
HDD	hard disk drives
HDTV	high definition television
HPC	high performance computing
IC	integrated circuit
ICT	Information and Communication Technologies
ICT	Institute of Computing Technology
IDP	Integrated product development
IFPMA	International Federation of Pharmaceutical Manufacturers Associations
IPR	intellectual property rights
ISM	Integrated supply management
IT	Information technology
IVA	Swedish Academy of Engineering Sciences
JEPZ	Jinqiao Export Processing Zone
JVs	joint ventures
KDI	Korea Development Institute
LCOS	liquid crystal on silicon
LDC	less developed country
LFTZ	Lujiazui Finance and Trade Zone
LII	labour-intensive industries
LME	large and medium enterprises
M&A	mergers and acquisitions
MDC	more developed country
MEMS	Micro-Electro-Mechanical Systems
MII	Ministry of Information Industry
MMT	million metric tonnes
MMTA	mobile multimedia technology alliance
MNCs	Multinational corporations
MOST	Ministry of Science and Technology
MPSR	Mega-projects of science research
NAFTA	North American Free Trade Area
NASA	National Aeronautics and Space Administration
NC	network computers
NERC	National Engineering Research Centres
NHTIDZ	New and High Technology Industry Development Zones
NIS	National innovation system
NRCSTD	National Research Centre for Science and Technology for Development

ODI	overseas direct investment
OECD	Organization for Economic Cooperation and Development
P&A	processing and assembly
PAS	personal access system
PBX	private branch exchange
PDA	personal digital assistant
PIRI	public interest related institutes
PLA	People's Liberation Army
PLA IEC	Information Engineering College of the People's Liberation Army
POS	point of sale
PPKIP	Pilot Project of Knowledge Innovation Programme
PRC	Peoples's Republic of China
PRD	Pearl River Delta
PVO	Russian missile defence system
QDI	Quantum Design Institute
R&D	research and development
RAN	radio access network
RCA	revealed comparative advantage
RF	radio frequency
RFID	radio frequency identification
RIS	Regional Innovation Systems
RMB	renminbi (Chinese internal currency)
RLV	rocket launch vehicle
RNC	Radio network controller
RPI	Research Policy Institute
S&T	science and technology
SAIC	Shanghai Automotive Industry Corporation
SASAC	State-owned Assets Supervision and Administration Commission
SAST	Shanghai Academy of Spaceflight Technology
SCIDZ	Shanghai Comprehensive Industrial Development Zone
SCIP	Shanghai Chemical Industrial Park
SCSTIND	State Commission of Science Technology and Industry for National Defence
SDI	Strategic Defence Initiative
SDRC	State Development and Reform Commission
SEI	Software Engineering Institute
SETC	State Economic and Trade Commission
SEZ	special economic zones
SHIB	Shenzhen High-Tech Industrial Belt
SHIP	Shenzhen High-Tech Industrial Park

SIA	Semiconductor Industry Association
SiGe	Silicon-Germanium
SIPA	Shanghai Intellectual Property Administration
SIPO	national patent office
SJEPZ	Songjiang Export Processing Zone
SJIZ	Songjiang Industrial Zone
SJTU	Shanghai Jiaotong University
SMG	Shanghai Municipal Government
SME	small and medium-sized enterprise
SMIC	Semiconductor Manufacturing International Corporation
SMT	surface mount technology
SOC	system on chip
SOE	state-owned enterprises
SPC	stored programme computer
SPGC	Shanghai Pharmaceutical Group Corporation
SPSP	Shanghai Pudong Software Park
SSL	Songshan Lake Sci. and Tech. Industrial Park (Dongguan)
STIC	Shanghai Technology Innovation Centre
STINT	Swedish Foundation for International Cooperation in Research and Higher Education
SVU	Shenzhen Virtual University
SZHTP	Zhangjiang Hi-Tech Park
SZSP	Shenzhen Software Park
TCL	TCL Holdings Corporation Ltd.
TCM	traditional Chinese medicine
TD-SCDMA	Time Division Synchronous Code Division Multiple Access (Chinese 3G system)
TII	technology-intensive industries
TRIPS	Trade Related Aspects of Intellectual Property Rights
TSMC	Taiwan Semiconductor Manufacturing Corporation
TTM	time-to-market
UAI	Universal Automotive Industries
UAVs	unmanned aerial vehicles
UCAVs	unmanned combat aerial vehicles
UMTS	Universal Mobile Telecommunications System
UNCTAD	United Nations Conference on Trade and Development
VINNOVA	Swedish Agency for Innovation Systems
VTE	village-and-township enterprises
WAPI	Chinese proposed system for Wireless Local Area Networks
W-CDMA	Wideband Code-Division Multiple-Access (one of the main technologies for the implementation of third-generation (3G) cellular systems)

WFTZ	Waigaoqiao Free Trade Zone
W-LAN	Wireless Local Area Network
WLL	wireless local loop
WTO	World Trade Organization
YRD	Yangtze River Delta region
ZGC	Zhongguancun (science park in Beijing)
ZSIB	Shanghai Zhangjiang Semiconductor Industry Base
ZTE	Zhongxing Technologies Corporation
ZWU	Zhejiang Wanli University

1. China becoming a technological superpower: a narrow window of opportunity

A 'NOBEL' THOUGHT

It is a dark winter evening in Stockholm. It is 10 December 2004 and it is time for the yearly Nobel Prize Festival that is always held in the Stockholm City Hall. More than 1300 people have assembled in the grand Blue Hall, and the Chinese Minister of Science and Technology, Mr Xu Guanhua, is seated at the centre table, which is reserved for guests of honour. Soon he will listen to the speeches of the Nobel laureates who have received the 2004 prizes in medicine, physics and economics.

His thoughts this evening may have included a number of reflections on China's future development of science and technology. In the coming year the Chinese government will make the final decision on the contents and direction of the 2020 Science and Technology Plan that has been the focus of intense deliberations for the past couple of years. He may wonder how soon China's scientists will enter the podium and describe how they were able to scale the scientific pinnacles that qualify for a presence in this forum. He may reflect on how institutions should be changed, what resources should be allocated and how scientists should be identified and selected to move beyond the present scientific frontiers.

Earlier in the week he has received the news that a Chinese company, Lenovo, an early spin-off from the Chinese Academy of Sciences, has acquired the personal computer division from IBM. In the same week he has also learned that China's leading company in telecommunications equipment, Huawei, has landed a first major contract in the Netherlands, which is the heartland of the next generation of mobile telecommunications. It has become obvious that China's high-technology companies are on the move in a globalized economy. The Chinese Minister for Science and Technology is likely to consider how to balance attention and resources between immediate industrial technology development and supporting advances at the scientific frontiers. They are not unrelated, but the time perspectives are very different and China's window of opportunity for rapid economic development may be quite narrow. China's technological advances are based on using foreign technology combined with its own manpower resources, and its integration of regional ambitions with national policies and programmes. By rapidly developing its economy China moves towards a future status as a potential technological superpower.

The US remains unchallenged in the early part of this century in defence, economics, politics and technology. Japan can be termed a superpower in economics and technology. The EU is conceived as a superpower in economics, politics and technology, while Russia has remaining strengths in defence and technology. Today China is already a superpower in politics; it is emerging in economics and has great ambitions in technology.

In any attempt to understand recent and future industrial and economic development in China, it becomes unavoidable to think about its various regions as the equivalents of major countries in other parts of the world. In several ways the regions of the Pearl River Delta, the Yangtze River Delta and the BoHai Rim, including Beijing, can be compared with France, Germany and the UK in Europe. By world standards the regions essentially have middle-income purchasing power.

China has dramatically increased the number of students in tertiary education and provided more funding for R&D during the past ten years, not only in absolute terms but also in relation to its GDP. Traditional indicators, such as patents, still suggest that China is far from reaching its goal of becoming a knowledge-based economy. However, monitoring signs of dynamic changes within industrial sectors and emerging competencies in a number of research fields presents a much more optimistic scenario.

China has set ambitious goals to become once more a nation that ranks among the world leaders in science and technology. China was a technology leader in many advanced technologies until the end of the 18th century, which is evident from the European 'East Asia' trade at the time. However, the country was trapped in a social system that, with its advanced administration and civil service examination system, had served very well during earlier dynasties. The slowness in changing attitudes and organization made it impossible for China to adjust to the onslaught of the Western powers, not only in military areas but in a wide range of industrial technologies. Japan was able to start a modernization process almost a century earlier than China.

Today there is a growing concern in Japan, indeed a realization that China will within the next couple of decades catch up in basically all industrial technologies. However, China was also in earlier days a pioneer in scientific discovery and its leadership and its scientists appear convinced that China will again play a leading role in most scientific fields by the middle of this century.

TECHNOLOGICAL ASPIRATIONS – NARROW WINDOW OF OPPORTUNITY

Lester Thurow noted in 1997 that the global economic system is simultaneously affected by movements of five tectonic plates: the end of communism; a technological shift to an era dominated by man-made brainpower industries; a demography never before seen; a global economy; and an era where there is no dominant economic, political or military power.

To a great extent China has become part of this aggravating shift and is left with a narrow window of opportunity to become a leading nation once more – a technological superpower. China has been living with a frustration since the industrial revolution took off in Europe and left the nation outside the mainstream of technological mastery; in the past Japan was the only country in Asia to be able to break out of a self-imposed straitjacket. Japan has in more recent times been followed by Korea and other countries while China was contained within a planned economy.

The industrialization of Communist China that started in the early 1950s, supported by the then Soviet Union, made a successful break with China's backwardness. A substantial number of strategic industrial complexes were built, many of them in the Northeast provinces, which had until 1945 been under Japanese occupation – formerly known as Manchuria. However, an ideological fixation and rashness to catch up with the outside world led to the disaster of the Great Leap Forward. China was eager to catch up with Great Britain in steel production, as the USSR announced its goal of overtaking the US in this sector. The successful collectivization of agriculture in its early stages made the leadership believe that rural labour could easily and quickly be transferred to industrial production. This proved to be a grim mistake – the expected rapid industrialization resulted in dismal failure and many industrial products of the Great Leap Forward were completely useless.

In an ensuing power struggle, with its roots in ideology, China at the time of the Cultural Revolution almost reached the stage of civil war. During this period China attempted major reforms of its innovation system and reorganized research institutes and universities, although with more attention to ideological dedication than results that would support the country's technological and scientific modernization. The situation was further complicated by the real or perceived threat of a military conflict, possibly involving nuclear weapons, with the US, which was at the time heavily involved in its war in Vietnam. This forced the planners in China to relocate major industries of importance for its defence capabilities to the interior of China in order to reduce their vulnerability.

It took until the end of the 1970s before China could establish anew its modernization goals, this time exploiting foreign capital and foreign technology. The process started modestly after the Open Door Policy had been announced in 1978. The size of potential markets in China, the seriousness of the leadership and rapidly changing economic conditions eventually persuaded the outside world that China was going to become part of the world economy. For almost 30 years until his death in 1976 Mao Zedong remained the dominant political leader, with at times a truly formidable power. He sought to change the character of the Chinese people by inspiring them to heroic deeds while in parallel tormenting them through his ambitious but unrealistic goals of quickly modernizing the nation. Still Mao remains the icon of a united China, where his party remains in power although almost all his policies have been abandoned.

In the meantime China in 1979 introduced a family planning programme that drastically reduced family size with its very strong emphasis on the one-child family and measures to ensure its success. The population would today have been 1600 million instead of 1300 if China had not followed this course. However China, like industrialized countries in Europe and Japan, will rapidly become an ageing society and the working-age population will already have begun to shrink around 2015.

The median age of the population will climb from 32 in 2003 to around 45 in 2040. As an ageing society China is facing the same financial predicament that is now developing in Europe and Japan. The leadership has to cope with at least two emerging challenges. First, the country must get rich hurriedly in order to be able finance the costs of an ageing population in terms of health care and other social security needs. Second, China being still in the stage of initial development, must move swiftly to utilize fully its still young people to, without any delay, raise all levels of knowledge which will be required in a more mature industrialized society.

China's narrow window of opportunity compels it to use youngish entrepreneurs, engineers and scientists vigorously and forcefully to move the country to the high pinnacles of economic development before the population gets old and tired and has to be cared for financially. Thus, brainpower industries have to replace the labour-intensive industries that still dominate most industrial sectors in China. Brainpower industries require talented people, technological ideas, advanced knowledge, sophisticated infrastructure, financial capital and other resources. This book offers facts and insights on how China is mobilizing such resources to become a technological superpower.

The process of achieving rapid technological progress and reaching the status of a technological superpower has been the substance of the deliberations of thousands of scientists, engineers and policy makers who have since early 2003 been preparing the 2020 Science and Technology Plan.[1] This covers altogether 20 different but also closely related features. Manufacturing is one of them, for which the Chinese Academy of Engineering (CAE) has been given the mandate of coordinating views on China's future course in industrial development. Basic research is naturally the responsibility of the Chinese Academy of Sciences (CAS), while the State Defence Science and Technology Commission is considering China's future in military technology. The National Research Centre for Science and Technology for Development (NRCSTD) is charged with the task of bringing all the various opinions and views into an overall strategy for China to reach its ambitious goals by 2020. The Centre has also been given the special task of charting the course of China's regional innovation systems (RIS).

In any country, and particularly in a country like China with its extraordinary size and diversity, technological innovation will take place in a number of its regions that are becoming spatial innovation systems. External connections to other businesses, to component suppliers and to researchers are as important for these regions as internal ones.

Oil – Achilles Heel of China

For a couple of decades we have taken China's entry into the world economy for granted although being surprised by its speed and its far-reaching effects. Before China initiated its Open Door Policy in late 1978 it was evident that the country had subscribed to a planned economy approach, although with a strong desire to utilize advanced technology from the outside. Inside China there was a realization at the time that reforms were necessary to deal with serious shortcomings caused mainly by the turmoil of the Cultural Revolution that had come to an end only recently. The supporters of the radical reforms that China was soon to implement – first in agriculture and then in industry, to be followed by a completely new foreign trade regime – greatly benefited from two important factors. First, those who wanted to return to China's earlier planning system lacked insights into China's real economic morass and did not have the qualified manpower to formulate realistic plans. Second, their assumptions were based on the expectation that China's oil reserves were abundant and could easily be exploited to meet not only an expanding domestic demand but also to pay for large-scale technology imports.

China's domestic oil consumption has more than doubled during the past decade to over six million barrels per day. As a consequence crude oil imports have increased from zero to 40 per cent of local consumption, and China has overtaken Japan to become the world's second-largest oil consumer.[2] China's fast-growing demand for oil was responsible in 2004 for much of the increase in worldwide prices. The East-West Centre in Honolulu estimates that the demand in China could double within a decade, while China's domestic production is expected to remain virtually stagnant.[3] China's attempt to diversify its sources of oil has prompted its oil companies to initiate exploration projects in countries like Sudan, Peru and Syria, which have been considered to be on the periphery of the global oil industry.

The cost of China's oil imports increased 65 per cent in the first ten months of 2004, with a total cost of more than US$26 billion, according to figures from the Customs General administration of China.[4] China's inability to produce enough oil domestically and its dependence on imports provide a stark contrast with the optimistic planning shortly before China embarked on its Open Door Policy. Barry Naughton argues convincingly that the inability to set realistic targets for the economy eroded the political basis for the leaders who wanted gradual reforms in China, and opened the doors for the far-reaching consequences of the Open Door policy.[5]

The traditional planners who took over after the Gang of Four had been ousted following the death of Mao Zedong in 1976, returned to a centrally planned economy approach. However, they lacked knowledge about the real situation in China and, among other things, set an extremely optimistic target for the oil sector, expecting it to pay for substantial and urgently needed technology imports to revitalize the downtrodden industry. Domestically produced oil also had to provide feedstock for the petrochemical industry and the necessary expansion of the fertilizer industry to improve productivity in agriculture. Naughton describes the situation in the following way:[6]

> China's petroleum sector had been one of the few success stories of the Cultural Revolution era: Output had grown by 20% annually between 1969 and 1977, and planners were projecting that rapid increases would continue. Moreover, the key technocrats drawing up the plan (most notably Yu Qiuli, the head of the Planning Commission) had come directly out of the petroleum sector. In projecting that petroleum output would reach 250 MMT (5 million barrels per day) by 1985, planners were stating that China would approach Saudi Arabia's position as the world's third largest petroleum producer (behind the Soviet Union and USA). Though ambitious, such plans could not be automatically dismissed. The Chinese were attempting what the Soviet Union had accomplished between 1957 and 1965. The question was whether they possessed the same vast resources that the Soviet Union enjoyed.

The long-term plan being formulated was based on a number of faulty assumptions. They included an assumption of national richness in oil resources and an ability to handle efficiently a massive programme of industrial plant import and construction. They planners also assumed that agricultural output was basically achieved with incentives and that reorganization was irrelevant. Naughton states that 'Everything was being gambled on a programme of economic acceleration. As it turned out, virtually all the assumptions on which that programme was based were wrong.' Planners had not made feasibility studies and were often unaware about total project costs. The situation gradually deteriorated throughout 1978, which made it impossible to achieve consistency in the plan for the following year.

The Baoshan Steel Project in Shanghai provides a good example of grand expansion from a modest five million ton smelter and port construction, approved in November 1977. A year later the project had metamorphosed after several changes into a fully integrated state-of-the-art steel combine with all equipment to be imported from Japan – with foreign exchange requirements tripling to US$5.7 billion.[7] By the end of 1978 Chinese buyers were greatly expanding their shopping lists and 'literally hundreds of billions of dollars worth of contracts were under discussions'. The buying rush was based on the belief that continued expansion of oil production would be the source of hard currency to pay for technology imports, but expanded drilling during 1977–1980 was to little avail.

Persistent and habitual optimism combined with lack of systematic investigation would quickly create disaster in planning not only for oil exploration but for the national economy, which had been re-centralized. The economy was decentralized in a manner that had hardly been practiced since the 1950s and training of managers and many categories of experts had been badly neglected during the Cultural Revolution, when colleges and universities were almost completely closed. Thus the Chinese leaders were forced to rethink their development strategy completely. Naughton says: 'When the central prop of that strategy, petroleum development, was removed, the entire edifice collapsed'. This set the stage for the beginning of China's reform era, which formally started at the Third Plenum of the 11th Communist Party Central Committee that was held in December 1978.

On this occasion the leadership realized that development of agriculture needed a drastically new strategy, as the planned acceleration of industry – eventually to have supported agriculture – had to be abolished. A new economic leadership moved away from the concept of central plans and decided on liberalizing rural economic policy and improving the terms of trade for agriculture. With serious constraints in the energy sector it became indispensable to move economic developments towards an industrialization strategy that would be less energy-intensive. This meant light industry with a focus on consumer goods, a change that would also improve the opportunities for handling the employment problem that had been neglected for a long time. Following the experience of neighbouring countries, China embarked on capital-saving and labour-intensive light manufacturing that became an important component of China's new development strategy. These policy changes quickly became irreversible and have ushered China on to a development path that has eventually led to an almost complete break with the earlier USSR-inspired planned economy approach, and its modifications after the Great Leap Forward and during the Cultural Revolution.

NATIONAL SCIENCE AND TECHNOLOGY STRATEGY

The People's Republic of China has throughout its history always given a great deal of attention to science and technology, which until the early 1980s remained completely within the government domain. Nowadays, the corporate sector has officially taken on a major portion of R&D in China, although most research-intensive companies are state-owned enterprises or maintain close links with the state sector. However, government programmes and initiatives remain important, while a private sector is growing in importance in the funding and implemention of R&D.

Since the beginning of the reform period China has launched five major science and technology programmes. The first one was the Key Technologies R&D Programme, started in 1982 to serve the mainstream of industrial development by concentrating resources on technologies that were urgently needed for industrial upgrading and economic development – its contents evolving as the economy advanced. Its present focus is on information technologies and biotechnology.

The next was the Spark Programme in 1986 to develop the rural economy through science and technology and to initiate technological changes in village-and-town enterprises (VTE). It supports a large number of technical projects that use rural resources, need small amounts of investment, give early benefits and use appropriate technologies. The programme is also creating demonstration zones to stimulate comprehensive regional development and furthermore develops regional industries that have their basis in a comparative advantage in regional resources. The Ministry of Science and Technology has the overall responsibility for the programme, while most management is decentralized to province, district and county levels. The Spark Programme has attracted attention in many countries and its experience has been utilized by various international organizations, including the World Bank.

Only a little later followed the High-Tech Research Development Programme (863) that was launched in March 1987. This programme coincided in time with initiatives in Japan and the EUREKA programme in Europe that were launched in response to the Strategic Defence Initiative (SDI) – Star Wars – in the US. The Chinese leadership realized that China and its manufacturing units would face increasingly fierce international competition. The origin of the 863 programme can be linked to a March 1986 report with the title *Suggestions on Tracing the Development of World Strategic High Technology* that gave the programme its name. The report, written by three researchers, was submitted to the State Council and immediately submitted to Deng Xiaoping. It then became the platform for in-depth and hurried deliberations and resulted in an operational report – *The Report on R&D Plan on*

High Technology. The main mission of the 863 Programme was monitoring the international situation in advanced technologies, and proposing relevant and reasonable national schemes. An intermediate aim was reducing the gap between China and developed countries in several important fields and achieving breakthroughs in fields where China holds a comparative advantage. The programme covers a substantial number of research topics which are selected from eight priority areas, which are the biotechnology, information, automation, energy, advanced materials, marine, space and laser fields. MOST is responsible for all areas except the last two. China has reported that its ability to join the international human genome sequencing project had its roots in an earlier 863 project.

Next came the Torch Programme that was launched in 1988 with the specific objective of developing new-technology industries in China. It was implemented by the Ministry of Science and Technology (MOST) after approval from the State Council. The Torch Programme included a number of activities, aside from a general mandate to provide a legal and organizational environment for the development of high-tech industries. First, it was given the responsibility for establishing new high-tech industrial development zones, where the objective was to convert R&D into successful industrial production, although foreign-dominated companies have become key actors. Second, the programme is also responsible for establishing service centres to support high technology development and for training and attracting talents which will raise the level of expertise and management of high-tech industries. Third, it was supposed to establish torch projects in high technology enterprises and enterprise groups that could successfully venture into areas such as new materials, biological engineering, electronics and information, optoelectronics, energy saving and environmental protection.

A more recent programme is the National Key Basic Research (973) Programme that was launched in June 1997 and was designed to stimulate research that would result in original innovations and to provide support for future development, with a perspective on 2010. Many of its projects have their roots in research activities which were initially supported by the National Natural Science Foundation or other funding agencies. The programme is supposed to establish a number of scientific engineering projects of significance for the long-term development of the Chinese economy and society. The programme has four major tasks. First, the 973 should conduct multidisciplinary research and provide scientific and theoretical foundations to solve important scientific issues that China will face in the intermediate and long-term future. Related to this the programme should engage in explorative research to advance the knowledge front. To meet the first two objectives the programme is supposed to cultivate outstanding scientists who have creative capability and who can tackle challenging research tasks. Finally the 973 programme is also responsible for the creation of interdisciplinary research centres responsible for carrying out projects of high national priority.

The 973 programme not only absorbs the largest investment ever made by the central government in research and development but also includes some of the country's largest R&D projects outside the defence and space sectors. MOST suggests that its implementation has had a remarkable impact on the scientific and technological community in China and greatly boosted the rapid improvement of China's international competitiveness. The programme has involved senior researchers belonging to the Chinese Academy of Science (CAS) and the Chinese Academy of Engineering (CAE) in a major way, and also encouraged overseas Chinese researchers to join many projects. It has also provided postgraduate training for a large number of students.

Previous projects have included material research on carbon nanotubes. Ongoing research includes the basic study of super-high-density, super-high-speed optical information storage and processing, and basic research into novel devices and novel processes such as systems on chip (SOC). The objective of this project is to solve the basic problems of device structure, device modelling and simulation, and key processes and reliability for semiconductors with features in the range of 20–50 nanometres. The long-term ambition and expectations are that China will take a leading position not only in the manufacture of advanced semiconductors but also in their further development.

All five programmes are still operational, although their nature has been modified and new projects have been introduced as earlier ones were completed. The 2020 Science and Technology Plan, to be announced in 2005, is likely to provide an overall framework for the programmes already mentioned. They aim to develop different parts of the Chinese economy and provide support and funds at various levels of scientific and technological sophistication. Such national programmes in China, as well as in any other country, undergo changes and naturally encounter obstacles and failures. However, there is little doubt that they have made a great contribution to the reform of the R&D system in China, which is still not completely transformed from the straitjacket of the earlier planned-economy times. In this process big programmes have created new interfaces among various sectors of the economy and between research institutes; thus CAS still plays a dominant role in major fields of basic research.

There can be little doubt that the development of science and technology will remain a concern of utmost importance in China for some time. China's science and technology (S&T) strategy was outlined in a speech by the Chinese Minister of Science and Technology in 2003.[8] At the time he stated that China should follow seven major development strategies to become successful in the 21st century.

1. China should adopt a leaping development strategy to accelerate its competitiveness in IT industries by developing new operational systems with associated software platforms and advanced CPU chips. In biotechnology China should focus on functional genomics, bioinformatics, bioengineering medicine, and biogenetic seed breeding – to be recognized in the international biomedical industry.
2. China should boost its capacity for original innovation, move away from the country's earlier emphasis on copying proprietary innovations and establish evaluation systems that support this objective.
3. China should improve its capability to integrate and manage national R&D resources, and national S&T programmes should have an explicit emphasis on disciplinary and interagency cooperation with a focus on the development of products and industries.
4. China should speed up its high-tech industrialization, to include the reform of national high-tech parks, supporting small and medium enterprises and providing needed support services.
5. China will use IT to support the industrialization process, which should develop and diffuse high performance computer environments and create common information technology systems, particularly in areas such as banking, insurance and manufacturing.
6. China will promote international S&T cooperation and provide support for Chinese scientists to participate actively in global large-scale science projects; it will encourage multinational companies to invest in R&D and use imported technology and personnel closely integrated with foreign direct investment (FDI), which has in the past played such an important role in China's economic development.
7. China will improve its human resources strategy to establish a system which is open and competitive, and gives more weight to personnel costs in the total R&D expenditure. China will pay more attention to the development of human resources and will enhance the import of top notch people from overseas on a selected basis.

CREATING AND USING BRAINPOWER – THE ROLE OF UNIVERSITIES

The university system in particular, as well as education in general, has received great attention as a source for China's future as a knowledge-based economy. Universities have undergone dramatic changes in recent years, although it is still too early to judge fully their performance with regard to undergraduate teaching, graduates studies and research. Universities in China have until recently not been involved in advanced research, and graduate studies were only introduced after major reforms started in the late 1970s.

It will be several years before the combined effects of a rapid increase of enrolment and a major restructuring of universities can be seen completely. The annual enrolment of students in regular institutions of higher education was only 400 000 in1978 and dropped to less than 300 000 the following year, which corresponded to about 1.5 per cent of those entering secondary schools in the same year. It increased rapidly after 1998 with 1 000 000 new students, and the intake expanded to 3.2 million in 2002, which equals 10 per cent of the intake in secondary schools in the same year. Total enrolment in Chinese universities had reached more than 9 million by 2002 and continues to increase. More than one third of all university students study engineering, and including science students, the share is close to 40 per cent, and increasing. Thus, Chinese universities will in the predictable future graduate at least one million students every year in science and engineering, the latter with a focus on electronics.

The number of students enrolling as postgraduates has undergone a similar and also very dynamic expansion from a total annual intake of some 10 000 in 1978 to more than 200 000 in 2002. The number of students having completed postgraduate degrees reached 80 000 in 2002, with a total enrolment in postgraduate studies of 500 000. At the same time 125 000 Chinese were enrolled in postgraduate studies at overseas universities, which shows that presently one out of every five Chinese postgraduate students are pursuing their studies abroad.

In 1997 China embarked on an ambitious reform plan that should bring a number of its universities into world leading positions during the coming century – the National 211 Project.[9] The selection of 100 universities is expected to create an essential focus on higher education by providing special attention and favourable funding. Simultaneously many universities have merged into more comprehensive entities and have recently begun to be governed only by the Ministry of Education. This is a reversal of the reform in 1952, when China adopted a model for higher education from the Soviet Union which led to dividing and dispersing a large number of units. For example, the School of Sciences and School of Law and Liberal Arts at Tsinghua University were transferred to Beijing University while an opposite transfer was made for the Engineering School of Beijing University. The result that Beijing University became focused on social sciences while Tsinghua only contained engineering departments after the reform.

In 1952, almost all ministries were given direct responsibility for higher-level training to meet their specialized needs, and they established their own universities and colleges. For example, the Ministry of Post and Telecommunications established its own university by merging two departments from

Tianjin University and another one from Chongqing University. Subsequently the ministry created four universities in different locations, of which Beijing remained prominent. The latter has now become Beijing University of Post and Telecommunications (BUPT) and, like its parallel UPTs in other locations, is under the direct control of the Ministry of Education.

The process has now been reversed and involves multiple objectives, aside from giving the Ministry of Education full control of higher education. First, major universities were given the task of upgrading the teaching and facilities of lower-tier colleges. Second, many mergers were undertaken to provide complementary strengths. Finally staff and overall administrative costs were reduced.

Project 211, the national programme for higher education, is mainly oriented to economic development in China and emphasis will be given to supporting institutions and key disciplinary areas which are closely related to 'pillar sectors of industry' where high-level professional manpower is needed. Priority will be given to some 25 universities which have a concentration of critical disciplines. These universities are expected to reach high international standards in both teaching and research and become models for other universities in China. An underlying objective is to break away from the narrow disciplinary orientation that existed in the former university system, to broaden the coverage of various disciplines and to foster the emergence of cross-disciplinary teaching and research.

The sudden increase of university training, if maintained and successful in terms of quality, will have two important effects – one inside China and the other one in the global economy. In 2002 alone, about 20 million new workers entered the employment market in China, but half of them had only received junior middle school education or less. Currently, a mere 5 per cent of China's whole population has received higher education, a figure that is far lower than the average level in developed countries. However, the upsurge in higher education has had the result that students after having completed their university studies can no longer expect to choose freely among job opportunities and will see salary levels shrink compared with only a few years ago. The second consequence is that China will be able to offer an increasing pool of young and educated university graduates who will be available in a job market that already has a well developed industrial, technological and logistics infrastructure in many places all over China. Assuming that universities continue their expansion in engineering and science at reasonably high quality levels, China will increasingly be competing in brainpower rather than in labour-intensive production.

An obsession with commercializing university research has apparently created a conflict of interest as campuses at many colleges have become cradles of entrepreneurship, and new technologies are being raised into business ventures at an increasing rate. A number of incubators have taken the form of campus-based science parks. Beijing University Science Park is one of these. In the past ten years, it has aided the growth of more and more start-ups. In 2000, 300 projects were evaluated, of which 30 were selected for incubation. Altogether, 400 businesses currently have operations in the park, 80 per cent of which are high-tech enterprises, although critics argue that many campus companies have little that is high-tech about them.

However, China's major universities have produced some of the best-known names in the technology sector. Beijing University Founder Group Corporation came into being in 1986 and now has total assets of RMB6000 million, with shares in 17 other companies and a controlling stake in four listed companies. Its core business has diversified from developing word processing software into hardware, Internet-related products and systems integration. A non-profit school system and its highly specialized teaching and research may not always resist the temptation of the business world, and research-obsessed scholars and market-oriented managers usually have very different perspectives.

Chinese Companies as Global Players

In late April 2004 a Chinese company, TCL Communication, which originated in 1981 in Guizhou in South China, and Alcatel in France signed an agreement to combine their mobile handset resources into a joint venture (JV) company. The JV will include research and development, sales and distribution of mobile phone handsets and related products and services. The formation of the JV involves contributions of 55 million euros by TCL Communication and 45 million euros by Alcatel, so TCL is the major shareholder that expects its partner to provide a platform to expand international business rapidly.

TCL started as a manufacturer of tape recorders and was reorganized after having coped with a financial crisis in 1986, it developed into an important trader in TV sets and was subsequently listed on the Shenzhen Stock Exchange in 1993. In 2000 TCL started new production lines for TV sets and entered into mobile handsets the following year. In 2002 the TCL Group invited strategic investors from abroad and made important changes in its management structure.

Alcatel has, through its acquisition of the ITT partnership in the Shanghai Bell telecommunications company, been a pioneer in providing advanced digital switches to China. The present Alcatel owes its origin to a company that was started in 1888 as a general electrical equipment factor, CGE, to follow, in the footsteps of Siemens in Germany and General Electric in the US. The company now specializes in telecommunications with a focus on solutions rather than manufacturing, but since the early 2000s it has experienced difficulties in making its mobile handsets business profitable.

In 1999 the Ministry of Information Industry in Beijing indicated that domestic makers captured a miniscule 2 per cent of the market for mobile handsets. This share increased to as high as 55 per cent in 2003, a development for which companies like TCL, Bird, Konka and Lenovo have been responsible. This dramatic expansion has its roots in China being able to enter into a global technological system. Many Chinese companies have entered into dynamically evolving learning processes where they have rapidly been acquiring manufacturing and design technologies. Such companies are now entering a stage where they are aggressively looking for resources that will enable them to become global players.

The TCL-Alcatel deal coincided with the visit of Premier Wen Jiabao to France, but is hardly related to any initiatives or direct support from the Chinese government. TCL had already signed an even more important agreement with Thomson, also in France, in late 2003, to form another joint venture in which TCL would moreover be the dominant partner. The new company will become the world's largest producer of colour TV. One year earlier, TCL also bought Schneider, an ailing electronics company in Germany. TCL is the majority partner in all these new ventures and in this way the Chinese company has gained access to technology and well-recognized brand names – Alcatel for handsets worldwide, RCA for TV sets in the US and Thomson in Europe. There is no doubt that liquid-crystal and plasma displays will eventually replace most TV sets using cathode ray tubes (CRT) and TCL has declared that it will be one of the major global suppliers of digital TV sets.

In the meantime another Chinese company has been carving out a major share in the manufacture of CRTs. Ancai Group, a state-owned producer of TV tubes based in Anyang in the North of Henan province, has done a series of aggressive merger-and-acquisition deals in China as well as in the US. Ancai acquired all plant and equipment from the joint venture between Asahi Glass and Corning in the US in the second half of 2003. When in operation in China the transformed plant will have an annual production of six million tubes for TV CRTs of 25 inches and larger. Ancai is making these daring moves when Japanese companies are hesitant to continue their investment in a technology that will eventually disappear. The company argues that TV sets with traditional CRT will secure market shares for quite some time, particularly in developing countries. Ancai is poised to become the world's largest and the major producer of CRT glass tubes; this includes a joint venture with LG-Philips, located in Zhengzhou, the capital of Henan province. Thus its annual production will amount to 60 million tubes.

Apparently, industrial circles in China are well aware that major technological changes are underway, and are looking for ways to speed up the process inside China. Creating brand names and acquiring technology for Chinese products can be speeded up by joining forces with foreign companies or buying foreign companies outright that have encountered serious economic difficulties. This was the case when Beijing Oriental Enterprise (BOE) purchased the flat panel division from Hynix, formerly Hyundai Electronics, in 2002, when creditors forced the Korean company to offload its non-core activities. BOE and another Shanghai company are already in advanced investment stages for creating major manufacturing capability for flat-screen panels.

BOE was an electronics company that was serving the military sector in the 1950s. It suffered serious financial problems in the early 1990s in its attempts to adapt to the new business climate that came into being after China announced its Open Door Policy and sped up the economic reform process. BOE was reorganized in 1993 – with employees also providing investment – and was listed on the Shenzhen Stock Exchange in 1997. The product range has shifted from traditional electronics components to high-tech communications and computer and digital systems, with a current focus on the new generation of display devices. In Beijing BOE is making a major investment in a production plant for flat panel displays, while SVA in Shanghai in cooperation with NEC in Japan, is also making a similar large-scale investment.

Although already successful in some sectors, Chinese companies are not always commonly recognized as global actors. This is still the situation for the China International Marine Container Group (CIMC), based in Shenzhen, that has become a major supplier in the global market for marine shipping containers. Its growth has been driven by the acquisition of container manufacturing facilities scattered along coastal China from Dalian in the North down to the Pearl River Delta, including a major manufacturing facility in Shenzhen. Like most large companies in China, CIMC is still state-owned but operates under stringent operational and financial control by group management.[10]

These moves are demonstrating a multi-pronged approach by companies and government to give China leverage and strengthen its technological position in the global market place which the country only re-entered in the late 1970s. China has become the industrial hub of the world and her products now reach every corner of the world. Foreign direct investment, booming export markets and a large domestic market have, together with abundant labour and improving infrastructure, been driving an explosive growth of manufacturing.

Meanwhile, while quality is constantly going up, the prices of China's manufactured goods are going down. China's terms of trade for its industrial products are much less favourable than was the case for Japan and Korea in their early industrialization efforts. China has yet to capture brainpower industries in a major way, without which it will be stuck with low levels of added value in its exported products.

Moving Up the Technological Ladder – Standards in Competition

China's industrial development, which fuelled its success in exports, has been based on abundant Chinese labour offered at low cost to foreign investors as well as domestic companies. Many and soon most of the world's advanced consumer electronics products will be manufactured in China but they rarely carry a Chinese brand name or contain advanced technology of Chinese origin.

Standardization has in recent years become a very important element of China's technology strategy. There are basically two trends to consider, in the light of the tectonic shift of electronics to East Asia. One is the increasing attempt to harmonize interests and development among three countries in East Asia – China, Japan and Korea. It has become obvious that the countries in East Asia have identified common interests, and a more formal approach was established at a ministerial meeting in Seoul in September 2003. The other and more important trend is that China wants to establish its own technological platforms, in as many areas as possible, in order to gain independence from foreign high-tech companies and drastically reduce the level of licence fees. Being able to establish or influence global standards has become vital for national technological efforts.

This situation can only be changed gradually and needs to involve both company forays and government initiatives. The recent emphasis on industrial standards should be seen in this context, although China earlier registered a success with the International Telecommunications Union by having its own 3G standard accepted as one of three ITU global standards. Today almost all DVD players, bought anywhere in the world, are manufactured in China using technology for which mainly Japanese companies hold the intellectual property rights (IPR). In 2003, Chinese companies reached an agreement with the Japanese IPR owners who request hefty licence fees. Currently the Japanese IPR holders of DVD technology have persuaded most of the Chinese DVD makers to pay licence fees at the level of 4 per cent of ex-factory prices.

China's IPR position is still feeble. The patent portfolio of Chinese companies in the US is very weak and the majority of patents granted in China – by the Chinese Patent Office – are still held by foreign-based companies. The Chinese government wants to change this situation and so do domestic companies. New approaches to standards will enable a greater share of Chinese IPR and fewer licence fees – and also greater possibilities for expanding globally. China has singled out a number of areas it which in wants to influence the process of setting international standards.

China wishes to establish IT standards and will use the combined forces of R&D and standard setting to create changes for its IT sector in order to raise the level of technological development and avoid paying high licence fees. Chinese companies have recently suffered as they have been faced with complaints from IPR holders in Japan and the US. Thus the Chinese government is today strongly supporting the development of various industrial technology standards in a number of areas, which include digital TV and the controversial Chinese (WAPI) protocol for W-LAN, which offers higher security but has greatly angered both the US and Japan.

TV services in China are still analogue and cable networks are very popular. Most likely digital services will be introduced soon and will be distributed both over cable and broadcasting networks. The US standard dominates in many countries, although China originally decided to base itself on the EU standard and was expected to pay licence fees. However, introduction has been delayed and at present there is no competition from the outside. China would be able to choose between two different domestic approaches from an original array of five. In December 2003 China made an announcement on its plans for WAPI and on 1 January 2004 announced that all production in China and all imports must conform to the Chinese WAPI standards. This made the US government angry and the US Trade Representative visited China for serious discussions in February 2004.

WAPI involves both security and economic concerns. A number of personal digital assistants (PDAs), notebook computers and mobile handsets already have built-in WiFi, based on the original US standard that has until recently been followed all over the world. Japan also shares the political and commercial concerns of the US as all Japanese companies buy WiFi chipsets from US companies, where Intel has a strong position.

China ruled that products to be sold in China by foreign companies must conform to the WAPI standard proposed as a Chinese standard. This would have required an agreement with domestic Chinese companies, including ZTE and Huawei, which hold strong IPRs for this technology in China. The US government was obviously very eager to delay the Chinese rules and persuade the Chinese government to change its requirements. The tentative agreement on this issue might indicate that China is buying time for its own 3G system to become fully operational and commercially viable before it offers licences to mobile operators.

Another sensitive area is openware software, where Japan and China have common interests. China wants a Linux version for the Chinese language. Although China, Japan and Korea have different interests, there are prospects for cooperation in East Asia that could reduce the market dominance of Microsoft.

Regional Innovation Systems

The dynamic changes that have erupted from economic policies since the late 1970s have formed new industrial structures along China's coastal areas and also significantly transformed the economic relations between the centre and the regions and provinces. The regionalization that has taken place prompts us to view China as a federation of economies rather than a completely integrated economic entity, and similar tendencies can also be identified in other countries. The result is that regions may interact more with agents from outside the country than within. Chinese regions have become innovative, with their own distinctive ambitions to foster their industrial dynamics and development which need to be supported by regional policies for institutional and structural changes – although also supported by national structures. Furthermore, it has also been practicable for the planners to create different approaches towards innovation in some technology sectors for which the size of China offers more local and regional possibilities than would be feasible in most other regions of the world.

It has for some time been possible to identify three regions in China where policies and structures reflect an ongoing regionalization that also includes science and technology as essential components. One is the Yangtze River Delta (YRD) region, which includes Shanghai and 14 other cities in the southern part of Jiangsu Province and the northern part of Zhejiang Province – close to the actual river delta. The population of the YRD region is presently 82 million, which corresponds to about 81 per cent of the population of the two provinces combined with Shanghai. A second distinctive region is the Pearl River Delta (PRD) region, which includes Guangzhou (Canton) and Shenzhen, Hong Kong and Macao and several closely linked cities in Guangdong province. The population of the PRD region is now 48 million. A third region, towards the north, still along China's coastal line, less clearly defined but also very important, is the BoHai Rim (BHR). This includes Beijing with its surrounding areas and Tianjin on the coast, parts of Hebei province and Qingdao, Weihai and Yantai Shandong province – with a population of around 80 million. These three regions share 3 per cent of China's land mass and have about 15 per cent of the country's population, while generating 45 per cent of gross domestic product and more than 70 per cent of international trade and investment.

The first two regions are important hubs in China, having become the manufacturing centre of the world. PRD and YRP have developed into the heaviest concentrations of industrial activities that can be found anywhere in the world. The Pearl River Delta started its race earlier and reached a GDP of US$267 billion compared with US$205 billion for the Yangtze River Delta region in 2001. PRD also dominates with its exports of US$195 billion against US$74 billion for YRD in 2001. Labour resources for the continued existence of such industries within the regions rely on large numbers of migrant workers, who in Chinese statistics are recognized as 'the floating population'.

With constant development in China's economy more and more people are leaving their registered permanent residences to seek better jobs.[11] The State Statistics Bureau states that China's floating population exceeded 120 million in 2002, although the official figure is likely to underestimate considerably the real numbers.

Until now, most migrant workers still move within provinces, while the trans-provincial floating population is only 42 million. Naturally, poor and populous provinces offer more migrant workers. Sichuan provides 16 per cent, Anhui and Hunan each 10 per cent, Jiangxi 9 per cent and Henan and Hubei each 7 per cent according to the official figures. The outflow from these six provinces makes up 59 per cent of the national trans-provincial floating population. Six provinces and two municipalities (Shanghai and Beijing) are the most attractive locations, with Guangdong province absorbing 36 per cent, Zhejiang 9 per cent, Shanghai 7 per cent, Jiangsu 6 per cent, Beijing 6 per cent and Fujian 5 per cent of trans-provincial migrants.

Although Guangdong province still remains the magnet for many migrant workers, some segments of the Pearl River Delta region, and some of the other more developed regions, are rapidly advancing. In the late 1970s at the location of the present industrial Shenzhen City, only a small town with some 50 000 inhabitants existed. Only one engineer was reported and no industrial base whatsoever existed. Shenzhen was at the time just the frontier of a booming Hong Kong. China's Open Door Policy in 1979 completely changed the city's conditions and over the past couple of decades Shenzhen has become a high-tech industrial city with a population of seven million.

Shenzhen is today the most multicultural city in China, developing at high speed with great flexibility. Forty per cent of the present inhabitants come from Guangdong. The rest come from Hong Kong, Taiwan and all other parts of China. The expansion of production has been followed by research activities in which IT companies dominate. Exports from Shenzhen have increased rapidly during the past ten years and today about one third of China's high-tech exports originate in Shenzhen; in this respect Shenzhen is becoming more important than Beijing.

Shenzhen's characteristics include prominence in IT, finance and logistics. IT products constitute 90 per cent of industrial output with an expected rapidly increasing share for biotechnology. Shenzhen companies have established a strong position in telecom switching equipment and in magnetic and optical reading heads. Two companies are looming on the horizon as serious competitors to the established European telecom companies like

Ericsson and Siemens, US Motorola and Japanese NEC – not only in domestic but also in global markets. The most prominent one is the still privately-owned Huawei Technologies, being closely followed by other domestic competitors. A second one is ZTE Corporation; when established in 1985 it was only a simple trading company, under the name Shingling Semiconductor, having been sponsored by the No. 691 Factory under the former Ministry of Aerospace Industry. ZTE entered the telecom sector in the early 1990s when domestic demand increased rapidly and management found that engineering competence inside the company provided a golden opportunity to enter a new and rapidly expanding industrial sector. In many other fields Shenzhen shows rapid advances in industrial technologies. For example an expansion of technological competence within innovative materials has driven rapid developments of production technology for new types of advanced high-tech batteries.

Foreign companies have located in Shenzhen in the past because of labour, which is not only attractive for cost reasons but also for its increasing competence levels. Recently, a new reason for the increasing attractiveness of Shenzhen has lain in its dominance of private companies while industry in many other Chinese cities is dominated by state-owned enterprises (SOEs). As a consequence more than 90 per cent of all R&D carried out in Shenzhen takes place inside the companies. Incubators have taken on a prominent role in recent years, for which private firms are mainly responsible. The government policy is focused on providing finance for incubators, and a favourable institutional and legal framework. The local government states that the industrial output of Shenzhen will increase seven-fold by 2020. The same optimism suggests that by 2020 semiconductors will become a very important part of the information technology structure in Shenzhen, and new materials and biotechnology will become leading industries.

Probably, one of the justifications for this optimism is University Town, which was designated in December 2003 to have 10 000 students of which 70 per cent are Shenzhen full-time students while the rest are 'local' part-time students who study in the evenings and weekends. The University Town will in particular draw on resources from Tsinghua University, Beijing University, Harbin Engineering University and Nankai University. Shenzhen does not have a well-developed university structure, although University Town is part of China's 211 Project whereby China should have the 100 best universities in the 21st century.

Halfway between Shenzhen and Guangzhou lies Dongguan, which has a total population of 6.5 million of which only 1.5 million are permanent residents. The remainder consists of migrant workers, some of whom have become semi-permanent residents. Similar to Shenzhen, having moved through several stages of manufacturing, technology development, venturing and high-technology development, Dongguan will in the future emphasize industrial development of biotechnology and other high-technology fields.

The first factory moved from Hong Kong (HK) to Dongguan in 1978 to start manufacturing handbags, soon followed by a large number of other HK firms that moved their labour-intensive manufacturing on to the Mainland. The first companies from Taiwan moved to Dongguan in the mid-1980s, soon to be followed by a substantial number of electronics firms, which created the foundation for Dongguan to evolve gradually into a high-tech industrial city. As a consequence Dongguan has attracted more and more high-tech industries – also from Europe, Japan and Singapore. Nokia started its production of telecom equipment there in the mid-1990s.

Dongguan authorities have realized that the land is overloaded with too many labour-intensive companies scattered throughout, which among other things created environmental problems. Thus, municipal authorities decided that industrial development, population and environment must be harmonized, and further industrial development should be concentrated only on high-tech industries with considerable restrictions on others. This has eventually created a Dongguan focus on industries in the electronics and telecommunications sectors that increased their share of industrial employment from 17 per cent in 1995 to 34 per cent in 2000. As a consequence industrial production has become more diversified and less labour-intensive. Dongguan has in parallel been expanding and reorganizing its higher education and is also establishing a science and industrial park. Published information on Songshan Lake Science and Technical Industrial Park indicates that:[12]

> Dongguan has set the strategic goal to become a modern industrial city noted for its manufacturing in the new millennium. It is a crucial decision to develop Songshan Lake Science and Technology Industrial Park (SSL) aiming at achieving the goal. In the future development of Dongguan's economy, as the economic and technological centres of this city, SSL will be the new platform for foreign economic and technical cooperation, the driver of upgrading the industrial structure, the centre of industrial supportive services.

Within the three regions mentioned above China is shaping regional competence blocks – in other contexts referred to as regional innovation systems – which are increasingly interacting on a global scale, whereby China's national innovation system sometimes takes on a subordinated role. A competence block would include a set of competencies that are essential to generate, identify, expand and exploit business ideas successfully. The overwhelming concentration of shoe manufacture in Dongguan City in Guangzhou province offers an example of a competence block that has evolved in less than 20 years. Embedded competencies exist in clusters which have become endowed with synergetic structures of customers, inventors and innovators, entrepreneurs and industrialists. A competence bloc becomes fully operational, or complete, when it has a critical mass of activities, which attract competent actors for all its functions. Such competence blocks are coming into existence all over China these days.

However, it would be reasonable to consider China's late industrialization in the light of what took place in Europe two hundred years earlier. The industrial revolution took off in England in the late 1700s and swiftly crossed over to the European continent in the following century, where it was quickly embraced by France and Germany, also to become industrial powers. The industrial revolution in Europe not only created dominance over the rest of the world but also dramatically changed power relations, which started intensive rivalries among the European nations and social struggles within. The ensuing instability, resentment and envy contributed to the two world wars that started in Europe.

China today is undergoing a dynamic industrialization and modernization process on a land area that constitutes a continent and is larger and more populous than Europe. Industrialization in China started earlier but the very rapid and dynamic process that we are now witnessing started only some twenty years ago. This recent industrial revolution took off in the south of China, in Guangdong with its now industrialized cities, and has been progressing northward along the coastal line and more recently to the interior of China. Pulsating industrial efforts have created great disparities between regions and social tensions, which in certain aspects replicate early industrialization in Europe, with the significant difference that China is one nation. However, uneven speed of modernization in various parts of the country challenges the central government to alleviate disparities and control the leading regions so that they do no become too autonomous. Ways of handling disparity issues in the ongoing industrialization process would reveal the aptitude of the central government for handling some of the country's most touchy problems.

Global Rivalry – Competition and Cooperation

China has become a major player in global trade and its comparative advantage in a number of industrial sectors has given its manufacturers a dominant position for a number of goods. China's industrial advantage lies in its almost unlimited supply of low-cost labour with increasingly higher technical skills that has become available in a large number of clusters, particularly in coastal areas. These clusters are served by good transportation, telecommunications and logistics services which enable close and efficient interaction with global markets.

Although China still enjoys a strong comparative advantage in the manufacture of more traditional labour-intensive products, more striking is the country's large-scale entry as the foremost supplier of almost all electronics goods. This includes not only traditional colour TV and household appliances, but also DVD players, notebook computers and mobile handsets. The role of China's manufacturers was initially limited to final assembly, for which low-cost high-skilled labour was a key factor while most components were imported, but it is likely to change in the quite foreseeable future.

The rapid entry of China into the world trade system to become a major supplier of shoes, garments and electronics is forcing other countries to change their export strategies. This has become very evident, particularly in developing countries that have realized that they must adjust rapidly to a situation that is undergoing constant change. However, effects are also reaching industrialized countries, where we can see an earlier comparative advantage in advanced components and subsystems being eroded. Chinese companies are moving up the ladder of industrial competencies in a growing number of technological fields.

The benefit is coming from being embedded in clusters, or competence blocks, that are not only closely linked locally, but also throughout the country and internationally. This phenomenon is also fuelled by the decisions of multinational companies that are lured to locate more value-added production in China, being attracted not only by manual labour but also by brainpower available at low costs. A further attraction lies in the rapidly expanding domestic market. As a result, increasingly large shares of production chains for advanced products are being transferred to China. This is causing serious concern in Taiwan, Korea and Japan, which fear that their economies will soon suffer from serious industrial hollowing-out, and recently such concerns have also been raised in the US.

China has, as a developing country, become a natural site not only for final assembly but also for manufacturing parts. However, large international companies usually control the final production into complete sets of advanced equipment. For most advanced and heavy machinery, this process would previously have been totally integrated in one physical location, almost always in an industrialized country. The dissolution of integrated production processes – not only for electronics products – has been of considerable assistance in enabling companies in China to get involved. Chinese companies are thereby acquiring skills and technology by expanded participation in the production of large-scale equipment for global markets.

At the same time China's rate of economic development has remained high year after year. A favourable budget situation has enabled the government to fund a number of farsighted programmes in science and technology. Although not always successful, they have in many fields acted as catalysts and contributed to lifting the country out of its earlier technological backwardness. More and more Chinese companies, although still few in numbers, have become research-intensive and will soon challenge competitors in global markets. As a consequence industrialized countries no longer see China as a challenger from the developing world but increasingly as a competitor. There are prospects, however, that China may become a potential collaborator in many fields of advanced science and technology.

Since the late 1990s the United States has been watching China's strategies and progress in both civilian and military science and technology with increasing concern. It appears that the European Union has come to view China strategically less as a competitor and more as a collaborator for a number of advanced technologies. Korea and Japan and other countries perceive their economies to become increasingly integrated with China and have been prompted with mixed feelings to further advance their scientific and technological prowess to stay ahead of China.

There is little doubt nowadays that US-China relations are undergoing structural and fundamental changes. China has been able to achieve much more rapid development of industrial capabilities than any other developing country. The apparent ability of the Chinese economy to integrate various strands of technological inputs quickly and efficiently is not only advancing its commercial success but also enables China to improve its defence technologies radically. Underlying this change is an expanding and open market for R&D required in high-tech industries and an increasing willingness of foreign companies to make substantial R&D investments in China.

At a US Congress hearing in the autumn of 2003 the following factors were highlighted.[13] First, the driving force behind the globalization of R&D is the need for commercial ventures to increase the value-added for goods and services. Thus, design as well as R&D is often most efficiently done in close proximity to production bases. This provides a strong impetus to move R&D to follow production, although advanced and basic research is generally maintained within the home base. Second, multinational companies no longer limit their R&D investments to industrialized countries in Europe and Asia. During the past couple of years China has in this respect taken on a prominent role, offering a number of attractive sites for R&D investment. Third, the ongoing revolution in information communications has made IT technologies all-pervasive. This has given global companies an almost unrestricted choice of locations for sophisticated technologies and drastically reduced the barriers to locating in China the manufacture of technologically complex products, including R&D. China has, impressively taken good advantage of these international dynamically evolving assets that have become available during the past few years. China has also benefited from its future enormous market potential and its policies and programmes to capture emerging global possibilities. By expanding its human resources and identifying 'pillar industries' for the future, China has become exceptionally well-positioned to exploit the global dynamics in research and development.

China successfully completed its first manned space flight in October 2003 and subsequently announced that the country will undertake major efforts in space research and development. This came as no surprise to the world as China had for a number of years given great publicity to its space efforts. However, these successes are likely to have given an additional boost to the US decision to announce in January 2004, ahead of the State of the Union address by the President, a moon colony plan. The objective is to establish a permanent human settlement on the moon and eventually to send Americans to Mars.

China announced that it aimed to launch 10 satellites in 2004 while preparing for its second manned spaced flight. The US apparently wants to maintain a lead, although it has not specifically targeted China, and a Senate hearing includes the following statement.[14] 'US policymakers must not sacrifice funding for science and technology to other priorities such as homeland security; both are essential to long-term US national security interests. Funding levels must also increase over time if the United States is to remain economically, technologically, and militarily competitive.' The hearing statement argues that China has greatly benefited from the globalization process, which has rapidly opened up new sources of technology and facilitated technology transfer to create an advanced industrial and knowledge-based economy, in considerably less time than could have been expected only a few decades ago. Thus, it is stated that 'the challenge it poses for US policymakers is not how to prevent China's technological advancement, but how to stay ahead of it'.[15]

CHINA – A KNOWLEDGE ECONOMY

Every month a book appears in China, Japan or the West that tries to portray China as a future knowledge-based economy. All of them testify that economic growth has been remarkable over the past decades with an annual average of close to 10 per cent. They also mention that half of China's high-technology exports include not only consumer electronics but also computers, telecommunications and office equipment, for which China is presently highly reliant on imported electronic components. This thriving trade in high technology has a strong basis in a large number of multinational enterprises that operate large-scale affiliates in China. An OECD report indicates that foreign affiliates of multinational enterprises located in China show low turnover per employee compared with foreign affiliates in Singapore or Hong Kong. This would indicate that multinationals are still using China for the production of low-technology goods.[16]

Any nation desiring to become a knowledge-based economy must develop its human resources in crucial ways and provide infrastructures to exploit fully the talents within its borders. China has dramatically increased the number of students in tertiary education and provided more funding for R&D during the past ten years, not only in absolute terms but also in relation to its GDP. Traditional indicators, such as patents, still suggest that China is far from reaching its goal of becoming a knowledge-based economy. However, using signs of dynamic changes within industrial sectors and emerging competencies in a number of research fields presents a more encouraging scene.

In many locations in China one can find an environment with young and mobile entrepreneurs, a multicultural mixture of people drawn from different parts of China, overseas Chinese, and visitors from a number of countries. Whatever direction China chooses for its national structure of S&T, its future in R&D management lies in its entrepreneurial urban middle class, for which the government must create good institutions, resources and infrastructure.

Participation in a global innovation system (GIS) has recently become highly valued. It shows itself for example in increasingly close relations with the European science community. China has reached an agreement to become a full member of the Galileo Project; this is carried out within the sixth Framework Programme and will involve close collaboration with the European Space Agency, giving China access to future advanced satellite technology and systems. Another example is China's decision in 2003 to become a full member of the ITER fusion project, which has created a geopolitical conflict as China, Russia and the EU opted for France as the location for the major ITER laboratory while the US, which joined the project after China decided to become a member, has strongly argued that Japan should be given this responsibility.

China has clearly demonstrated its capability in advanced technology industries (ATI) in a limited number of areas, which reflects spillover effects from defence programmes, although now supported through nationwide civilian initiatives. The nation's strength in labour-intensive industries (LII) is continuously demonstrated to the rest of the world. However, China still remains weak in industrial sectors such as semiconductors – not only in advanced manufacturing facilities but also in the development of, for example, devices and many other advanced components that are needed in all kinds of electronics products. China's share of the world's software engineers is still miniscule although it will soon deliver some 50 per cent of the world's consumer electronics products.

Thus China needs a much stronger basis in technology-intensive industries (TII) and aims to achieve this through its manpower development combined with TII infrastructure and strong national efforts to develop innovative technologies. If this is realized, we will in the not too distant future recognize a China that might simultaneously show extraordinary strength in ATI, LII and TII. This would categorically challenge the 'flying geese pattern of development' where latecomer nations just follow the leaders and only gradually move up the chain of technological sophistication. China might upset our concept of structural transformation by being able to compete concurrently on all fronts.

NOTES

1. Author interviews and Beijing conference participation, November 2003.
2. Ng (2004).
3. Romero (2004).
4. Ibid.
5. Naughton (1995).
6. Ibid., p. 69.
7. Ibid.
8. Xu Guanhua, 'Speech on technology strategy', Closing Ceremony of the Ministerial Forum on Industrial Policies of China, Beijing International S&T Industries Fairs, 15 September 2003.
9. 211 stands for the objective of bringing a number of Chinese universities into a Global One position in the 21st century.
10. Meyer and Lu Xiaohui (2004).
11. Perrins (2003).
12. Songshan Lake Science and Technology Industrial Park – Innovation Chapter; Administrative Committee Building, Dongguan City, Guangdong, PRC, Tel. +86 – 769 2891 228, fax: +86 – 769 2822 822; www.ssl.gov.cn.
13. Walsh (2003).
14. Ibid.
15. Ibid.
16. OECD (2004).

2. National reform programmes and human resources development[1]

In its science and technology policy, China has moved from one extreme of total national control of all resources to a flexible and open system which allows a great amount of freedom for foreign companies and domestic actors. However, since the early 1950s, national programmes for science and technology (S&T) development have received great attention, together with raising the level of human resources. An overview of the current innovation structure in China can be seen in Figure 2.1.

EVOLUTION OF TECHNOLOGICAL POLICY IN CHINA

A first major educational reform was implemented in 1952 when all universities were basically transformed into teaching universities, while most research was organized into research institutes directly controlled by line ministries. Most teaching universities were controlled by relevant ministries, such as the Ministry of Railways and the Ministry of Post and Telecommunications. Many ministries organized their own research institutes and research activities, which were normally referred to as academies, for example the Academy of Telecommunications. Following the Soviet model, the Chinese Academy of Sciences (CAS) was established in 1949 and soon became the main bastion for advanced scientific research, with a total employment that eventually reached more than 120 000, including also a large number of service functions.

While industrial and technological development was formulated and implemented under formal five-year plans, specific long-term plans were formulated on several occasions for the development of science and technology.

During a short period, 1958–59, the planned economy was completely disrupted by China's frantic effort to industrialize under the slogan The Great Leap Forward. A similar disruption occurred when the Cultural Revolution began in 1966. This led at the time to an almost complete closure of all universities, which did not start normal operation until China announced its Open Door Policy in 1978. China had already established the Chinese Academy of Social Sciences in 1976, partly from institutions within the CAS, and

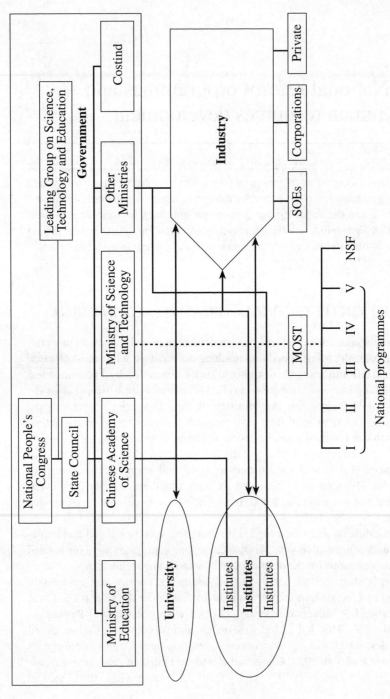

Note: NSF = National Natural Science Foundation of China.

Figure 2.1 National innovation system – key institutions in China

this had come to serve in many ways as a cluster of think tanks for the Government.

The new era after 1978 involved a number of far-reaching reforms, although the concept of five-year plans remained and still influenced significant sectors of the economy like the production of grain and mineral resources. However, a number of important reforms were implemented. First, the line ministries shed direct control of a majority of their enterprises; they were corporatized and often introduced on to stock exchanges in Shenzhen, Shanghai and Hong Kong, although with major equity shares still controlled by the state at national, provincial or municipal levels.

Second, the university system underwent a complete reform that reversed most of the 1952 changes. Universities were merged in whole or part to create comprehensive higher education units that were also given major responsibilities for research and were instructed to develop new curricula for key disciplines, particularly in new and emerging technologies. Undergraduate teaching expanded rapidly while China initially relied on foreign universities, primarily in the US, to provide Masters and PhD training. One hundred universities were included in the 211 Education Project[2] and selected for special attention, while some ten, among 30 major comprehensive universities, were given the objective of becoming internationally recognized.

Third, R&D that had until the late 1980s been completely controlled and carried out within state institutions, underwent gradual and very substantial changes. This was partly a reflection of the state giving up its direct control of state enterprises. Many industrial research institutes were simultaneously transferred to manufacturing plants. See Chapters 4 and 5 for further details.

In other instances high-performing research institutes of the Chinese Academy of Sciences (CAS), and similar institutions, established commercial high-technology companies. One of the most striking recent examples is Lenovo, formerly Legend, which is a spin-off company from CAS that acquired control of the personal computer division from IBM in early December 2004. Several well-known universities like Tsinghua University, Beijing University and Fudan University in Shanghai have successfully established their own high-tech companies. A recent rise in China's capability in high-performance computers can be traced substantially to knowledge transfer from the Institute of Computer Technology of the CAS. However, in recent years technology commercialization activities have changed as leading universities have become more involved in fundamental research, and the role of incubators has become more important than outright commercialization of existing technologies.

With line ministries having a declining role in the direct support of the development of R&D, and mainly responsible for formulating and implementing policies, the Ministry of Science and Technology (MOST) – formerly a Commission – has come to play an increasingly important role. MOST has established five major national programmes (see Chapter 1). Direct funding is limited, with major financial resources coming from banks, local agencies, companies and research institutions. MOST receives funding based on decisions taken by the State Council that also allocates funds to the CAS, the Ministry of Education, and COSTIND for the military sector. The National Natural Science Foundation of China has come to play an increasingly important role in its support of individual projects in basic research.

China's funding of research and development has increased more rapidly than economic growth, and in 2003 constituted 1.32 per cent of GDP. However, this figure is likely to exaggerate the present real potential of China's R&D capability as almost two-thirds is reported to be carried out in the corporate sector, with many state-owned enterprises that are still too poorly organized to exploit R&D results efficiently. Shanghai and Beijing are the major research centres in China, and Shanghai has set the objective that 2.5 per cent of its GDP should be used for R&D by 2005. The 973 Programme under MOST has a focus on agriculture, energy, information, natural resources, life sciences and material research. This programme, together with the Chinese Academy of China, provides most of China's national major thrust into basic research. MOST suggested in late 2004 that 'The State mid and long-term research and development strategy will allocate 15 per cent of total R&D expenditure from 2010 to 2020 to the state's basic research, or 973 projects',[3] which apparently does not include basic research within CAS institutes.

In March 2005 China was due to decide on its long-term science and technology structure to cover the coming 15 years until 2020. The IT sector has already been defined as one of the pillar industries and the semiconductor industry will receive special support. Biotechnology will also get substantial support. The development of these and other sectors will greatly benefit from an expanded talent pool resulting from a continued rapid expansion of higher education.

China's technological prowess will require efficient use of its R&D resources in the engineering and scientific fields to capture substantial intellectual property rights (IPR). Chinese companies remain weak in mature or maturing technologies as they lack IPR strength and recognized brand names. However, the size of the domestic market and the fact that China has become the world workshop gives it strong influence in setting standards, for example in future communication technologies.

China will gain major benefit from attracting returnees who have received education and advanced training abroad, primarily in the US. It is simultaneously attracting substantial R&D activities from industrialized countries. China may become the first developing country that is able to capture the entire range of R&D functions from multinational companies – from rudimentary upgrading of manufacturing technology to basic research at the frontiers of new knowledge. Thus, global IT companies are possibly embarking on a process of creating a global innovation system (GIS) in which China is going to play an increasingly important role.

Stages in Science and Technology Policy Formulation

The formulation and implementation of programmes for the development of science and technology represents a common feature and trend in the modern world. The programmes in China are the basic ways that the Chinese government organizes R&D activities, and are an effective tool for the government to use in rationally allocating resources into S&T activities; they will promote the development of the economy and society, and the prosperity of scientific and technological undertakings.

China's expenditure in 2003 for research and development (R&D) reached RMB152 billion, up 18.1 per cent from 2002, accounting for 1.3 per cent of GDP. Of this total, RMB8.6 billion was used for basic research, which amounts to 5.6 per cent of total R&D. This would correspond to some 17 per cent of total government spending, on the assumption that two thirds of all R&D is carried out in the corporate sector. At the time China had 28 million professionals and technicians of all specializations working in State-owned enterprises and institutions. There were 1573 projects under the National Key Technologies R&D Programme with another 4479 projects under the High-Tech Research and Development Programme (863 Programme). Another nine national engineering research centres were established. The initiation of 46 projects on updating key national laboratories and the designation of 302 national technical centres established in groups also took place in 2003.[4]

Most funds for research flow through the Ministry of Science and Technology (MOST), although some funds flow from MOST directly to other Ministries. The Chinese Academy of Sciences and the Chinese Academy of Social Sciences have ministerial status, and receive their funds directly from MOST and the Ministry of Finance. Some funds for research also come from research funding bodies, of which the most important one is the National Natural Science Foundation of China, established in 1986.

It is increasingly becoming evident that the forms of interaction among actors in a system of knowledge creation are more important with respect to performance, than the formal structure of universities, research institutes and companies. However, their actions and accompanying results are strongly influenced by the incentive structures and general policies that would govern the National Innovation System (NIS), for example the rate and course of learning and performance in scientific, technological and industrial fields.

Studies on the NIS were initiated by Freeman's analysis of the Japanese system in 1987 and have been followed by numerous studies which have provided a framework for ongoing research.[5] The OECD has developed and shaped the understanding of the NIS through various studies, although not fully accepting that the word 'national' in NIS has begun to lose much of its relevance. Subsequently a number of analytical methodologies have evolved which focus on institutions, relations and strategies, the latter primarily performed by companies. Regions in China, for reasons of size and history, play a very important role and regional innovation systems have evolved that, in combination with the NIS and an emerging global innovation system (GIS), constitute China's innovation system. Selected details will be given below about the structures, resources and major actors in China's innovation system, with a focus on the changing role of government initiatives and regional development to strengthen the performance of the enterprises, while at the same time providing the foundation for advanced scientific research.

Rapid changes in the economy, science and technology, prompted the Chinese Government to identify ways of designing a National Innovation System. In 1998, the Chinese Academy of Sciences (CAS) was given approval to initiate the Pilot Project of a Knowledge Innovation Programme (PPKIP), to be completed in 2010. It included three phases: an initial phase from 1998 to 2000; an implementation phase from 2001 to 2005; and the concluding phase from 2006 to 2010. The CAS will undergo major changes under PPKIP, which include the following measures. First, about 80 national institutes that already have substantial capabilities which can be sustained will be restructured. Second, 30 of these institutes will be supported and given resources to become internationally acknowledged, high-level research institutions. Among these, three to five would be expected to rank among the first-class research institutes in the world. As a result of this transformation process China will reach, in respect of basic science, international research frontiers in some important and strategic areas, thus making a contribution to the world's modern science development.

China has previously experienced acute struggles over how to organize science and technology. The model which we know from industrialized countries and also from the earlier Soviet Union may be called the professional or 'bureaucratic' model. The other is the Chinese mobilization model, which we know from the Great Leap Forward period in the late 1950s and its revival during the Cultural Revolution, when large numbers of people were expected to make their contributions to advance the technological front. The mobilization approach remained in place until 1978. Before that mobilization teams in agriculture and amateur scientists in meteorology made contributions, during a time when science and technology sources in China were very scarce. However, this did not conceal a realization that the thrust in all technological sectors would have to be based on a professional approach stressing a high degree of specialization – as is the case in highly industrialized countries. At the time, there were two objections to the professional model. First, it was seen as corrupting the socialist development within the country. Second, it was seen as opening the doors to foreign influence, directly as well as through exchange of goods. As a result, until the end of the Cultural Revolution there was a commitment to narrowly defined self-reliance or even self-sufficiency in order to counter such influences.

The realization dawned in China that technology transfer during the last quarter of the 20th century would have to differ from earlier periods. First, the demands of technology are increasing as countries and regions are drawn into much closer interaction and as differences in the standard of living among nations become much more apparent. Secondly, technology has become much more intimately related to science than in the past. Thus, developing countries have been forced not only to approach, by adaptation, a much higher level of science and technology, but also to reach for a science and technology that is continually and rapidly moving to higher and higher levels. Practically all of this advancement has taken place until recently in the industrialized countries.

The well-known Chinese scientist Qian Xuesen, who after his return to China made a major contribution to the country's missile programme, discussed in 1977 the gap between China's science and technology and the advanced countries of the world.[6] He talked about four issues related to the gap. First, he wanted to know if there exists a disparity between the level of China's science and technology and advanced world levels. He stated that there are some areas in which China has come close or even surpassed advanced world levels. However, they represent only a relatively small part of the whole, while in most areas China is backward. Furthermore, he mentioned that even in areas of success China has not been able to use advanced machinery and equipment.

Second, he said that speed and direction are important and that the gap has to be closed in order to meet requirements for (socialist) economic development and national defence. Relating to a third question, Qian suggested that China would be able to catch up with and surpass the industrialized world because of China's socialist system. He pointed out the contradiction existing in capitalist economies between private ownership and 'socialization of science'. However, he answered this question affirmatively and said that in the final analysis the interests of the individual, the collective and the State were in accordance with each other in China. Finally, he discussed how to bring the superiority of the socialist systems into full play in order to enable China's science and technology to catch up and surpass advanced world levels, although without providing any suggestions for specific reforms and measures.

Later on in 1977 the State Council announced that a national science conference would be held in early 1978. A planning conference was held in October the same year and subsequently a draft plan was prepared for China's science and technology that covered needs and tasks for the period to the end of 1985, with an involvement of more than 20 000 people. Regional units of the People's Liberation Army and provincial science and technology agencies were meeting all over the country to deliberate on the Plan and make suggestions. A plan was presented to the delegates of the Science and Technology Conference that was held in late March and early April 1978, after which followed a new series of provincial meetings all over the country. A number of specialized conferences were also held from July 1977 to July 1978. This resulted in a growing awareness and eventually an in-depth knowledge of the problems confronting China in its modernization programmes. An apparent outcome from all this activity was an increasing awareness of the need to improve the educational system, engage in technology imports, and collaborate with scientists abroad.

The Eight-Year Plan for Science and Technology, 1978–85, gave special attention to eight comprehensive areas – energy, materials, electronic computers, lasers, space, high-energy physics, genetics and naturally agriculture – for which 108 items were specifically identified. The Vice-President of the Chinese Academy of Sciences, Fang Yi, mentioned at the time that China expected the fulfilment of the plan to result in China approaching or reaching the advanced world levels of the 1970s in a number of important branches of science and technology, leaving a gap of about 10 years.

China's science and technology policy cannot be fully understood without realizing the underlying dual-economy development strategy. This gives rise to a variety of technology demands which have to be met with distinctly different technology policies. It is necessary to differentiate among various sectors where R&D results are required. Long-term projects, with substantial demands on resources, would generally require specialized researchers with specialized education. The control of such resources in terms of manpower and funding also requires adequate organizational structure. Such projects would normally be found within defence and the modern industrial sectors. Priorities and execution are handled by national agencies, with the consequence that the earlier popular participation in the decision-making process has basically disappeared. Thus the intensive political debate raging in scientific and technological circles in China came to an end. An understanding was reached that scientific research must precede production and this outcome was strongly influenced by an acute awareness that China's material base must rapidly expand and that science and technology with all their requisites are important and efficient instruments for achieving this.

The formulation of China's science and technology programmes started in 1956, and they have gone through a number of phases since then. The first phase was the implementation of the Perspective Programme of S&T Development from 1956 to 1967, whose mission was to strengthen China's science and technology system quickly, to solve a number of complicated S&T problems in industrial development using its own national resources, and to try to reach world advanced level in certain fields in 12 years.

The second phase was the application of the Programme of S&T Development from 1963 to 1972, where the stated goals were to provide S&T results for the development of agricultural and industrial industries so as to enhance the living quality of the people, to develop S&T resources quickly and to train S&T talents.

The third phase was the National Programme of S&T Development from 1978 to 1985, which included goals of achieving mechanization and electrification by the end of the 20th century, cultivating first-rate S&T experts and acquiring advanced equipment, establishing many modern S&T experiment bases and creating a national system of science research.

The fourth phase was the implementation of the Ten Year Programme of S&T Development, 1991–2000, where important objectives included the development of the economy and society, orientation to mainstream fields of economic construction, the development of vigorous new high-technology industries, and the continuous development of basic research.

A fifth phase is the Ninth Five Year Plan and Long-term Programme on S&T Development to end in 2010. The strategic goals include: to establish the groundwork for a market-oriented S&T system; to enhance the contribution of S&T progress to economic growth; to construct an economy in which social development substantially depends on the progress of S&T and improvement of the quality of labour; to realize the integration of S&T and the economy; and to become one of the ten strongest nations in the world by 2010.

A sixth phase is the implementation of the Tenth Five Year Plan on S&T Development, whose objectives include creating a base by which the country is thriving through science and education; reforming the current S&T system and establishing a market-oriented national innovation system; enhancing the international competitive capabilities of China's industries; stimulating the sustainable development of the economy; improving people's living quality; and improving considerably the level of Chinese S&T and the capability for autonomous innovation. According to China's Long and Medium-term Programme on S&T Development, China is attempting to reach by 2020 S&T levels that developed countries had reached in the early 21st century, and considerably shorten the gap between China and advanced countries in the world.

A seventh phase is presently underway in laying the groundwork for reaching an advanced global position in science and technology by 2020. The process of achieving rapid technological progress and reaching the status of a technological superpower has been underlying the deliberations of thousands of scientists, engineers and policy makers who have been preparing the 2020 Science and Technology Plan since early 2003. As in Japan, the science community in China is eager to lay the foundation for acquiring Nobel prizes for its scientists to boost their image and attract more international collaboration. This may contradict the interests of the engineering community, which recognizes that Chinese intermediate success lies in boosting industrial technological capability rather than advances in scientific research. The 2020 S&T Plan is likely to reveal some compromises when it is completed and finally released.

After years of experimentation China has been able to establish a relatively complete system of science and technology programmes, which include both mission-orientated programmes mainly for R&D activities, and government-guided programmes mainly for the commercialization of R&D results and the industrialization of high-technology products. They are all closely linked and often include joint development projects. Five of them illustrate in a major way the recent evolution of the Chinese science and technology system (see Figure 2.2).

Key Technologies Research and Development Programme
The National Key Technologies R&D Programme started in 1982, and is China's first National S&T Programme to serve main sectors of the economy in order to provide key results to upgrade industries. Having been implemented for more than twenty years, the programme has provided a batch of key technologies and equipment and has played a significant role in the rapid development of the economy and the swift progress of science and technology. The programme is readjusted by the Ministry of Science and Technology every five years – within the national Five-year plans.

The programme was approved by the Fifth Plenary Session of the Fifth National People's Congress in 1982, and was designed to serve the mainstream sectors of the national economy with concentrated resources of key and common technologies that were desperately needed for industrial upgrading and sustainable social development. It has been implemented during five five-year plans, including the one that is to be completed by 2005, and has provided a batch of key technologies and equipment for the economic and social development of the country. The programme included the three major components of agriculture, new high-technologies and social development.

As the Key Technology R&D Programme was evolving it shouldered the tasks of application to engineering and development of new/high technologi-

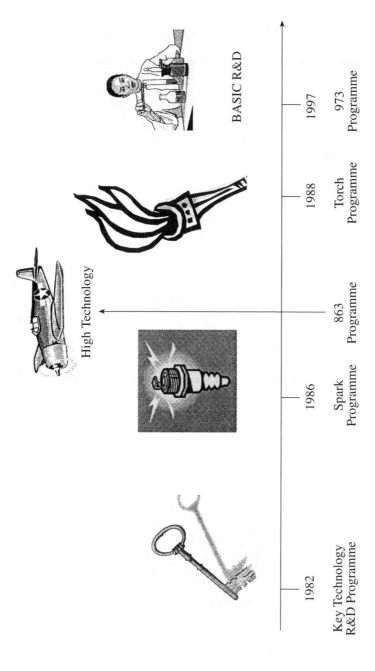

Figure 2.2 National science and technology programmes

cal achievements from the '863' and Key Technologies Research and Development Programmes. The objectives of the 863 Programme (High-Tech Research and Development) include breakthroughs in the key technologies of the new generation of large SPC switching systems, high performance computers, information technology-based commercial and trade equipment, high definition TV (HDTV) digital audio broadcasting (DAB), Chinese information software platforms, and so on. The research on advanced manufacturing technologies includes cars, integrated CAD, industrial robots, laser-processing and so on. The programme should also carry out coordinated development of the economy and society through science and technology with the following five aspects, in order to promote the implementation of the Chinese 21st-Century Agenda:[7] medical treatment and health, the exploration of resources in the western part of China, environmental protection technologies, the utilization of marine water and marine biological resources, and the establishment of national comprehensive pilot zones.

During the period of the ninth five-year plan (1996–2000), the central government allocated RMB5.3 billion, with another RMB17.6 billion financed by local governments. The programme has involved more than 70 000 researchers in some 1000 R&D institutions, 700 universities and colleges and 5000 enterprises. The Programme has also trained more than 20 000 researchers, including some 5000 researchers who have become young academic leaders in their respective fields.[8]

A recently reported successful project is the high-quality large container inspection system that is being used by customs in Beijing, Shanghai, Tianjin, Dalian and 11 other cities, which has greatly improved the checking efficiency and accuracy. The system was developed by Tsinghua Tongfang Co., a company attached to Tsinghua University.[9]

Spark Programme
In the late 1970s, the reform of economic institutions first started in rural areas in China. The establishment of a system of household family contract responsibilities had aroused the farmers' enthusiasm for getting rich through hard work, which has brought about a significant change to production relations in rural areas. From the early1980s, the township-and-villages industries developed rapidly, and the rural economy became unprecedentedly active. The specialization and commercialization of agricultural production have made farmers aware of the importance of science and technology, and hence demands on sciences have emerged widely in the rural areas.

The Spark Programme was the first major programme approved by the Chinese government to promote the development of the rural economy by relying on science and technology and is an important component of plans for national economic, scientific and technological development. The goal of the programme is to introduce advanced and appropriate technologies into rural areas, and lead the farmers to develop the rural economy by relying on science and technology, to pilot technological progress in town-and-village enterprises, to stimulate expertise of rural labourers, and to drive the sustainable, rapid and healthy development of agriculture and the rural economy.

The background of the implementation of the Spark Programme is as follows: in March 1986, the Chinese government started the reform of scientific and technological institutions in the whole country, and further clarified the S&T policy: 'Economic construction must rely on science and technology and scientific and technological workers to be oriented toward rural economy, extend research achievements of science and technology to rural areas and help develop the rural economy. China is a large agricultural country with the rural population occupying 80 per cent of the national total and it is necessary to solve the problems of agricultural, rural areas and farmers by realizing the modernization of China and promoting the development of the rural productive force'.[10] This provided good opportunities for S&T personnel to orient to the rural economy, to extend the research achievements of science and technology to rural areas and to help develop the rural economy.

In May 1985, the Ministry of Science and Technology submitted to the State Council a suggestion to 'implement a batch of short-term scientific and technological projects so as to promote the rapid development of the local economy', in which the proverb 'A single spark can start a prairie fire' was quoted, hence came the name of Spark Programme, meaning that the spark of science and technology would extend over the vast rural areas of China. The Chinese government approved the Programme in early 1986.

Implementation included the following approaches. First, to establish demonstration S&T-guided enterprises in order to guide township enterprises to develop healthily, and to stimulate the structural adjustment of rural products and industries; to train rural technicians, managerial talents and farmer entrepreneurs; to develop high-yield, high-quality and high-efficiency agriculture; and to promote the construction of rural social service systems. Second, to establish technology-intensive zones, that is, demonstration zones for the comprehensive regional economic development of rural areas, so that the growth of the economy is mainly based on the progress of science and technology in certain economic areas. Third, to develop regional pillar industries, which would develop in certain economic areas, step by step, and be based on the progress of science and technology. They would play an important role in extending science and technology to rural areas, producing dominant products based on regional resource superiority, and occupying a certain market share in the regional economy.

The implementation of the Spark Programme has come to depend mainly on enterprises. The construction of technology-intensive zones and the development of regional pillar industries is an important task, and similar projects for regional pillar industries at provincial and county level have also come to play an important role in terms of generated revenue. In 2000, the Spark Programme invested RMB153.7 million for training workers and held 105 800 terms of training classes, and close to ten million people were involved in training activities. The training programmes have their focus on the counties.

High-Tech Research and Development (863) Programme
The 863 Programme was initiated in March 1986, and is focused on advanced technologies in areas such as IT, biotechnology and new materials. The programme was launched in March 1987 with the aim of enhancing China's international competitiveness and improving China's overall capability of R&D in high technology including the six fields of information technology, biological and advanced agricultural technology, advanced materials, advanced manufacturing and automatic technology, energy technology and resources and environment technology. After 16 years, the 863 Programme has achieved considerable success in new and high-technologies, which has strongly stimulated the rapid development of S&T and the economy.

Approximately RMB10 billion was granted to researchers and research groups up to the end of 2001. The grants were widely distributed to expert scientists with the consequence that there was not sufficient concentration of resources in critical areas, after the committees were overwhelmed by the number of applications from universities and research institutes. There had been too little involvement of firms, a situation that is likely to change in the future.

The start of the 863 programmes may be attributed to the response of the Chinese government to the fierce international competition in new and high-technology. For example, in 1983, the United States first put forward the Strategic Defence Initiative (that is, the Star Wars Initiative), then came Europe's EUREKA, Japan's Policies for the Promotion of Science and Technology in the Next Ten Years and so on, which are all strategic plans aimed at the development of high technology. In view of the impacts on China of competition in the fields of high technologies in the world, four scientists submitted a report to the central government to suggest that China adopt appropriate policy measures to reach the world advanced level and reduce the gap between China and the developed countries.

On 3 March 1986, a report called *Suggestions on Tracing the Development of World Strategic High Technology*, written by Wang Daheng, Wang Ganchang, Yang Jiachi and Chen Fanyun, was sent to Zhongnanhai – the location of the State Council and Party headquarters in Beijing. It is reported to have reached Deng Xiaoping almost immediately. He considered the report very important, and prepared instructions within the next few days requiring headquarters to invite experts and relevant responsible comrades to discuss the report and make suggestions for a decision. This resulted in the formation of the 863 Programme, based on the above report. Deng Xiaoping is reported to have told his comrades that a 'Quick decision should be made on this matter without any delay'.

The strategic goals of the 863 Programme are as follows: working towards the international advanced level and reducing the gap between China and developed countries in several of the most important fields, striving to achieve some breakthroughs in the fields where China holds advantages so as to provide a good environment for economic development and social security; training a new generation with high-level scientific and technological knowledge; and enhancing the development of science and technology in related fields.

The 863 Programme, like the Key Technologies R&D Programme, was regularly updated with respect to ongoing Five-year plans. Revisions have included emphasis on the construction of a national information infrastructure, stimulating the development of key biological, agricultural and pharmaceutical technologies for the improvement of quality of life, and boosting the development of key new materials and advanced manufacturing technologies in order to enhance the competitiveness of industry.

In 2002, out of the six fields mentioned earlier, the focus was strongly on biochemical industries and agriculture (see Table 2.1). The projects were mainly and almost equally distributed among higher learning institutions, R&D institutions and enterprises and applied development and applied basic research received most of the resources. Funds for experiments and soft science were limited. Funds for new 863 projects were allocated to institutions all over China, but some ten provinces received 81 per cent of the projects and 87 per cent of the funds. Furthermore, Beijing was absolutely dominant followed by a trailing Shanghai. In 2002, the projects and expenditures in Beijing and Shanghai totalled 59 per cent (see Table 2.2).

Table 2.1   Distribution of 863 Programmes by field in 2002

	Projects	**Expenditures**	**Personnel**
Information	22%	20%	26%
Biochemical and Agriculture	25%	33%	28%
Advanced Materials	22%	17%	11%
Manufacturing and Automation	14%	10%	15%
Energy	5%	14%	6%
Resources and Environment	12%	6%	8%

Source: http://www.most.gov.cn (25 September 2003).

Table 2.2 Regional distribution of projects of 863 Programmes in 2002

	Projects	Expenditures
Beijing	34.9%	48.6%
Shanghai	10.1%	10.4%
Hubei	5.1%	4.0%
Jiangsu	4.9%	3.8%
Shanxi	4.9%	3.7%
Hunan	4.5%	3.5%
Liaoning	4.2%	3.3%
Guangdong	3.8%	3.2%
Sichuan	3.8%	2.7%
Shandong	3.5%	2.5%

Source: http://www.most.gov.cn (25 September 2003).

The importance of Beijing is evident in the news from 2003 that mentioned that Beijing 'would develop etchers and injecting machines for 0.10-micron semiconductor manufacturing technology, which has been on the list of the State's high-tech 863 Programme and will get capital support from the central government.[11] Since the building of the Beijing Integrated Circuit Design Park and the launch of Semiconductor Manufacturing International Corp's (SMIC) Beijing project in 2003 the Chinese capital has become an important national focal point for China's future development of the semiconductor industry. However, Beijing aims not only to develop semiconductor technology but also to become a major centre for production. To that end the municipal government wants to develop the market for six major projects: integrated circuit (IC) cards; shared point of sale (POS) systems for banks and taxation bureaus; digital cameras; high-definition TV (HDTV); network computers (NCs); and mobile communication devices.[12]

Beijing has an important advantage because most domestic central processing unit (CPU) designers are based there; Beijing is also considered to be the national leader in chip design and manufacture. Naturally procurement policies play an important role. Xu Qin, deputy director of the department of high-tech industry development with the State Development and Reform Commission, has indicated that the central government would promote the adoption of network computers in Beijing Municipality and western regions, and initiate major projects in education and government sectors.[13] A Vice-Mayor of Beijing, Fan Boyuan, said the city would launch pilot projects and allocate special funds for the procurement of network computers in government departments.

Beijing is also the pilot city for shared point-of-sales (POS) systems between banks and taxation bureaux. In this development Beijing is favoured by the presence of a number of government research institutes including the Chinese Academy of Science, Beijing University, Tsinghua University and SMIC Beijing. They all participate in 863 projects that are critical for the future of the semiconductor industry. Three developers of central processor units, Arca Technology, Godson (BLX IC Design) and Zhongzhi, have initiated three industrial alliances – comprising IC designers, hardware manufacturers and software firms – for the development of network computers to meet the demand in Beijing.

The Arca company was the first Chinese chip vendor to develop its own CPU in 2001, followed by competitor chip vendor BLX IC Design, which is also based in Beijing.[14] Arca has signed an agreement with the US company Wyse Technology of San Jose, California, to provide its new generation processor for an upcoming version of Wyse's Winterm thin clients that will be sold globally. Thin clients are computers that do not have a hard disk drive and access applications that run on a central server. The agreement between Arca and Wyse marks the first time that a Chinese-developed CPU will be used in computers sold by a multinational vendor outside China and it gives Arca an important entry into the global market for thin clients. Wyse is the world's largest vendor of thin clients, accounting for some 40 per cent of global thin client shipments during the third quarter of 2003, and Arca expects to deliver 'hundreds of thousands' of its processors.[15]

ARCA Technology has been supported by the Chinese government and widely recognized by Chinese industry. ARCA Technology Corporation, together with colleagues from the information industry, is attempting to establish China's microprocessor industry with the expectation of a presence in the field of CPU core technologies. Many Asian analysts seem to think China is poised to take a potentially competitive microprocessor market position within just a few years.[16]

The Dragon microprocessor is known in international markets as 'Godson', although the project name has changed on several occasions. BLX IC Design is considered to have been the first to develop a processor based on internally engineered design. The dragon architecture is being offered as a generally compatible 'MIPS-like' solution as the Chinese Academy of Sciences (CAS) has licenced the graphic synthesizer (GS) instruction set from MIPS Technologies in the US. However, China has chosen to implement a completely independent and proprietary architecture. Subsequently, the companies that are associated with the CAS microprocessor project have been opting for an implementation based on Linux.

Development is now underway for the next evolution of the Godson processor platform, with more features to make it attractive in the market for computer servers. BLX IC Design has moved quickly to gain Chinese industry support around the architecture and aims to create an alliance of some 100 members that will be supplied with a variety of designs over the next few years.[17] The Godson chips are largely based on the MIPS instruction set, but are not fully compatible because the Chinese companies want to avoid the use of key instructions that would intrude on MIPS patents.

Torch Plan
The Torch Programme started in 1988 and has been designed to develop new and high technology industry under the guidance of the Ministry of Science and Technology. Promotion includes grants as well as loans together with self-financing. Since 1996 several new measures have supported the development of specific Torch Programme Projects.[18] Financial resources are mainly raised from the public sector with the support of start-up funds from the state. The creation of high-technology development zones, most of them started in the early 1990s, represents a major thrust of the Torch Programme.

The programme includes four main aspects. First, Torch creates an environment for the development of high-technology industries, providing equipment, laboratories and other required inputs. This includes a system and mechanisms adapted to the development of new high-technology industries. Torch creates channels to meet the needs for funding and establishes risk investment mechanisms. It also develops information channels and networks domestically and abroad and is responsible for a practical long- and medium-term development programme and implementation plans. Second, Torch has the responsibility to establish Hi-New industries development zones. This is a very important component of the Torch Plan and has become a central area of development for Chinese high-technology industries. The aim is to transform the production of R&D results. Third, Torch is responsible for creating service centres for what in China are referred to as 'high technology pioneering undertakings' – more commonly known as incubator systems. Fourth, the plan is also responsible for creating specific Torch projects. These should be profitable high technology projects based on key technologies and R&D programmes from the national, local and industrial level. They should utilize high-technology R&D results with the aim of developing high-technology products into full industrialization. The expected outcome is high-technology enterprises and enterprise groups in areas such as new materials, biological engineering, electronics and information, new types of energy, high efficiency and energy saving, and environmental protection.

Through the support of state and local preferential policies, use of multiple funding sources such as bank loans, self-financing funds, private funds and foreign funds, the Torch Plan should stimulate the commercialization, industrialization and internalization of the Chinese high-technology R&D products.

There exist a number of preferential policies. First, when an enterprise in a development zone is confirmed as a high-technology enterprise the income tax rate is reduced by 15 per cent. Second, when exports from a confirmed high-technology enterprise in a development zone exceed 70 per cent of the total output value, income tax rate is reduced by 10 per cent. Third, a confirmed newly established high-technology enterprise in the development zone could be exempted from income tax within two years. Furthermore, projects of the Torch Programme also have access to preferential financing sources, such as special loans for S&T development, the national new products programme of the Chinese Torch Foundation, the Chinese S&T Development Foundation, special construction loans, and so on.

Having been in existence since 1988, the Torch Plan has made a major impact on the industrial and technological landscape in China. In many places the supported industries have become pillars of the local economy, which is the case for the information industry in Beijing, Guangzhou and Shenzhen, in which the Hi-New technology development zones have played an important role. By now there are a sufficient number of such zones, but many may need to expand in scale and improve in quality in order to have a real impact on China's technological development and economic growth.

The number of large companies within the zones has continued to grow, although this expansion has taken place in a limited number of zones. The number of enterprises with annual revenues higher than RMB100 has reached 2000, with high growth rates for many of them. The output value of enterprises in national high-technology development zones has increased rapidly – in 2002, it reached RMB1293.7 billion (see Figure 2.3). The exports of enterprises in national Hi-New technology development zones has continued to rise rapidly and reached US$32.9 billion dollars in 2002 (see Figure 2.4).

National Basic R&D (973) Programme

The National Key Basic R&D Programme was launched in June 1997, and is designed to stimulate the activities of original innovation, to solve significant scientific issues in the development process of the economy and society with the aim of providing scientific support for the future development of China. The establishment and implementation of the 973 Programme is an important policy measure for the improvement of capabilities in scientific and technological innovation, and the realization of the Chinese targets for the economy, science and technology and society in 2010 and by the middle of 21st century.

On 4 June 1997, the third meeting of the national leader group of science and technology decided to establish and implement a National Basic R&D Programme and the then Ministry of Science and Technology was given the

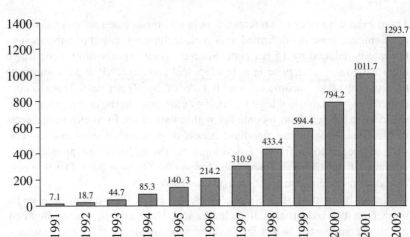

Source: http://www.most.gov.cn (27 September 2003).

Figure 2.3 Output value of enterprises in national hi-new technology development zones (billion RMB)

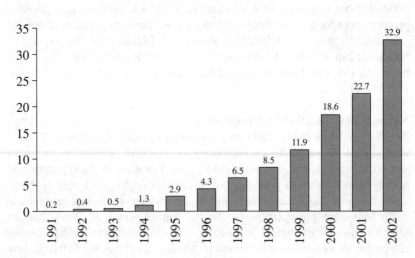

Source: http://www.most.gov.cn (27 September 2003).

Figure 2.4 Exports of enterprises in national hi-new technology development zones (billion dollars)

responsibility of organizing and implementing it. Under the broad direction of national goals, the organization and implementation of the 973 Programme was implemented according to the following procedures:

1. Make overall rational structures to provide impetus for technological innovation, while at the same time providing support for sustainable development of the economy and society.
2. Strengthen the strategic research and in-depth analyses into the major demands of China, such as the development of new high-technology; the adjustment of the industrial structure; increasing the importance of information in the economy and society; the improvement of the people's standard of living and health level; the conservation and efficient exploration of natural resources; the coordinated development of ecology, the environment and society; the development of the western part of China, and so on.
3. Implement the research into key scientific issues in terms of the future, aiming to be at the forefront of science.

The Key Basic R&D Programme has the following five tasks: first, to carry out multidisciplinary and comprehensive research in scientific issues relating to the development of the economy, science and technology and society, which are closely linked to the fields of agriculture, energy, information, resources and the environment, population and health, and materials, and provide theoretical foundations and scientific bases for solving problems. Second, to establish a set of significant scientific engineering projects, which reflect the current level of development in science and the overall strength of science and technology, and which will produce far-reaching impacts on the long-term and sustainable development of the Chinese economy and society. Third, to ensure basic research that is relevant, important and explorable. Fourth, to train high-quality talent with innovation capabilities and excellent scientific qualities, who can adapt to the demands of the 21st century. Fifth, to construct a set of high quality science and research bases that can undertake key scientific tasks for China, and establish some cross-disciplinary, comprehensive centres of science and research.

The objective of the 973 Programme is to support basic research; this can solve important issues in the long and medium-term development of the economy, can address important issues in the forefront of science, and can reflect characteristics of Chinese environment, geography and human resources, which may enable China to position itself at the forefront of international science. In the period of the ninth five-year plan, the expenditure of the programme reached RMB2500 million. From 1998 to 2002, the 973 Programme established 132 projects, distributed across agriculture, energy, information, resources and the environment, population and health, materials, and issues at the leading edge of science. The share of projects in resources and the environment is currently the highest, which indicates that the Chinese government has realized the seriousness and urgency of these issues.

Forty-four new projects were established in 1999, while the number dropped to 26 in 2002. The allocation of projects to various areas has undergone changes since the start of the programme (see Table 2.3).

Table 2.3 The distribution of projects of the 973 Programme by field from 1998 to 2002

	1998	1999	2000	2001	2002
Number of projects	15	44	27	20	26
Agriculture	13%	14%	11%	15%	12%
Energy	7%	14%	11%	10%	12%
Information	20%	11%	7%	15%	15%
Resources and Environment	20%	17%	15%	15%	22%
Population and Health	20%	16%	7%	25%	15%
Materials	20%	14%	11%	20%	12%
Synthesis and Intersection		14%	38%		12%

Source: http://www.most.gov.cn (26 September 2003).

Among the 973 projects started in 1999 was a basic research project on Digital Signal Processing (DSP) and Central Processing Unit (CPU) chips with the aim of developing reconfigurable architecture to function in software systems. This was part of a broader objective of creating a new generation of processor chips to integrate the capabilities of various chips, such as CPU, DSP and custom chips, thereby realizing lower costs and higher performance. Research was carried out by seven groups from the Institute of Acoustics of the Chinese Academy of Sciences (CAS), Tsinghua University and the Beijing Broadcasting Institute. The project included basic research centred on the study of integrated chips, such as algorithms and structures, logic and electronic circuits, layout, global optimization methods, simulation and verification. A successful completion of the project could provide chips that could be applied to a wide range of areas, including multimedia information processing, telecommunications, information electronic appliances for family use and mobile computation – from which might follow strong intellectual property rights.[19]

Other National Programmes – Examples

National Engineering Research Centres
National Engineering Technology Research Centres (NERCs) were first started during the eighth Five-year Plan period (1991–1995), as outlined in the Ten-Year Plan for the Development of Science and Technology, and have several major objectives. First, NERCs will integrate science and technology with the economy and shorten the commercialization period, while at the same time accelerating the renewal of products. Second, the centres will have research institutes in key scientific and technological fields of importance for national economic and social development.

According to its originally formulated objectives, China intends to establish about 200 Engineering Centres which will employ a total of some 40 000 staff for research and development in engineering technology. The centres should meet the following requirements:

- Possess first-class engineering technology; have outstanding professional teams for research and development, design and experimentation; have relatively complete facilities for comprehensive experiments in engineering technology; and have the strength to participate in international competition.
- Have considerable capabilities in experimentation and engineering design; be able to transform scientific and technological achievements rapidly into commercial goods; and be able to promote the formation of industries.
- Be capable of self-management and of developing financial resources.
- Train and stabilize high-level teams for the research and development of engineering technology and provide training services for the technical personnel of enterprises.

The establishment of nine national engineering research centres, the initiation of 46 projects on updating key national laboratories and the designation of 302 national technical centres established in groups took place in 2003.

The Ministry of Science and Technology is responsible for the overall planning, rational structure and comprehensive arrangements for the formation of the Engineering Centres; this lasts for about three years. According to the Task Document for the Plan of Formation of the State Engineering Technology Research Centre, these tasks are approved by the state. After being examined and accepted a NERC can start its operation.

State Key Laboratories
Until recently there existed 161 State Key Laboratories in China, although in late 2004, 12 of them were deprived of their status. They were established early on in the reform process and have improved the country's research in many areas of the fundamental sciences and technologies over the past two decades. However, Vice-Minister of Science and Technology Cheng Jianpei announced in December 2004 that some laboratories had lost their status because they were found to be badly managed, had come up with few results, or had even lost some of their best staff.[20] MOST says that freed resources could be used for those that are running well, while poor performers may find an incentive to improve themselves.

National New Products Programme
The National New Products Programme is supported by the Ministry of Science and Technology with the purpose of guiding and encouraging enterprises and research institutes to accelerate technical progress and promote the capacity of technical innovation. The priority fields to be supported by the programme include:

- New high-tech products in areas such as space science and aviation and spaceflight technologies; photoelectric science and opto-mechano-electronics integration technologies; biological science and bio-engineering technologies; materials science and bioengineering technologies.
- New products with self-owned intellectual property rights;
- New products that are aimed at export or can replace import products, or can be produced with imported technologies with over 80 per cent of the parts domestically made.
- New products that adopt international standards.

Mega-Projects
Mega-projects of science research (MPSR) are projects relating to the development of large-sized modernized key instruments and equipment (or ones in which a large amount of investment is needed), necessary to the process of scientific research. During the seventh Five-year Plan period, China constructed ten such projects. During the ninth Five-year Plan period, the state implemented a rolling programme of several major scientific engineering projects based on in-depth study and widespread consultation of opinions in various circles; projects will be established only when conditions are ripe in the relevant discipline. The list of mega-projects includes among others the second phase project of the national synchronous radiation laboratory and the HT-7U Superconducting Tokamak Fusion Experimental Equipment. The

latter has a direct relevance to China's participation in the ITER international nuclear fusion energy project.

The brief examples above have indicated not only that China is supporting a strong national ambition to develop its scientific and technological capability, but also that it is probing the use of various institutional mechanisms to achieve its goals. With restricted financial resources, the planners and administrator will eventually reduce or delete some programmes while emphasizing others. However, the broad panorama of approaches has been justified in the present situation when the corporate sector has not fully identified the most suitable roles. The success of the national programmes must be judged by the extent to which the firms are benefiting and involved in R&D. China has successfully initiated a number of programmes in basic research and in technological development but they are still poorly linked to the country's economic development, and continued institutional changes are needed.

At the general level it is possible to trace the major flows of financial resources and other assets within the Chinese science and technology system. However, given the rapid evolution and expansion of various programmes it remains difficult to judge the efficiency of resource allocation, and even more difficult to measure the real outcome in terms of contributions to the national economy or advancing frontiers of new knowledge. Nevertheless, there is abundant evidence that China has sown the seeds that have the potential of yielding a rich harvest. However, the ongoing transformation of the state-owned enterprises poses a great challenge as many domestic companies have not yet fully realized their need to engage in efficient R&D activities to survive. Thus the competitive climate has to be strengthened and MOST is exploring policies to boost efficient R&D spending by providing tax incentives and expanding programmes for small and medium-sized enterprises. In the meantime advanced production sites in China depend heavily on foreign direct investment, where the foreign companies provide much of the needed technology; at present, this is poorly integrated with the rest of the economy. Most likely the solution to the problem lies less in MOST policies than in a major reform of corporate governance, rules for access to credit and new ways of controlling the substantial amount of assets in state-owned enterprises.

TECHNOLOGICAL AND SCIENTIFIC MANPOWER

China has one the largest global pools of scientific and technological manpower, for which a few figures and comments will be provided. The growth of educated manpower, in particular people with college diplomas and university degrees, will transform China in a major way during the first half of the 21st century. The tradition of paying great respect to learning and higher education has received a tremendous boost from the widespread existence of one-child families. Parents and grandparents will consider it their mission in life to bring the youngsters into institutions of higher learning. Ningbo, south of Shanghai, is already recruiting more than 30 per cent of its senior high school students to colleges and universities. The level of recruitment has already reached 50 per cent in Shanghai and current policy is that it should soon reach 60 per cent.

Thus, China has the opportunity to become eventually the most highly educated country in the world. The creation of this extensive knowledge pool will during the next couple of decades get a substantial injection from Chinese knowledge workers now residing in the US or Europe, where they remained after completing university degrees. Many will be attracted to return to take up key positions in research or industrial development as the Chinese economy moves into more advanced phases of development. Highly educated people and very qualified returnees will become the catalytic agents in national research programmes within all industrial segments.

However, like other developing countries, China faces the problem of scarcity of skills in a number of sectors. Thus developing human resources and making full use of existing ones is the greatest challenge that the current Chinese government has to handle. The legacy of the Cultural Revolution caused great destruction to the Chinese economy and human resources; intellectuals, particularly many scientists and engineers, suffered unjust treatment and were deprived of their working environment. The policy direction fundamentally changed soon afterwards; this naturally included the changing of scientific policy after the statement by Deng Xiaoping that 'Science and technology is the first production power'.[21] Subsequently, Chinese S&T and human resources have experienced fast development.

In recent years, China's universities have seen a series of structural reforms which have integrated them into overall science and technology policy. China's domestic universities have undergone major changes with the objective of strengthening their disciplines and making many of them into comprehensive universities, rather than being the specialized training schools of the earlier period. Subjects such as information science, life sciences, materials sciences and environmental sciences have become priorities.

Simultaneously, China's universities have undergone a dramatic expansion since the late 1970s, and the growth period has not yet come to an end. One of the results is that university graduates need to reduce their expectations and design reasonable career development plans to meet a new employment situation. Statistics show that the annual starting salary of university graduates in 2003 dropped by 40 per cent from the previous year.[22] China only had 400 000 university graduates in 1978. At that time, every university graduate was regarded as a 'treasure' by the government and state-owned enterprises. Today, millions of students graduate from universities every year, making university graduates a less scarce resource. The cost of higher education began to increase dramatically in the mid-1990s. The Government stopped covering it completely in the State budget as part of the reform of China's higher education system.[23] Tuition fees rose from several hundred yuan a year in the 1980s to 3500 yuan or 5000 yuan, with the highest fees reaching 8000 yuan, not counting boarding expenses.

Table 2.4 Number of student enrolments by level and type of school (millions)

Year	Institutions of Higher Education	Secondary Schools
1978	0.86	66.37
1980	1.14	56.78
1985	1.70	50.93
1990	2.06	51.05
1995	2.91	61.92
2000	5.56	85.19
2001	7.19	89.01
2002	9.03	94.15

Source: China Statistical Yearbook 2003.

Higher education is mainly run by the government in China. However, since the early 1990s, private colleges and universities have sprung up all over the country, especially in economically and culturally developed areas such as Beijing, Shanghai, and provinces like Jiangsu, Shandong and Zhejiang, contributing to the rapid growth of higher education in China.[24] Chinese formal education has achieved fast development during the past couple of decades, with the most extraordinary expansion in higher education (see Table 2.4). The yearly intake shows great variations. Enrolment in institutions of higher education actually dropped slightly in the early 1990s, and a similar reduction also took place in the high schools, in particular the senior high schools – reflecting policy changes at the time. The expansion of enrolment in higher learning institutions originated from policies in 1997 onwards, while the contraction may partly be attributed to the 'Tiananmen Square Incident' (1989) when government decreased enrolment to safeguard the stability of higher learning institutions.

Chinese institutions of higher learning are already playing an important role and will increasingly do so. Enrolment of undergraduates reached 7.19

Table 2.5 Enrolled students by field of study in regular institutions of higher learning (thousands)

	1996	1997	1998	1999	2000	2001
Postgraduate	162.3	163.3	**185.1**	218.7	283.9	371.6
Undergraduate	3021.1	3174.4	**3408.8**	4085.9	5560.9	7190.7
-Philosophy	5.1	4.9	**4.8**	4.9	5.6	5.4
-Economics	461.0	483.5	**508.4**	614.0	876.5	359.9
-Law	104.7	118.4	**136.5**	174.5	272.0	387.9
-Education	121.8	128.9	**138.7**	167.1	236.2	374.5
-Literature	384.0	412.0	**453.6**	563.7	818.6	1059.3
-History	47.5	48.8	**50.3**	55.7	62.2	53.4
-Science	315.4	332.2	**359.5**	421.0	536.8	716.3
-Engineering	1212.6	1262.7	**1354.6**	1613.3	2148.3	2491.2
-Agriculture	106.4	111.9	**119.0**	142.4	181.8	186.0
-Medicine	262.7	271.1	**283.3**	329.2	422.9	529.4

Source: http://www.sts.org.cn (28 December 2003).

Table 2.6 Graduates by field of study in regular institutions of higher learning (thousands)

	1996	1997	1998	1999	2000	2001
Postgraduates	39.7	43.2	**43.2**	50.8	54.8	63.7
Undergraduates	838.6	829.1	**829.1**	847.6	949.8	1036.3
-Philosophy	2.0	1.2	**1.2**	1.1	0.9	0.9
-Economics	127.0	133.0	**133.0**	134.3	159.3	57.3
-Law	25.9	28.3	**28.3**	31.5	44.1	61.5
-Education	40.6	39.6	**39.6**	40.3	42.1	52.6
-Literature	120.0	116.1	**116.1**	121.0	147.0	157.8
-History	16.4	14.6	**14.6**	13.4	13.7	10.2
-Science	97.3	90.5	**90.5**	90.4	98.2	115.8
-Engineering	315.0	314.4	**314.4**	326.2	354.3	349.1
-Agriculture	33.0	30.2	**30.2**	28.1	30.4	28.5
-Medicine	61.4	61.2	**61.2**	61.5	59.9	62.6

Source: http://www.sts.org.cn (28 December 2003).

million in 2001, an increase of more than 4 million from 1996. The scale of enrolment of postgraduates has also increased dramatically. Enrolment in engineering fields constitutes more than one third of the total (see Table 2.5). The second and the third largest fields are Literature and Science respectively, each of whose shares in total are less than one sixth. The number of annual graduates from various disciplines is given in Table 2.6.

With the reform of the Chinese scientific and technical system, the distribution of human resources by sector has changed substantially. In 1999, nearly half of total human resources were allocated to large- and middle-size industrial enterprises, and the remaining human resources were mainly in R&D and higher education institutions, and primarily in East China, particularly in Jiangsu, Shandong, Sichuan and Beijing.

The number of R&D personnel employed by enterprises, has increased substantially during the past ten years, reaching 412 thousand in 1999, while R&D personnel in research institutions decreased. The result shows that the reform of the Chinese S&T system was successful in formally transferring human resources from R&D institutions to enterprises. The absolute number of R&D personnel in higher learning institutions had grown only slowly. Staff employed in large- and middle-size enterprises in 2001 had increased, while the number of S&T personnel in R&D institutions had decreased. However, project-related employment had only changed slowly (see Tables 2.7 and 2.8). In 2000, the number of personnel involved in S&T activities in higher learning institutions reached 613 000 (Table 2.9).

Table 2.7 Employment in large- and middle-size enterprises (thousands)

	1995	1996	1997	1998	1999	2000	2001
Total staff	38 930	38 740	37 800	34 280	31 369	29 020	28 040
S&T personnel	1 234	1 455	1 474	1 410	1 454	1 387	1 368
-Share in total staff (%)	3.2	3.8	3.9	4.1	4.6	4.8	4.9
R&D personnel	260	428	443	383	428	543	558
-Share in S&T personnel (%)	21.1	29.4	30.1	27.2	29.4	39.1	40.8
Scientists and engineers	710	796	802	637	669	769	791
-Share in S&T personnel	57.5	54.7	54.4	45.2	46.0	55.4	57.8

Source: Management and Economic Institute of Beijing University of Science and Engineering: *Research Report on Conditions of Chinese Human Resources* (2000) – Basic Conditions', p. 30; *China Statistical Yearbook on Science and Technology* (2001, 2002).

Table 2.8 Employment of S&T personnel in R&D institutions (thousands)

	1996	1997	1998	1999	2000	2001
Total staff	1040	1020	986	870	740	620
S&T personnel	653	633	607	553	491	427
Scientists and engineers	391	381	369	342	303	276

Source: China Statistical Yearbook on Science and Technology (2001, 2002).

Table 2.9 S&T personnel of higher learning institutions (thousands)

	1995	1996	1997	1998	1999	2000
S&T personnel	598	602	600	598	607	613
S&T scientists and engineers	522	527	530	531	542	582
R&D personnel	237	242	241	238	251	228
R&D scientists and engineers	224	229	229	228	241	223

Source: http://www.sts.org.cn (17 February 2003).

Universities in China

Universities in China have a history of little more than 100 years; Beijing University, the oldest, was created in 1897. The university system was dramatically reformed and expanded in 1952 following the communist takeover a few years earlier. University education came to an almost complete standstill with the onslaught of the Cultural Revolution in 1966 and it was not until 1978 that universities again started to provide tertiary education. A subsequent rapid expansion has also included a major structural reform that reversed the specialization trend of 1952. In 2004 the number of colleges and universities exceeded 1000, of which more than several hundred also maintain graduate schools. The total number of students in colleges and universities has reached 14 million and with an expected entrance rate of 35 per cent around 2010 the enrolment will more than double to exceed 30 million.

Change and development of the higher education system, and the National 211 Project, have already been described. Here we look at a few institutions affected by the 211 Project, in order to gain a deeper understanding of the changes that have taken place. Developments at Jiaotong University in Shanghai are described, followed by Tianjin University, Zhejiang University in Hangzhou, and Tsinghua University in Beijing. The China University of

Mining and Technology (CUMT) provides a further illustration of the complex changes that have taken place within the Chinese university system.

Shanghai Jiaotong University
Jiaotong University[25] was founded in 1896 as Nanyang Public School. When the Republic of China was founded in 1911 it became the Government Institute of Technology of the Communications Ministry. A school of management was founded in 1918 and is considered to be one of the oldest university departments in China. A graduate school was founded in 1943. When the People's Republic of China was established in 1949 the university was divided into several units, with subsequent transfer to other places in China; the main part remained as Shanghai Jiaotong University. As one of the key universities in China, the university is jointly run by the Ministry of Education and Shanghai Municipal Government.

Since China's opening up policy, Jiaotong University has been among the leading universities involved in the reform of management systems in the institutions of higher learning. It has entered into a number of new disciplines such as naval architecture, ocean engineering and metal plasticity processing, and has established important national positions in areas such as large-scale integrated circuits and optical fibre technology.

Today Jiaotong University has 21 academic schools, with 60 undergraduate programs, 152 masters degree programs, 93 PhD programs, 16 postdoctorate programs, 16 State key doctorate programmes and 14 State key laboratories and National engineering centres.[26] Its total enrolment of full-time students amounts to more than 24 000, with more than 13 000 enrolled in Master's Degree and Doctor's Degree programs.

Jiaotong University has five campuses, in Xuhui, Minhang, Qibao, Shangzhong Road and Fahuazheng Road. It is the nodal point of the 'China Education Science and Research Networks' in the East China region, and through computer networks, Jiaotong University has faster and closer connection with universities, scientific research institutions and corporations both at home and abroad.

Nankai University
Nankai University[27] in Tianjin was founded in 1919 on a very modest scale with less than 100 students in three departments – liberal arts, science and business. By 1937 Nankai had expanded into a university of three colleges, 13 departments, and two research institutes – with some 400 students and more than 100 faculty and staff members. Nankai Institute of Economics was established in 1927 and soon recognized as a national economic research focus in China. After 1949 Nankai became a comprehensive university with an emphasis on arts and science. It is now one of the leading universities in

China and its programmes in mathematics, chemistry, history, and economics are seen as the best in China. Total enrolment has reached more than 10 000 undergraduate students and 6000 graduate students in 18 colleges and schools.

Zhejiang University

The present Zhejiang University was created in 1928 from the former National Chekiang University. It is included in the list of universities in the 211 Project and is expected to rank among leading universities in the world by 2017. In 2003 there were 41 000 full-time students, including 30 000 undergraduates, 8000 master degree students and 3200 PhD candidates. There are also nearly 38 000 students enrolled in vocational education courses.

Tsinghua University

Tsinghua University has its origin in Tsinghua Xuetang, a preparatory school for students who were to be sent by the government to study in universities in the United States. This was funded with part of the 'Boxer Indemnity'. The school began its first term on 29 April 29 1911 and was renamed Tsinghua School after the 1911 Revolution. It is located in the north-western suburbs of Beijing on the site of Qing Hua Yuan, a former royal garden of the Qing Dynasty.

University undergraduate students were first enrolled in 1925; the name National Tsinghua University was adopted in 1928 and in 1929 the Research Institute for graduate students was set up.[28] In 1952, the university was moulded into a polytechnic institution focusing on engineering, to become the national centre for training engineers and scientists. Since 1978 Tsinghua University has re-established departments in sciences, economics, management and the humanities. The university consists of 11 schools containing 44 departments, with its graduate school taking on a very important role. A medical school is currently being established. In 1999, Tsinghua opened the School of Arts and Design by merging with the Central Academy of Arts and Design. It has now become a leading comprehensive university with engineering as its focus, and has over 20 000 students, including 12 000 undergraduates, 6200 master's degrees candidates and 2800 doctoral candidates.[29] In 1984, Tsinghua became the first university in China to establish a graduate school.

China University of Mining and Technology (CUMT)

CUMT is a leading multidisciplinary polytechnic university with an emphasis on mining. CUMT-Beijing, which is an important part of CUMT, has a focus on engineering and also includes other disciplines such as liberal arts and management. It was approved at an early stage as one of the universities for the national 211 Project. The predecessor of CUMT was Jiaozuo School of Railroad and Mines founded in 1909. It was formally named the China University of Mining and Technology (CUMT) in 1988. CUMT was officially approved as one of the key universities in China by the government in 1960 and again in 1978. In February 2000 CUMT was transferred to the administration of the China Ministry of Education (CME) and became one of the state universities directly supervised by CME. In June 2000 the Graduate School of CUMT was officially approved by the China Ministry of Education.

China University of Mining and Technology (Beijing), or CUMT-Beijing, was established from the Beijing Graduate School of CUMT. In 1978, with the approval of the State Council, CUMT Beijing Graduate School started to enrol postgraduate students and doctoral candidates. In 1997, with the approval of the China Education Commission (now the China Ministry of Education) and the Beijing Municipal Government, CUMT-Beijing was developed into a multidisciplinary polytechnic university in Beijing.

CUMT-Beijing consists of nine academic schools and one graduate school and is responsible for seven national key disciplines. It has one key laboratory under the China Ministry of Education, one key laboratory under the Beijing Municipal Government, and more than 20 laboratories in various disciplines.[30] CUMT-Beijing has established a multi-level education pattern emphasizing simultaneously the development of graduate, undergraduate and adult education. It enrolment has reached some 7000, including 2400 graduate students and 500 students receiving adult education; it now has a total of 665 faculty and staff members, including 300 full-time teachers. Among the academic staff, there are seven academicians of the Chinese Academy of Sciences or the Chinese Academy of Engineering.[31] In 2002, CUMT-Beijing established cooperative education programmes with both the University of Nottingham and the University of Leeds by sending excellent undergraduates to these two universities to pursue masters and doctoral degrees.

Zhejiang Wanli University
Zhejiang Wanli University (ZWU) provides an innovative approach. ZWU was created by restructuring a former provincial institution of higher learning – the Zhejiang Junior College of Training Teachers for Agricultural Technology – which was established in the early 1950s. ZWU is supervised by the Zhejiang Education Department and is operated by the Zhejiang Wanli Education Group (ZWEG). It has two campuses – Huilong and Qianhu. The Junior College of the university is located in the Huilong campus, 'where the management of freshmen is paramilitarised'.[32]

ZWU also has a College of Commerce, and includes departments such as foreign languages, law, culture and media, artistic designing, computer, electronic information engineering and life sciences with altogether 38 specialities in junior college programmes and eight specialities in undergraduate programmes. The number of full-time students was more than 8000 in 2003 and an Adult Education Programme has more than 1000 students. ZWU has seven laboratory centres and three municipal key laboratories – with installed laboratory equipment to the value of RMB45 million. It gives much attention to research and national science endeavour as well as economics and management issues. It has three research institutes – the Institute of Culture, the Institute of Biological Technology and the Ningbo Industrial and Economic Development Research Centre.

The university encourages its staff to look for new models in China's economic development. ZWU is so far the only market-economy university in China. It is a private university but the land and buildings belong to the local government in Ningbo. It follows the university regulations put forward by the Ministry of Education. However, ZWU receives no support for its teaching activities, which have to be covered from much higher tuition fees than those levied at nationally-supported universities. The student fees at Fudan University in Shanghai are around RMB4000 per year while ZWU charges their students RMB16 000.

ZWU started with 1200 students in a three-year programme and has now reached total enrolment of 16 000 of which 10 000 are studying for four-year university degrees while the rest are studying in three-year diploma programs. The planned intake is more than 4000 students in four-year programmes and another 1000 in diploma programs. Total enrolment will stabilize around 20 000.

Creating a new style university like ZWU has its origins in the reform policy of the Ministry of Education, which has encouraged experimental models as society desired more higher education; it has also responded to a growing demand in Eastern China. However, private universities have in the past only offered diploma training. ZWU is in the forefront of private universities in offering a full range of four-year programmes for university degrees.

ZWU is actively pursuing funds for its laboratories and research activities and is exploring ways of collaborating with firms. An important objective is to train students and staff. Zhejiang Wanli University is involved in a major project to establish the Nottingham-Ningbo University, for which premises have been built on a new campus, presently under construction. The Chinese Ministry of Education approved it in March 2004 as China's first Sino-foreign university.[33] The new university will be founded jointly by the University of Nottingham of Britain and the Zhejiang Wanli University in Ningbo with an investment of RMB600 million.

The Nottingham-Ningbo University will have about 4000 students who will study business and social sciences. The Vice-Chancellor of the new university is professor Yang Fujia, currently dean of the University of Nottingham and academician of the Chinese Academy of Sciences. Doctor Ian Gow, current deputy president of the University of Nottingham and dean of its business school, will be the managing deputy president of Ningbo-Nottingham University. Professor Yang has been a guest of Wanli University and played an important role in the creation of the new university, after Nottingham failed to establish desired partnerships with other universities in China, having had discussions with Fudan and Tong universities in Shanghai and with Tsinghua University in Beijing.

The new university is supported by both Ningbo City and the Zhejiang province government. The Ministry of Education will assess the results after three years and promote a number of similar universities if it is found to be successful. The annual tuition fee will be RMB50 000, which will cover the cost of teaching; this will be carried out by staff from Nottingham University, who will initially have full control of teaching. Students will spend an extra initial year to acquaint themselves with English and the international character of the university. The final degree will be a BA according to UK requirements.

The Beijing Geely University
Another interesting university development originates in Ningbo although located in Beijing. Geely University is mostly known for its investor – the Geely Group. In 1994, the Group became aware of the need to train its own workers, and decided to invest RMB50 million in a college. A deal to do the training through an existing college in Hangzhou came to naught. Its application to set up a school in another Zhejiang city was rebuffed when the local authority liked its idea so much that they decided to open their own school. Geely University was finally opened in Beijing in 2000. The first year it enrolled 3000–4000 students, and the second year saw a big jump to 7000–8000.

Beijing Geely University (BGU), approved by Beijing Municipal Government and registered by the Ministry of Education of China in 2001, is an international and new type of higher education institute and it is qualified to issue graduation certification and diplomas. It is located in Beijing Zhongguancun Chang Ping Science and Technology Park, close to the Olympic Village. Meeting the requirements for social economic development, BGU has established 14 colleges and schools, including a Business School, Information and Technology College, Journalism and Communication College, and Finance and Security College.

The brief university profiles above clearly indicate a number of important factors that have shaped the landscape of higher education in China. First, universities in China have a short tradition, with the first one having been established in 1897. Second, all universities that were created prior to 1949 suffered from the turmoil of the revolution in 1911 followed by civil war, the Sino-Japanese War (1937–1945), and again civil war. Third, existing and new universities were shaped into a planned-economy model of Soviet origin in 1952. Fourth, the Cultural Revolution and the university reform of 1978 again completely reshaped the structure of higher education.

The situation of the past decades and projections into the future suggest difficulties and also great prospects for higher education. A commitment to higher education, combined with a demographic window of opportunity gives China the opportunity to bring higher education to a very large share of its young population. The universities are receiving substantial resources and are seen as critical assets for national modernization. However, the exploding enrolment will create heavy pressures on teaching and administrative staff not only to expand in numbers but also to improve quality. Also relevant is the necessity of maintaining the best teachers and making domestic universities a first choice for Chinese students, and increasingly drawing on staff from the overseas talent pool.

FUTURE BRAIN GAIN

From the beginning of its decline in the 19th century, China has had a long tradition of having its best students and professionals go overseas. Many of them excelled, and notable personalities include Nobel Laureates Professor C.N. Yang and Professor C.T. Lee. Surprisingly it may be noted that it was Deng Xiaoping, having spent his early years in France, who created the Open Door Policy, which allowed tens of thousands of Chinese students to depart for overseas education and training. It is occasionally argued that this single action perhaps ensured that China's doors would never be closed again as its young minds were exposed to new and modern technical know-how and management techniques.[34]

Modernization in China can draw on two important human resources outside its territory. One is the entrepreneurial business-oriented Chinese community in South-East Asia and in other parts of the world – the original Chinese diasporas. These people of Chinese ancestry have played an important role in China's recent modernization by bringing capital, development projects and ideas to China. They have, together with entrepreneurs and businessmen in Hong Kong and Taiwan, made great contributions to this first stage of China's real Great Leap Forward.

China is now drawing on its second diaspora – the large numbers of university graduates and PhDs who were sent or went abroad for advanced studies after 1978. A majority of them, primarily now living in the US, have remained abroad and constitute an important brain-pool for China's next development stage, which is to become a knowledge-based economy. The number of Chinese graduates now living in the US is estimated to be in the region of 200 000–300 000. Among them are ten to fifteen thousand who are scientists and engineers of world-class standard. Another fifty thousand are highly-qualified professionals and generally better than their colleagues who have graduated from domestic universities.

When the Cultural Revolution came to an end the university system was in total disarray and there was a dire need for trained and qualified personnel throughout the country. Thus it was natural to look for advanced training possibilities at foreign universities, and the most attractive were found in the US. This started a stream that developed into a flood of young and talented people who left China for overseas studies. Some were sent by government agencies and others went on their own, with families often providing financial support. However, a majority in both categories decided not to return to China, which has been perceived as a serious brain drain (see Table 2.10). It is reported that some 700 000 Chinese have left China since the late 1970s for studies or research and that most of them have been reluctant to return and directly contribute to their home country's development. Available statistics indicate that only about one quarter of them have returned and Professor Cao Cong suggests that the situation is aggravated by the fact that the best and brightest are most likely to remain overseas.[35]

A number of these scientists, engineers and other professionals have been attracted to work with institutions and colleagues in China. They constitute a broadening intellectual bridge between the scientifically and technologically most advanced country, and the economically and industrially most rapidly developing country in the world. They continue to pursue their own professional and career ambitions while contributing to China's expansion in advanced science and technology. They have at the same time given the US government and its security agencies a headache for which no remedy may exist.

The Chinese Government decided in the early 2000s to create a more favourable climate to attract more overseas Chinese scholars to return home, to run laboratories, technological firms or scientific parks. The Ministry of Education started to implement a series of talent-oriented plans, such as the Hong Kong-based Cheung Kong Scholars Programme[36] and the Cross-Century Talent Programme. These are just two of various programmes that are also supported by other agencies such as the Chinese Academy of Sciences, The National Natural Science Foundation and the Ministry of Personnel.[37]

Furthermore, universities are simultaneously encouraged to establish science parks or engineering research centres to speed up the commercialization of new technologies. To encourage college teachers to excel at academic

Table 2.10 Number of postgraduates and students studying abroad

Year	Number of Students Studying Abroad	Number of Returned Students
1978	860	248
1980	2 124	162
1985	4 888	1 424
1986	4 676	1 388
1987	4 703	1 605
1988	3 786	3 000
1989	3 329	1 753
1990	2 950	1 593
1991	2 900	2 069
1992	6 540	3 611
1993	10 742	5 128
1994	19 071	4 230
1995	20 381	5 750
1996	20 905	6 570
1997	22 410	7 130
1998	17 622	7 379
1999	23 749	7 748
2000	38 989	9 121
2001	83 973	12 243
2002	125 179	17 945

Source: China Statistical Yearbook 2003.

programmes, the Ministry of Education proposed that major universities should invite 500 to 1000 overseas or domestic professors within the next five years, with a primary obligation to carry out research. The Ministry had noted with concern that that researchers and university teachers were quitting their jobs because of low incomes and moving into business to make more money or going overseas. Under a new programme each professor will get RMB100 000 annually in subsidies to support research programmes. Those who make breakthroughs will be awarded prizes ranging from RMB500 000 to one million each year. This comes in addition to substantially increasing the incomes of teachers and researchers.[38]

A most striking example of contributions made by an overseas Chinese is the case of the physicist Dr Qian Xuesen, who returned from the US in 1955 and made outstanding contributions to the development of China's aerospace industry. He played a leading role in the research, manufacture and testing of carrier rockets, guided missiles and satellites. A similar role could be played by Dr Steve Chen, who used to be the chief architect of Cray supercomputers in the US. Since late 2004 he has been CEO of Galactic Computing in Shenzhen and sees the possibility of China leap-frogging into blade supercomputing technology, and making its presence felt in global markets.[39]

NOTES

1. This chapter incorporates two contributions from Dr Yongzhong Wang prepared for the project 'Emergence of New Knowledge Systems in China and their Global Interaction': *Government Programmes and Human Resources in Science and Technology*.
2. The '211 Educational Project' focuses on transforming approximately 100 Chinese colleges and universities into research bases for the country's economic, technological and social problems. It will also reinforce support for new academic fields and for cross-field research. As part of the project, an information service system combining the Internet, data banks and digitized libraries of Chinese colleges and universities will be established in order to make more efficient use of the existing infrastructure and software of each school. The project, initiated in 1996, was China's largest educational project. Since 1996, China has put RMB18 billion into the project, which has greatly improved the teaching and research conditions of some universities. Source: 'China to put six billion yuan into higher education projects', *China Daily*, 17 September 2002.
3. 'China channels funds into 32 RD projects', *China Daily*, 15 October 2004.
4. 'Nation witnesses progress in many fields', *China Daily*, 27 February 2004.
5. Freeman (1987).
6. Qian Xuesen, 'Science and technology must catch up with and surpass advanced world levels before the end of century', *Red Flag*; BBC FE/5563/B11/6; referred in Naughton (1999), pp. 70–71.
7. The 21st-Century Agenda was approved by the Environment and Development Convention of the United Nations in Brazil in June 1992, which represents the new idea of sustainable development of human society, and worldwide common understanding on the issues of sustainable development.
8. In this period, the revenue of the programme attained 430 billion yuan; and the direct economic revenue was 95 billion yuan and the exports were 2.35 billion dollars.
9. Cui Ning (2004).
10. Ministry of Science and Technology (MOST), 'Spark Programme – background and purposes', available at http://www.most.gov.cn.
11. Liu Baijia (2003).
12. Ibid. Mr Liang Sheng, chief of the technological innovation and development division of the Beijing Municipal Economic and Trade Commission and Vice-Chairman of the Beijing Semiconductor Industry Association stated that 'the focus of the development of the semiconductor industry for us this year (2003) is to foster market demand'.
13. Ibid.
14. 'China entering global CPU market', *PC World Malta News* (IDG News Service), 17 February 2004, http://www.pcworldmalta.com/news/2004/Feb/171.htm.
15. Ibid.
16. Richmond (2003).
17. Clendenin (2003).
18. MOST Brochure 2001.
19. 'New Chip studies make progress', *China Daily*, 4 September 2003.
20. Chen Hong (2004).
21. Mr Deng Xiaoping said 'Science and technology is the first production power' twice in 1988. The first time was on 5 September, when he said, 'Marx had said that science and

22. technology is production power, of course, this view is very right. However, this view is not perfectly right under current circumstances, science and technology is possibly the first production power'. The second time was on 12 September, he said perfectly clearly, 'In my view, science and technology is the first production power'. These sayings are cited from Deng Xiaoping (1993), pp. 274–6.
22. Zeng Xiangquan, head of the School of Labour and Personnel at People's University and China's prominent specialist in labour economics, reported in 'Expert: university graduates need to drop job expectations', *Xinhua*, 11 December 2003 (*China Daily*).
23. 'Poor students struggle to fund higher education', *China Daily HK Edition*, 26 August 2003.
24. So far, there are over 200 such institutions in the country, taking in about 200 000 students every year. Song (2003).
25. Of all the academicians of China's Academy of Sciences and Academy of Engineering, more than 200 are the alumni of (Shanghai) Jiaotong University. Jiang Zemin is among its famous alumni.
26. Their staff includes 22 academicians of the Chinese Academy of Sciences and the Chinese Academy of Engineering, as well as 23 'Changjiang Chair Professors'.
27. The founders were the renowned patriotic educationist Zhang Boling (1876–1951) and Yan Fansun (1860–1920). Nankai University is the alma mater of the late Premier Zhou Enlai.
28. The faculty greatly valued the interaction between the Chinese and Western cultures, the sciences and humanities, the ancient and modern. Tsinghua scholars Wang Guowei, Liang Qichao, Chen Yinque and Zhao Yuanren, renowned as the 'Four Tutors' in the Institute of Chinese Classics, advocated this belief and had a profound impact on Tsinghua's later development.
29. The university has over 7100 faculty and staff, with over 900 full professors and 1200 associate professors, including 24 members of the Chinese Academy of Sciences and 24 members of the Chinese Academy of Engineering.
30. During the period of the Ninth Five-Year-Plan, CUMT-Beijing undertook 44 National Key Projects, including 6 National Hi-Tech 863 Projects, 10 National 973 Projects and 4 National Key Projects for the Tenth Five Year Plan.
31. Over 400 research projects are now being undertaken by CUMT-Beijing, including National Key Projects for the Tenth Five-Year-Plan, National Hi-Tech 863 Projects, 973 Projects and projects for the Innovative Scientific Research Group of the National Natural Science Foundation of China; also 35 projects for the National Natural Science Foundation, and over 70 scientific research projects of provincial or ministerial level. A number of national key projects, such as technology, coal water fuel preparation technology, silicon carbide whisker technology and industrial coal briquetting technology, have been transferred into productivity, achieving favourable economic and social profits.
32. Zhejiang Wanli University brochure, 2003, states that the Huilong campus attempts to 'paramilitarize administration to students by strictly regulating their behaviour while stimulating their study potentials'.
33. http://english.people.com.cn/200403/25/eng20040325_138422.shtml.
34. Yip, Vincent, 'Prodigal Sons', manuscript.
35. Cao (2004a).
36. This was jointly launched in August 1998 by the Ministry of Education and Hong Kong entrepreneur Li Ka-shing and his Cheung Kong Infrastructure Holdings Ltd. Source: 'Profound thinking on stopping brain drain', *China Daily*, 23 February 2002.
37. See for example Cao (2004a, 2004b).
38. Ibid.
39. Zhu Boru (2004).

3. Technology access through FDI and technology transfer[1]

China is eager to make the most of new technology to modernize the country rapidly and lay a strong foundation for its sustained future industrialization. Simultaneously, the country must provide employment for its huge and still expanding labour force and improve its living conditions. To this end China has again, since its Open Door Policy, embarked on a twofold strategy. On the one hand its has opened its markets and industries to foreign direct investment (FDI), initially to boost exports, although at the same time laying the foundation for industrial and technological upgrading that will facilitate China's technological catch-up with the advanced countries. On the other hand the leadership has systematically promoted FDI in labour-intensive processing and assembly (P&A) industries that have the potential of absorbing significant portions of unemployed or underemployed people in rural areas. One can find apparent similarities between this strategy and the Great Leap Forward in 1958. However, there are also very important differences which have made it possible to reap great benefits continuously from the industrialization strategy that was initiated in the early 1980s.

EVOLUTION OF INDUSTRIAL POLICY IN CHINA

China at the time of the creation of the People's Republic in 1949 was almost desolate of industries outside a few selected places, of which Shanghai was one of the most important. With assistance from the Soviet Union a substantial number of heavy industry kombinats[2] were created primarily in the Northeast provinces, using large coal and mineral resources which had earlier been developed under Japanese occupation. The USSR provided substantial on-the-spot assistance and trained a large number of engineers in its universities. This large-scale industrial development assistance came to an end in the late 1950s when political views seriously diverged. This was also the time when China attempted a forced industrialization almost completely based on domestic resources and indigenous technologies – The Great Leap Forward. These efforts led to an economic catastrophe, particularly in the agricultural sector, where farmers had been forced to leave their fields to engage in industrial pursuits.

A subsequent recovery was disrupted by the Cultural Revolution in 1966. After that a major relocation and expansion of industrial activities was carried out, to interior mountainous regions which would shield them if China came under military attack. Many of the new industrial locations lacked good infrastructure and were far away from China's major economic centres. Although self-reliance remained an honoured slogan, China continued to buy major industrial equipment, such as large fertilizer plants, although it was not able to update the imported technology continuously. Major industries were under complete state control, although rural industrial and small-scale processing plants flourished during this period, thus providing a broad diffusion of manufacturing technologies.

A self-imposed autarky stifled technological and industrial development, although China was able to maintain a certain closeness to the industrial landscape in advanced countries at the prototype level. However, China's industrial structure was in a miserable shape at the time when the Open Door Policy was announced in 1978. In principle, although only gradually implemented, China took the bold decision to enter the global economy in a major way, for which a final seal of approval was awarded in 2001 when China became a member of The World Trade Organization.

In the meantime China first attracted a stream of industrial foreign direct investment directed to special economic zones – initially in specific sectors and often organized under a joint venture formula in collaboration with state-owned enterprises. The stream of industrial FDI had already expanded into a flood by the late 1980s and in particular since the mid-1990s, although mainly targeting China's coastal zones. The joint venture requirement was gradually relaxed and wholly-owned subsidiaries have now become prevalent. Most industrial FDI has been directed to special zones in which China has created a favourable environment for meeting these specific needs. Industrial FDI was given various incentives such as preferred access to land at low cost, tax holidays and easy access to bank credits.

While encouraging industrial FDI, China's traditional industrial structure has undergone an almost complete transformation. The state sector of China's economy had been reduced from some 80 per cent to less than 40 per cent of GDP by 2003. Most of the earlier state-owned companies have been incorporated and listed on the stock exchanges in Shanghai, Shenzhen, Hong Kong and occasionally New York. Manager buy-out schemes have happened in a number of companies, although China's share of privately owned companies is still limited.

The earlier 'kombinat' character of state companies has gradually been abolished and a prevalent feature has been the merger of closely or not so closely related companies into industrial groups – a development that was partly inspired by the earlier successful *chaebols* in Korea. A number of companies and industrial groups – altogether 189 – are directly controlled by the State-owned Assets Supervision and Administration Commission (SASAC). Most of these enterprises are considered to be pillar industries of the Chinese economy and will remain under state influence for a number of years, and some could possibly emerge as global companies.

China's industrial development is concentrated in three major coastal regions. The first is the Pearl River Delta (PRD), where the first special economic zones of Shenzhen and Guangzhou (Canton) have become prominent industrial areas. The second is the Yangtze River Delta (YRD), with Shanghai and nearby cities such as Ningbo, Hangzhou, Wuxi and Suzhou. The third region is the BoHai Rim Region (BHR), which includes Qingdao, Dalian, Tianjin and Beijing further inland. These three regions are competing for FDI and national support programmes, a competition that takes place at provincial and municipal levels. Thus the economy of China could be regarded as a multitude of economies that are loosely coupled together.

Until recently, significant areas and their population in the Northeast and Western China benefited only marginally from China's rapid economic development. Policies are now being formulated to redress this imbalance. However, an important feature of industrial development in China is the artificially controlled pattern of urbanization. Thus, FDI and domestic enterprises have a privileged access to low-cost migrant labour that comes from less developed regions and provinces. They are generally not allowed to settle in their new workplaces and stay only for a limited period of time, and are willing to work for considerably lower wages than local workers. An ample supply of industrial labour at low costs is still an important feature of China's comparative advantage for many industrial products.

China's Second Great Leap Forward[3]

In 1958, China under the leadership of its at the time unquestioned ruler, Mao Zedong, mobilized its masses, predominantly in rural areas, to bring the country quickly into a new era of industrialization. Hundreds of millions would build small plants all over the country and start manufacturing all kinds of products, very simple as well as more sophisticated ones. Production of steel, with the stated goal of overtaking Great Britain within a few years, was the mainstay of these hasty efforts, that have become known to the world as the Great Leap Forward. It failed totally and created misery that lasted for several years.

Twenty years later China embarked on a second Great Leap Forward, although China's ongoing extremely rapid industrialization has not yet been recognized under this label. There are distinct similarities between the first and second GLFs, as they are both based on political approaches to mobilizing China's huge manpower resources for rapid industrialization. During the first event collectivized citizens, almost completely taken from the agricultural sector, were mobilized to replace machinery and advanced equipment with their manual labour to industrialize the country. Today, China is again mobilizing its manpower to substitute machinery and advanced equipment – this time to offer the industrialized world wide access to low-cost processing and assembly production for almost any product.

Dongguan, an industrial city located halfway between Hong Kong and Guangzhou, gives an illuminating insight into the way China's rural masses are mobilized in the Second Great Leap Forward. Four million of the city's some six million are only temporary residents. They come from the interior and stay only for a limited number of years, while sending home their wages and returning to get married. They live in huge dormitories that are owned and managed by local factories. This arrangement is multiplied all over China. It started in the South and has spread northwards along the coastline and gradually into the interior. Several hundred million Chinese are today engaged in labour-intensive industrial production – making products that are sold in global markets.

There are three significant similarities between the first and second Great Leaps Forward. First, both have their basis in mass mobilization, although today's is less controlled. Second, both are aiming for a great variety of industrial products. Third, they have their political basis in a twofold strategy which combines primitive or traditional industrial technology at one end of the scale with advanced technology and research at the other end.

There also exist three equally important differences between today's industrialization efforts and those of the first Great Leap Forward. First, the earlier approach was based on a fundamental ideology of equality throughout the society, while China today strongly supports the notion of individuals being successful and becoming rich. Second, China at the time of the first GLF lacked management skills to carry out the gigantic task of organizing hundreds of millions who at the time were completely unaware of industrial production. Today, China has reached a high level of management in many sectors and can also draw on resources and information from any part of the world. Third, China in the earlier period was seriously lacking in capital, with little patience to wait for its accumulation. Today China is at the receiving end of a torrent of foreign direct investment while at the same time benefiting from high domestic savings, and accumulating huge reserves of foreign currency.

Aside from the similarities and differences, China's first Great Leap Forward suffered from a fatal error. Mao Zedong, and in all likelihood many of his supporters, assumed that ample manpower resources could be withdrawn from the agricultural sector once the collectivization into communes had been formally accomplished in rural areas. This created an immediate economic catastrophe and political instability that later on resulted in the turmoil of the Cultural Revolution. Furthermore, during the first Great Leap Forward, China was still directed by a planned economy belief that gave only limited scope for trade and interaction with the rest of the world.

Mass mobilization of rural manpower was a cornerstone of the first Great Leap Forward, as it still is in the second. An overriding distinction is that the first one involved a self-contained autarchic approach while the second one is undertaken in an open global economy at a time when rural manpower is amply available for non-agricultural tasks. China's second Great Leap Forward has provided China, through foreign direct investment, with a massive infusion not only of capital as such but also direct access to technology and management skills that were so scarce in China's first attempt to leap forward.

In China's present Great Leap Forward the process of industrialization and technology is losing the traditional characteristics of industrial production. Companies in China are presently removing machinery and equipment in some processes and replacing them by manual skills, and handling more work on the plant floor rather than on automatic transfer lines. What is being done is that companies operating in China today are introducing what has previously been referred to as appropriate or intermediate technology. This offers more efficient combinations of technology and labour than would have been the case for processing and assembly operations that are completely based on production concepts prevalent outside China.

EXPLOITING FOREIGN TECHNOLOGY

The People's Republic of China was able to build up a huge research and development system employing a large number of scientists and engineers during its first 30 years of existence. Many of them initially received training and education in the USSR or its associated countries. At times the system served China well and the country was able to establish a credible military capability, primarily based on its technological development. A few civilian sectors worked well although the turmoil of the Cultural Revolution and the almost complete isolation from the global system of rapid technological development and advanced scientific discovery made China's domestic R&D obsolete in many ways.

First, many research institutes, embedded in vertical structures, were out-of-date by the late 1970s as they were overstaffed with over-aged people because too little fresh recruitment had taken place. Second, many industrial research institutes were running inefficient operations as equipment had become outmoded and research practices out-of-date. More seriously, the planned-economy approach, still remaining in place, did not offer incentives and mechanisms for effective direct contacts between clients and consumers of S&T results and their suppliers within the R&D system. This set the stage for a two-pronged policy approach – a more-or-less complete restructuring of the existing national innovation system and an initially heavy reliance on technology transfer from abroad. The decision to promote technology transfer into China has four distinct elements, some of which are closely related, while operating within different time frames.

First and temporarily foremost is foreign direct investment, for which the central government and local governments have provided an attractive environment through a number of policies and instruments. FDI brings capital, management skills and technologies. The latter two are important not only in processing and assembly industries where most of industrial FDI is to be found, but also across a wide industrial spectrum. The real challenge for China, its provinces and industrial cities with their industrial clusters and industrial enterprises, is to create a local environment that will facilitate a rapid diffusion of technologies and capture what the economists refer to as the spillover effects of FDI.

Second, China has seen a substantial inflow of FDI in R&D activities in the early 2000s, which fall into two separate categories. One is design and product development activities which adjust the production process to Chinese conditions, and products to meet customer preferences in China. Furthermore, since 2000, China has seen an influx of investment in design centres which exploit the availability of low-cost highly skilled labour (brainpower) to tailor certain designs to global market requirements. The second activity, related to low-cost brainpower, is an increasing number of research relations that have been established between foreign companies and a number of well-recognized research universities and R&D institutions in China. These relations also include contracts with various government research institutes including those of the Chinese Academy of Sciences. Classified as applied basic research, they are benefiting China in opening up new channels of communication and access to advanced research methods – although not necessarily seen in this perspective by the foreign investor.

Third, access to new or necessary technology can be obtained through mergers and acquisitions outside China. Recent events indicate that a number of Chinese high-tech companies, supported by the local governments, are increasingly looking for companies in Korea and Japan, and also in the US and Europe, that offer them the opportunity to leap forward. Information on already completed deals indicates a strong interest within the IT sector in acquiring enterprises completely or entering into joint ventures. Such deals are generally being carried out with ailing companies that want to rid themselves of divisions that hinder their restructuring and recovery. To a Chinese firm, such acquisitions offer access to advanced technologies and often marketing channels as well. However, given the limited experience of Chinese firms, the deals pose a number of challenges in terms of handling foreign labour issues and complex legal matters.

Fourth, a number of Chinese high-tech companies have already established listening posts in Silicon Valley and other high-tech locations to collect intelligence on technological and advanced consumer trends. A small number of them have already established overseas R&D centres to support not only their marketing abroad but also the continued development of their product portfolios in their home bases in China.

These FDI streams – two of them into China and another two out of China – are integrated in several ways with the National Innovation System (NIS), where the different parts are funded by the state in various ways. The interaction between China's NIS and the four FDI streams will bring about the participation of enterprises, research institutes and their staff into a global innovation system.

Inward Foreign Direct Investment

Foreign direct investment (FDI) is now regarded as the most significant source of technology inflow into China. During 2002–03 it hovered around an annual amount of US$50 billion and remained at this level during 2004.[4] At the same time, foreign multinational corporations, as the purveyors of FDI, play an important role in promoting the transfer of advanced technology through three channels, namely technology diffusion, intensified competition and personnel intercommunication. Because of the obviously positive effects of FDI on developing countries, especially the inflow of high technology, many governments adjust their policies to attract more FDI, as did China.

The Chinese history of attracting FDI started in the early 1950s, when the Communist Party established the People's Republic of China. The Chinese government began to adopt policies intended to encourage investments from other countries. Academically, there are different approaches to analysing the historical phases of FDI in China, such as the scale of FDI, the sources of FDI, and the regional distribution of FDI.

The chronology of FDI in China shows four distinctive phases. The first phase lasted from the establishment of the new China to the implementation of open policy reform, from 1949 to 1978. During this period, the most important task for China was to achieve modern productivity and to ensure a basic standard of living for its people. These basic tasks were to 'dress warmly and feed decently' and to set up an industrial system which would be appropriate for a country as large as China. Under such circumstances, the introduction of advanced technology, production experience, even social systems and ways of policy making from other developed countries, was considered the best solution for China. However, China's ability to get access to advanced industrial equipment was almost immediately influenced by the Korean War that broke out in the summer of 1950. This limited the main sources of foreign investment in China to the former Soviet Union for a number of years, until the political fallout with the USSR in the late 1950s.

A second period started on the heels of the failed Great Leap Forward in 1959 and lasted until the major economic reform policies were initiated in 1978. The results were unsatisfactory and far from sufficient, especially during 1963–66, and were to remain precarious during the Cultural Revolution period.

The third phase lasted from 1979 to 1992. Before 1978, imported technology was mainly obtained through the acquisition of advanced equipment. In the 1980s, most of the new technology introduced also included technical know-how, technical licences, technical consultancy and collaboration in production and design. The improvement of imported technological structure was regarded as a result of both the reform policy implemented by the Chinese government, and as a consequence of domestic achievements during the first and second phases.

It is widely agreed that 1993 was a turning point for a new, fourth phase in China's FDI policy. Only from this year did multinational corporations (MNCs) begin to invest in China, and many of them were among the world's largest corporations. The advanced technology and experiences they brought to China had a profound effect on both local enterprises and different industries. A new element of FDI was introduced in 1997 with investments in foreign research and development (R&D) in China. This was the beginning of a new stage in attracting FDI which introduced a qualitative improvement in types of technology imports. The first foreign R&D institute in China, the R&D Centre of Northern Telecom (now Nortel) at Beijing University of Posts and Telecommunications, came into existence in 1994, but the real quantitative increase in R&D investments in China only occurred in 1997. Many MNCs chose to establish their own independent R&D institutes in China that year, among them IBM, SUN, Ericsson, Dupont, Unilever, Rhone-Poulenc and Bayer, although most of them on a modest scale.

Geographically ten countries or regions were still the main source of FDI into China. According to actual value, up to the end of 2001, these were: Hong Kong, USA, Japan, Taiwan, Singapore, Virgin Islands, Korea, UK, Germany and France. The Eastern region is the main recipient of FDI from other countries and regions.

The pattern of foreign direct investment in China underwent a significant change in the early 1990s and MNCs have since come to play an increasingly important role. In many sectors they have out-competed their domestic rivals and captured significant shares in a small number of key industries. Large amounts of foreign direct investment flowed into China before the country's entry into the WTO in late 2001 and the amount has surged since then. A heavy inflow of capital has been directed to high-technology sectors, with the result that foreign-owned companies today completely dominate China's high-technology exports.

There are three main issues that have to be addressed. First, what are the benefits and opportunities of using FDI to promote economic and industrial development and what are the associated risks and threats? Second, how well do international markets function to supply FDI into the development process? Third, will international markets offer appropriate FDI and how should it be directed?[5] However, government policy measures to influence the inflow of FDI are strongly opposed by some economists. For example, Moran argues strongly for a new policy agenda which has the following components: 1) integrating world-scale manufacturing subsidiaries into the global-regional sourcing network, and policies for reinforcing the longer-term stability of investment; 2) no local content rules; 3) no joint venture requirements; 4) no insistence on technology licensing agreements in place of FDI. On joint ventures Moran argues strongly and provides the following argument.

> Compared to foreign firms with no constraints on ownership, those with constraints exhibit older technology and business practices and lag in introducing upgrades in technology and business practices. Constrained firms are less likely to export, and their backward linkages into the local economy are less sophisticated and dynamic.[6]

Development stages in outsourcing

Outsourcing has become the prevailing mode for manufacturing by MNCs. Boston Consulting Group (BCG) has identified five different strategies by which companies approach the outsourcing challenge, with an increasing influence from supplier countries. First, companies recognize at an early stage that sourcing is important for some basic commodities and are gaining incremental benefits from a selective use of subcontractors in low-cost countries. Second, as a next stage many companies have taken the approach of buying a large range of components and even complete products from low-cost countries. Third, a few firms have reached the stage of having developed a comprehensive sourcing methodology, often by integrating themselves into the local economy, and Motorola in China is seen as such an example.[7]

Fourth, a next stage would be an integrated sourcing strategy where the activities in a country targeted for outsourcing serve global markets as well as the domestic market. In the past the large car manufacturers only saw China as a location for making cars that were sold in the domestic market. The rapidly expanding car industry in China clearly indicates that the same companies now view China not only as an important market but also as a major source for car components and complete cars to be sold worldwide. The mobile phone industry in China gives another example of global companies having adopted an integrated sourcing strategy.

Fifth, a few companies have moved one step further by using an approach that BCG refers to as 'capturing global advantage',[8] and provides the example of Toyota, which outsources whole vehicle sub-assemblies from many Asian countries, allowing them to keep costs low and achieve just-in-time delivery. BCG stresses that a major hurdle in leveraging the benefits of outsourcing in low-cost countries is not necessarily the availability of suppliers or labour but barriers within the MNC which is trying capture the benefits of outsourcing. Far-reaching decentralization of procurement functions combined with natural inertia will considerably delay the kind of full-scale sourcing that China could offer in any large organization.

The two last strategies differ significantly in that the latter assumes that the MNC has become a networked company with a high degree of decentralization. In the meantime several Chinese companies will on their own capture the inherent capabilities of the Chinese industrial structure by acquiring advanced technologies and marketing channels on their own.

Contrary to conventional wisdom it often makes more economic sense to transfer a new product line to China than an old one. This becomes obvious when you consider that a distinctive new product line will realize more cost savings not only in the manufacturing itself but also in design and product development, where today an abundant supply of Chinese engineers are available at no more than one third of the costs in the US or Europe. Transfer of an old price-pressured production line gives a much lower payoff as there are a number of one-off costs.[9] These include changes in product and process design and establishing a new network of local suppliers. The above provides a solid justification for Nokia, Siemens and Motorola to outsource the production lines for a substantial proportion of their new mobile phone models from the beginning, including design and product development.

Another positive factor relating to sourcing in China arises from the utilization of a different mix of labour and capital. Workers in capital-intensive factories in the West are often several times more productive than their Chinese counterparts. The reason is that many factory workers have been replaced by complex flexible-automation and material-handling systems which reduce labour costs, but at the same time increase expenses for capital and support systems. It has been observed in many factories in China that the process is being reversed by removing certain equipment from the production lines and expanding the role of labour. This could reduce total capital by as much as a third, and a combination of lower wages and less capital would raise the return on capital to above the levels in factories found in highly industrialized countries.

FDI dominance in high-technology trade

The world is not yet a completely integrated economic system that follows the principles of perfect markets in which politicians have a limited or no role at all too play. However, it becomes immediately obvious that such an ideal world does not exist as the EU and the US economic policies clearly demonstrate an ability and willingness to protect not only specific product markets but also their markets for manual and highly-skilled labour. Furthermore, they have the political power to prevent China from gaining access to advanced technology that can be used in the military sector.

For a deeper understanding of the role of high-technology trade in China's industrialization one should look into its regional aspects. The coastal provinces have been able to establish dynamic industrial structures that are presently fuelled by two strong market forces. One is the availability of low-cost skilled labour, which will remain abundant in the intermediate future as the coastal areas and other favoured locations can constantly rely on labour pools in the hinterland.

The second force is the hunger for foreign investors to be active participants in this great bastion of a developing country that will ultimately become the major market for almost all industrial products and services. Attracting FDI by offering access to the huge Chinese market has lured more and more foreign companies to establish bases for high-technology production, although still dominated by final-assembly production. One characteristic of FDI-based industrialization becomes evident when studying the statistics of China's trade in high-technology products. During the first seven months of 2004 this trade amounted to US$112 billion, of which imports accounted for 59 billion.[10] Both imports and exports of high-tech products have increased rapidly over the years and the balance has improved, although the statistics shows that imports still prevail, which would be normal for any developing country, which China is at its present stage of development. However, the statistics reveal highly significant trends when looking into the origin of high-technology trade by type of company.

Foreign-owned enterprises in the first seven months of 2003 were accountable for roughly US$45 billion out of 53 billion in China's export of high-tech products, and the proportion is similar in imports, with $43 billion out of a total of 59 billion. Interesting is the fact that wholly-owned foreign enterprises have become dominant over equity joint ventures in China's high-tech international trade, with a ratio of approximately 3:1. Equally noteworthy is the import of high-tech products by state-owned enterprises that amounted to US$15 billion, while its export of high-tech products only registered 6 billion. However, private enterprises accounted for $2.3 billion for imports against 1.1 billion in exports. It is also striking that the role of private enterprises is expanding more rapidly than those of other types of companies, although still from a very low level.

Obviously China has been amazingly successful in attracting foreign investors to support its rapid industrialization drive. However, at the same time the same companies have become the dominant actors not only in high-tech trade but also as carriers of high-tech industrial development on China's territory. Most assessments suggest that expansion of high-tech industry in China will be controlled by foreign-owned enterprises as mastery of technology, control of intellectual property rights and access to advanced components lie in the hands of the foreign investors.

A very large share of high-tech parts is made up of a wide array of semiconductors, for which China still has very limited production capacity. A significant share of high-tech imports comprises parts for wireless mobile phones, for which China is rapidly becoming the production centre for the world – showing up in export statistics. Other important export categories include DVD players, digital cameras, laser jet printers and various equipment using liquid crystal displays. China has also become a big exporter of CRT displays and is taking on a world-leading position, although CRTs will eventually be almost completely displaced by flat panel displays.

Value-added in China's exports

Products resulting from processing and assembly (P&A) constitute a substantial portion of China's exports. P&A refers to a situation in which foreign firms provide raw materials, components, parts and sometimes machinery and equipment as well as design under a contractual agreement to enterprises in China, which in turn manufacture products using the imported inputs and reexport them back to the foreign firms. The proportion of Chinese domestic value-added in Chinese P&A exports has been roughly estimated to be 10 per cent in 1992, rising to 20 per cent in 1995 and 35 per cent in 1999.[11]

Before joining the WTO, which has accelerated China's reform, including the reform of the state-owned enterprises, there was a fear that China's unemployment problem would worsen. However, China's exports have continued to increase, which in turn generates employment. The authors conclude that the direct effect on domestic employment is 130 person-years per US$1 million of aggregate exports with a total average effect of 375 person-years. The highest domestic employment effect, either direct or total, occurs in agriculture – 1261 and 1671 man-years respectively – which reflects the high labour intensity of agriculture in China. However, the lowest employment effect occurs in the manufacture of electronic and communication equipment – 29 and 72 person-years respectively. Analysis of the textile sector shows employment effects of

84 and 558 person-years respectively for each US$1 million of exports to the US. For P&A exports, the largest direct and total employment effects occur in agriculture – 171 and 181 person-years respectively.

Calculated on data for 1995, Chinese exports to the US would directly increase China's GDP by 19 per cent of the export value, and taking into account the indirect effects in the rest of the economy, the total share would increase to 48 per cent. A significant proportion of Chinese exports in the mid-1990s were first shipped to Hong Kong and then re-exported to their ultimate destinations. Such exports in the past accounted for a very significant percentage of Chinese exports to the US.[12] For P&A exports the average direct value added is 15.3 per cent, while the average total value-added is only a little higher, 17.6 per cent. For non-P&A exports the value-added is much higher – 32.9 per cent and 92.5 per cent respectively.

China's Outward Foreign Direct Investment

Foreign investment is not only flowing into China but also out of China. At the end of 2001 China's direct investment abroad had reached an accumulated value of US$35 billion, according to estimates made by the IMF.[13] A small number of companies in petroleum, iron and steel, transportation and construction make up a major chunk of overseas direct investment (ODI).[14] Haier Group in household appliances and Zhong Telecommunications (ZTE) are the only other companies mentioned in UNCTAD's ranking of 12 Chinese companies with the largest foreign assets in 2001. This reflects China's need to expand overseas to get hold of natural resources in countries like Australia and Canada. Outward FDI to acquire technologies is a new trend for Chinese companies seeking to acquire Japanese firms as a result of their increasing financial strength and the Chinese government's policies encouraging overseas expansion.

The evolving structure of Chinese overseas investment now reflects a need to get access to advanced technology as the country attempts to integrate into the global economy. China's ODI has gone through four stages since the Open Door Policy was formulated. During the first period, 1979–1985, foreign investment was completely controlled by state agencies and was mainly possible only for state-owned enterprises. During this phase total foreign investment amounted to less than US$200 million for some 200 projects. The next period, 1986–1991, saw a considerable relaxation of government control and non-state firms were allowed to establish overseas subsidiaries if certain conditions were fulfilled. The average size of projects increased considerably and the total amount of foreign investment reached US$1200 million, according to official statistics. The following stage, 1992–1998, experienced a great outpouring when local and provincial enterprises started to invest overseas, which led to a number of failures. Officially, the level of foreign investment did not increase during this stage. In the next stage, in the late 1990s, as a result of the Asian financial crisis, a consolidation took place under government guidance that sought to ensure that investment projects went to genuinely productive targets.

A directive of 1999 encouraged enterprises to engage in trade overseas with the following emphasis. 'Enterprises producing light industrial goods like textiles, machinery and electrical equipment were specifically encouraged to establish overseas manufacturing projects that could process Chinese raw materials or assemble Chinese-made parts that could eventually spur China's exports.'[15]

There are a number of factors which prompt Chinese companies to make overseas investment. First, overseas investment has been driven by the country's need to secure access to natural resources and various raw materials, for example, to guarantee market access to critical resources such as iron ore, aluminium, oil and a number of other resources in countries like Brazil, Indonesia, India, Papua New Guinea and so on. Chinese firms have established joint ventures or wholly-owned subsidiaries in a number of countries that have the desired resources. Second, saturation of the domestic market has prompted companies to exploit foreign markets by setting up manufacturing plants, which is illustrated by the situation of the household appliances sector in the late 1990s. Third, Chinese companies, like those of other countries, also attempt to secure markets through investment to circumvent trade barriers.

Fourth, overseas direct investment is also an important channel for obtaining advanced manufacturing know-how and up-to-date technology, which is often crucial for future advances in both domestic and foreign markets. An early good example is provided by Shougang Iron and Steel Corporation, which bought 70 per cent of equity in Masta Engineering and Design Inc. in California in 1988. It thereby obtained access to the American company's high-tech design capability in steel rolling and casting equipment.

Access to technologies and marketing channels for branding and distributing high technology products with good profit margins may increasingly become more important. A striking example is Thomson Multimedia of France, which produces television sets and DVD players for a global market. The new company will have a network of plants located in China, Vietnam, Germany, Thailand, Poland and Mexico. Expected annual revenues are expected to be close to US$4 billion, mainly based on its production of TV sets that will reach 18 million annually thus far exceeding Sony and Matsushita with their annual production in the region of 12 million units.

Another daring venture was the decision by China National Bluestar[16] – a major chemical company – in December 2003 to sign an initial agreement to buy a controlling stake in the Korean car maker Ssangyong Motors, which would require an investment of US$1000 million. This ailing car company was offered for sale after it was spun off from the Daewoo Group, which was dissolved due to heavy debts in 1999. Creditors took over through debt-for-equity swaps. The Chinese company, with a chain of car repair shops, did not expect to experience any problems.[17] However, the deal immediately raised serious concerns in trade union circles and also raised concern in business circles in South Korea. This arises from the fact that they could see no spillover effects from Chinese investments and also feared that such Chinese ventures would further erode an already shrinking technological edge over China. Subsequently the deal fell through. Although this early attempt to acquire the Ssangyong car maker was not accomplished, another Chinese company appeared as a bidder for the Korean car maker.

Shanghai Automotive Industry Corporation (SAIC) was named as China's preferred bidder for Ssangyong Motor in July 2004. SAIC is China's largest car maker and is the joint venture partner of Germany's Volkswagen and US General Motors (GM), which are the two biggest foreign car makers in China. If successful in acquiring Ssangyong, this will be the first Chinese acquisition of a foreign car group, which could help the state-owned Chinese company to advance its technology. Local automakers are encouraged by the State Development and Reform Commission (SDRC) to develop into global players with their own brand names. The purchase of an overseas plant gives SAIC much greater opportunities for developing internationally.

China's ambitions in upgrading industrial technologies will trigger a number of other large Chinese high-technology investments overseas in the future, where major investments have in the past been concentrated on resources for which China is eager to secure supplies for energy, and a number of critical minerals and metals for its booming economy. A first example of new approaches came in 2002 when Beijing Oriental Enterprise (BOE) acquired the LCD panel division of South Korea's Hynix Semiconductor Inc. (see below).

In 2003 the state-owned enterprise China National Chemicals Import and Export Corporation – usually known as Sinochem – submitted the best bid in a public tender to take over another ailing company, Inchon Oil Refinery, also in Korea. The company was established in 1968 as a part of the Hanwha Group – then Korea Explosives – and almost went bankrupt during the Asian financial crisis in 1997. Under government guidance the firm was spun off from Hanwha and subsequently acquired by Hyundai Oilbank Corporation. However, Inchon was again spun off and placed into court receivership. Sinochem was selected by South Korea's bankruptcy court to purchase the company and will pay around $550 million, with final agreement to be reached by the end of August 2004. Inchon had sales of 1.95 trillion Won in

2003, with substantial increase in profits, and the company is expected to continue to progress after restructuring.

Korea has not experienced many major acquisitions by Chinese companies until recently. The surprise came in 2002 when BOE Technology Group acquired the LCD panel division of Hynix Semiconductor for $380 billion, renaming the firm BOE Hydis Technology Co. The company produces one category of LCD panel, referred to as 3.5 generation, in small and medium sizes, with IBM among its customers. A major reason for the sale of the division was that Hynix Semiconductor was under pressure from its main lending banks to reduce its outstanding debt. Discussions are underway by a number of Chinese companies to make bids for counterparts in South Korea's information technology field. Several Chinese firms in the electronics field have shown an interest in acquiring the plasma panel division of South Korea's Orion Electric. This recent interest and activity by Chinese companies has aroused mixed feelings. Many observers in Korea sound the warning that what the Chinese enterprises are really after is the technological capabilities of the firms they acquire, and that once these are obtained, manufacturing will be shifted to bases in China. Concern is mounting in industrial circles that channels are forming for an outflow of technologies to firms in China that will grow into future business rivals.

In the past, only big state-owned enterprises were encouraged to venture overseas. Shougang Group, the national iron and steel producer, built several plants abroad, but it was unusual for a private enterprise to be allowed to operate overseas. However other companies are moving into overseas operations.

Another example from Japan shows that FDI is used to integrate access to markets with technology upgrading. The Wanxiang Group from Hangzhou is the first major Chinese maker of car components to establish itself in Japan.[18] The company is perceived as a private company under the Chinese township enterprise category. It specializes in the manufacture of brake-related components and other parts and is a proven supplier to the Toyota and Mitsubishi car companies in Japan. Wangxiang was established in 1969 and employs 31 000 people in China with sales of RMB15.2 billion in 2003. It has already acquired a number of overseas companies and has offices in a handful of countries including the UK, Germany and the US. In an early move in 2001, the Wanxiang Group bought a 21 per cent stake in Universal Automotive Industries (UAI) Inc., a NASDAQ-listed company based in Chicago, and became the biggest shareholder under an agreement that entitles it to 58.8 per cent of the voting rights and guarantees its control over the company.[19] UAI mainly produces and sells automotive braking equipment and components.

The Chinese company plans to establish a wholly-owned subsidiary in Japan towards the end of 2004 and is in negotiations to acquire Japanese manufacturers of car components. One company in question is making auto body parts and suffers from a capital shortage. The Chairman of Wanxiang, Mr Liu Guanqiu, says that his planned expansion will come in the wake of constitutional changes that were approved by the National People's Congress in March 2004 that will guarantee the protection of private property.

An important reason for the move to Japan is the concern that car makers have expressed about the quality of Chinese car components, although they are generally considerably cheaper than Japanese parts. With its new subsidiary in Japan, Wanxiang aims to upgrade its brand image and improve product quality with the technologies of the Japanese auto parts supplier it hopes to acquire. Wanxiang and other Chinese companies look forward to absorbing the technologies held by Japanese companies and using them to improve production in China, and such alliances are likely to increase in the future.[20] Aside from gaining access to advanced technologies, Chinese companies are eager to utilize Japanese marketing channels and brand names that would otherwise take a long time to establish.[21]

There are a number of reasons behind the increasing Chinese investments in operations abroad.[22] Lagging domestic demand often provides the drive for moving manufacturing overseas. This is exemplified by Haier in electric consumer goods, and Changhong and Konka, which established production bases for TV sets outside China. Another powerful incentive is the need to secure market access by evading trade barriers abroad and taking advantage of free trade areas. Production was shifted to Mexico and Canada when the US imposed restrictions on the imports of certain textiles from China. Furthermore, overseas Chinese encourage companies on the Mainland to make investment in host countries such as Thailand and Indonesia. In recent years China's outward FDI has included mergers and acquisitions with the aim of securing access to advanced technology, global marketing channels and recognized brand names. The recent trend is likely to gather further strength during the rest of the decade.

Entering the WTO and its Impact on Foreign Technology Absorption

China became a member of the World Trade Organization in December 2001 after lengthy discussions and tedious negotiations that had been dragging on for a number of years. At the same time Taiwan was also admitted. Before entry a number of adjustments and policy changes were introduced, but China has faced a number of challenges that were not clearly foreseen at the time, as it is no longer feasible to support state-owned enterprises (SOEs) directly, although they still play an important role in many high-technology sectors.

The entry had been preceded by a number of policy changes in science and technology development, some of which date back as far as the early 1980s.

Funds for R&D have increased not only in absolute terms but as a share of GDP. Manpower resources for R&D have substantially increased. However, the WTO entry is a two-edged sword for China's future industrial development. Given the legal framework that WTO provides, more and more foreign companies are willing to make high-technology investments in China, but Chinese companies will, even in domestic markets, face formidable competitors armed with heavy Intellectual Property Rights (IPR) portfolios which very few if any Chinese companies can match. This compels institutions, researchers and policy makers to restructure the innovation landscape in China. Indeed, the new situation will affect Chinese economic development in many aspects. High-technology industries are key pillars for S&T development and the effect on high-technology industry from China's entry into the WTO will have a heavy impact on the general development of S&T.

There are a number of treaties directly relating to high-technology industries, four of them of fundamental importance: the Agreement on Information Technology Products, the Agreement on Trade Related Aspects of Intellectual Property Rights (TRIPS), Trade Related Investment Measures, Agreement on Subsidies and Countervailing Measures and the Agreement on Basic Telecommunications.[23] They will all be implemented and require changes in Chinese institutions. In some areas the outcome is remarkable with its direct effects on the direction, scale and pace of the development of Chinese high-technology industries.

The Agreement on Information Technology Products includes a comprehensive mechanism for tariff reduction. Accordingly, the Chinese government should eliminate customs duties on IT products from a list that includes 200 categories. The agreement states that the price of products from foreign countries will not increase when coming into the Chinese market. As a consequence domestic IT products will face tougher competition. The apparent effect may be less than expected as many IT products from foreign countries have been entering China without paying full tariffs, and actual tariff levels may have been around 4 per cent. Secondly, since most Chinese IT products are made of materials imported from other countries, tariff reduction means that input costs are going down, which will benefit Chinese high-technology enterprises. This would strengthen the competitive power of domestic IT products in the international market, and could also attract more foreign investment.

The Agreement on Basic Telecommunications requires openness that forces Chinese telecom services companies to compete with foreign telecom enterprises, and to reduce prices considerably. The intensive competition in the foreseeable future is a huge challenge for the Chinese telecommunications industry. For a long time, the development of the telecom industry in China was protected by government policy. Management by government agencies contributed greatly to low efficiency in telecommunications and related industries. The Agreement on Basic Telecommunications stipulates that all member countries have to abolish all government monopolization, and open the market for local services, which includes data transfer, private lines rented by enterprises and so on. This agreement means that the market for telecommunications services must be open to all WTO member countries within a period of three to five years after China's entry, and the tariff on telecommunication equipment should be reduced to zero within three years.

In the evolving situation after China joined the WTO, the state has been playing a reduced role in directly financing high technology industries. Enterprises, universities, independent institutions and so on are gradually financing R&D activities for themselves. Thus, the role of the Chinese government has become more explicit in other areas within the WTO framework, for example, in the support of industrial innovations, which among other things is exemplified by incubators that have come into existence to commercialize R&D results.

The WTO has created a more competitive environment. Many state-owned enterprises (SOEs) and other enterprises fall short of competitiveness and could disappear. Those which want to survive must quickly improve their management of technology capabilities. The creation of S&T enterprises out of former ministry-controlled research laboratories has been regarded as one of the strengths in the development of the Chinese high-technology industry. Such enterprises have plenty of experience from domestic market competition. However, the technical capability of these enterprises is not strong enough to carry out R&D activities related to long-term technology projects.

Weaknesses of developing countries in global S&T performance are not only embodied in their innovative capability, but also in the level of the domestic S&T demand. The absorptive capacity is weak. Therefore, developing countries depend on the knowledge provided by developed countries. In this sense, developing countries are in an inferior position in the world S&T platform relative to those advanced countries. The challenges include primarily four aspects – deficiency in R&D investment, limited involvement in strategic alliances, being outside the mainstream of technology flows, and the large-scale control of R&D resources by global MNCs.

More and more technology and knowledge is being transferred from home countries to host countries, and critically, these activities are controlled by enterprises from investing countries. In order to exploit local materials and low-cost human resources more efficiently, R&D investment has followed in order to satisfy better the demands of local markets. Thus, enterprises in host countries could benefit from R&D activities providing that technology diffuses.

Strategic alliance is a protocol for technology cooperation between enterprises, which includes R&D cooperation or other innovative activities. The main forms of strategic alliances include R&D cooperation, R&D joint ventures, sharing technology information and so on. Over the last couple of decades technology agreements have become very prevalent. International technology networks are mainly made up of MNCs from developed countries. As a result, developed countries and MNCs were able to control international S&T and knowledge resources in greater depth. China and its enterprises rarely have enough competitive technology to enter into such networks.

FOREIGN R&D INVESTMENT IN CHINA

China is attracting foreign R&D activities from the industrialized countries, while still a developing country with a GDP per capita in the region of US$1000. China may be considered the first developing country that is able to capture the entire range of R&D functions of multinational companies – from rudimentary upgrading of manufacturing technology to basic research at the frontiers of new knowledge.

There were more than 600 R&D centres established in China by foreign enterprises by the end of June 2004, according to statistics published by the Ministry of Commerce. These centres represent a cumulative investment of about US$4.0 billion.[24] Major investments have been made by Microsoft, IBM, Motorola, Siemens, Nortel, Dupont, GE, General Motors, Honda, Hitachi and Toshiba to mention only a few.

A large share of foreign R&D centres has been established in Shanghai and Beijing. The number is given as around 140 for Shanghai and close to 200 in Beijing.[25] These figures do not include the large number of agreements between university departments and foreign companies, although such activities are also handled by the officially registered foreign R&D centres. A rough estimate by the author suggests a total number of R&D staff working for foreign companies in the region of 50 000, which would correspond to 6 per cent of China's officially recorded number of R&D personnel.

During recent years China has become increasingly attractive as a location for R&D, for which cost considerations are only one factor and not necessarily the dominant one. IBM was not the first major foreign company to establish an R&D centre in Beijing, although its wholly-owned IBM Research China launched in 1995 was the starting point for major IT companies to follow suit. Available documentation shows that out of 167 registered foreign R&D centres in Beijing, the majority of them are active in developing and researching information technologies. All major IT companies have directly or indirectly outsourced significant shares of their manufacturing operations to China since 1985. Similarly, there has been a growing relocation to China of R&D activities spearheaded by the IT sector since 1995, now followed by biotechnology and pharmaceutical companies.

The attraction of having R&D in China is driven by four considerations: 1) availability of a large low-cost and highly skilled talent pool; 2) being inside a large, expanding and increasingly sophisticated market; 3) supporting product and process development in China; and 4) capturing new technology possibilities by being in touch with China's research frontier. The decision to locate R&D in China is seldom based on only one factor. A combination of factors is illustrated by the example of a Danish pharmaceutical company (Box 3.1).

BOX 3.1 R&D WORK OF NOVOZYMES IN CHINA

Novozymes supplies biological solutions to industrial problems. The company produces and markets more than 600 different products which are used by industrial companies in the production of a wide variety of everyday products, including food and beverages, clothing and detergents. The turnover in 2003 was DK6800 million.

Novozymes was one of the early entrants to establish a R&D foothold in China. Its R&D Centre was inaugurated in 1997 and at the same time the company initiated a fungal screening programme in South China. A detergent laboratory was established in 1999, and a basis for application developments for Asian markets was initiated in 2001. R&D for textile applications will have been transferred from the US by early 2005.[26]

Novozymes is a knowledge-based company. It employs some 4000 people worldwide with 760 full-time employees in R&D, of whom 395 hold science degrees.[27] A growing proportion of sales originates from sales of new enzyme solutions – launched during the past five years. New products increased their share of turnover to more than 30 per cent in 2003. In the same year Novozymes invested about 13 per cent of turnover in R&D. The company has around 100 separate product development projects in various stages of completion, of which one third are in the closing stages. Around 10 per cent of the company's investment in R&D is in projects which are outside existing business areas.

The company will expand its R&D activities in China, the major reason being that the company wants to be close to its customers. One example is the processing industry for textile technology now moving to China, for which Novozymes is an important

> supplier. Another important reason for being in China is the availability of skilled microbiologists. An attractive recent phenomenon is the increasing availability of returnee microbiologists – primarily from the US.
>
> The global expansion of many R&D functions will take place in China and new production plants will also be located in China. Novozymes benefits from being in the country in a number of ways. First, the company has close access to an attractive market with a variety of requirements and possibilities. Subsequently Novozymes can work closely with customers and universities and institutes in developing new applications as well as product adaptations. Second, the activities in China are strongly supported by business development within other parts of the company by providing technical services which provide a competitive edge. Third, some activities can be performed at lower cost in China than anywhere else in the world.

As mentioned above, 167 R&D centres have been established in Beijing by foreign enterprises. Their presence is dominated by US companies with 70 units, followed by Japan with 33 units. Taiwan, Korea and Hong Kong are present with ten, nine and eight units respectively. Companies from Germany, France and the UK have only registered six, five and three units respectively. A number of companies have registered multiple units which appear as separate entities in the Beijing official registry. This is the case for 18 enterprises that have jointly established 62 units. This includes Ericsson with six units and Lucent Technologies with nine. Motorola is actually listed as a single R&D centre in Beijing as it has consolidated 18 centres, located in various places in China, into the Motorola (China) Research Institution. Lucent in Beijing provides a good example of various approaches in carrying out research in China. Three of its units are collaboration agreements with Tsinghua University in software technology, communications and opto-systems. Two other units include collaboration with the Software Institute of the Chinese Academy of Sciences (CAS), and Beijing University on software development. Two other units are listed as the Beijing R&D Branch of Lucent-Bell Laboratories and Lucent Technology (China) Co., Ltd. The remaining two are listed as the Bell Laboratories Advanced Technology Research Institute, and the China Basic Research Institution of Bell Laboratories. Outside Beijing, Lucent Technologies in Qingdao has become one of the company's four worldwide system integration centres. In addition, Lucent opened a new research and development centre in Nanjing in September 2003, and is also present with R&D functions in Shanghai.

A dominant majority of foreign R&D centres in Beijing have been registered as being engaged in information technologies – altogether 122 out of 167. The next prevalent category of centres is doing research in biotechnology and pharmaceuticals, with a total of 21 units. Delphi from the US has two units for automobile research, of which one is in collaboration with the Tsinghua University Automobile Research Institute.

The Example of Motorola R&D in China

The R&D activities of Motorola provide an excellent illustration of the interaction of a global company that has a global network of R&D centres, and a wide-ranging R&D structure in China that includes both corporate and government R&D units. Motorola, like other IT companies in China, originally had its focus in manufacturing operations. However a shift in the characteristics of the Chinese market from the beginning of the 2000s prompted Motorola to pay attention to value-added services and innovations, when China started to lead the world market for handsets in terms of fashion and new design features. Thus Motorola moved R&D functions to China to become closer to the market and more cost-efficient, while at the same time having access to an excellent talent pool of engineering graduates from Chinese universities.

Motorola has major R&D centres in the US, Canada, France, Germany, Japan, Israel, India, Singapore, Hong Kong and China, although some core research remains in the US, such as CPU R&D, which includes the PowerPC used in Apple computers.[28] The company's global innovation network has evolved over a period of some 30 years since labour-intensive manufacturing facilities – integrated circuit (IC) testing and packaging – were moved to Hong Kong and Malaysia in the 1970s. Substantial decentralization of R&D started in the 1980s, although low-end process R&D was moved to off-shore locations from the very beginning.

France and Germany were the first locations of Motorola's major overseas laboratories, followed by R&D centres in Canada, Japan and Israel. At a later stage Motorola dispersed further R&D activities to India and China in order not only to have access to low-cost engineers but also to establish strategic sites in emerging markets. Motorola R&D centres in China cannot yet take on advanced assignments that are presently carried out in France (Crolles) but are very competitive in various technology domains within the telecommunication sector, where 'locational windows of opportunity are still open'.[29]

Motorola in China has 12 000 employees, one holding company, one wholly-owned company and nine joint ventures. Motorola's major manufacturing complex in China is located in various science and industrial parks in Tianjin. There are five major factories manufacturing products which include network equipment, GSM cell phones and various kinds of telecom equipment. However, Semiconductor Manufacturing International Corporation (SMIC) has taken over the Motorola IC production plant in Tianjin, after lengthy negotiations that were completed on 1 January 2004.

Motorola employed 1400 staff in its R&D centres in China in mid-2003; of these, five centres in Beijing employed 740. The centres in Beijing include the following:[30]

1. Beijing R&D Centre for the Personal Communications Sector (PCS): product research on software, and electrical, mechanical and industrial design development, established in 1999.
2. China Design Centre for the Global Telecom Solutions Sector (GTSS): research and development of mobile communication systems, established in 1995 (Software Engineering Institute Capability Maturity Model – CMM – Level 5 in 2000).
3. Global Software Group (GSG) China, established in 1993 (First CMM level 5 organization in China).
4. Digital DNA Labs China (DDL): research on predictive technology of semiconductor devices and processes, systems on chip, advanced materials and relevant software development).
5. ESDL Metroworks: focus on the Internet and PowerPC.

Major manufacturing facilities are located in Tianjin, which also includes a major plant for integrated circuits, established in 1993 shortly after Motorola decided that China should be one important focal point in its global business strategy.[31] Aside from manufacturing, the company also has one of its advanced centres located in the city – the Motorola Advanced Technology Centre (MATC-Asia). Other centres include the Motorola China Research Centre in Shanghai, Suzhou Technology Centre, and centres belonging to the Global Software Group in Nanjing and Chengdu. Furthermore, Motorola has entered into a number of collaborative research agreements with Beijing University, Tsinghua University, the Chinese Academy of Sciences, Fudan University in Shanghai, Suzhou University, Nanjing University, Chengdu University and Zhejiang University in Hangzhou.

Motorola is present in almost all important R&D locations in China and has established collaborative relations with major universities. The pattern of Motorola R&D activities in China indicates what other IT companies are already doing or will be doing to capture market possibilities, exploit the Chinese talent pool, explore the knowledge frontiers in China and support product and process development. In doing so global IT companies are possibly embarking on a process of creating a global innovation system (GIS) in which China is going to play an increasingly important role.

FOREIGN RESEARCH IN CHINA – DOMESTIC AND GLOBAL ISSUES

There is little doubt that China is trying to pull its weight through the combined influences of its huge consumer markets and being the world workshop. This could influence technology standards significantly in many fields, including wireless communications. China is today the location of around 12 per cent of global manufacturing and is number three trader in the world, behind the US and Germany and ahead of Japan. It already has a highly educated population and is attracting major foreign investments in research and development, as has been amply illustrated.

The large-scale presence of foreign R&D institutes poses a number of serious issues – for Chinese policy makers, for policy makers in industrialized countries and for the companies themselves. Let us briefly look at the last issue confronting companies. They are now relocating and expanding their R&D activities in China following on the heels of relocating their manufacturing activities to capture low-cost options. As in foreign investment in manufacturing, substantial spillover effects are likely to occur whenever the Chinese absorptive capacity goes beyond a certain threshold. Intellectual capital is very mobile and the IPR regime in China is still not commensurate with the situation in advanced industrialized countries. However, locating R&D in China is being carried out for long-term survival and to capture market possibilities and profits, which have to be balanced with the risks of intellectual capital seeping through porous corporate boundaries. This has for the time being prompted foreign companies to retain the control of core technologies in home countries.

The Chinese planners and leadership had expected that FDI in advanced manufacturing would more-or-less automatically be followed by substantial technology transfer to benefit the domestic industry in its upgrading efforts. This has hardly happened and the policy makers may now voice concern that China is offering its intellectual resources to foreign companies that will gain in competitive advantage for their own companies in highly knowledge-intensive industries. Furthermore, China may feel threatened by a gradual 'colonization' of its intellectual landscape.

How does the situation appear from scrutinizing potential consequences in the US and Europe? There is little doubt that policy makers in the US feel threatened by a shift in gravity of R&D activities towards Asia – in particular towards China and in the intermediate perspective towards India. This will eventually lead to reduced dominance of the US – by its research institutes, universities and companies – in a number of important fields. This will in the longer term also affect the present dominance of the US in almost all space and military technologies, which is in all likelihood the cause of the most serious concern among the US policy makers watching the changing science and technology landscape in China.

The perspective in Europe is different, although the sense of concern is also growing as is apparent in the 2004 OECD *Science, Technology and Industry Outlook*. At the March 2002 meeting of the European Council in Barcelona,[32] European Ministers announced a goal of '… turning the EU into the most competitive knowledge-based economy in the world'. One identified objective for achieving this status was to raise spending on R&D and innovation in the EU from its 2002 level of 1.9 per cent of GDP so that it approaches 3 per cent by 2010. This development is based on the expectation that approximately two-thirds of the increased R&D spending is to come from the private sector.

The OECD notes that China continues to deepen the reform of R&D institutes to enhance their innovative capacity, it is developing a regional innovation system and it encourages the return of overseas Chinese graduates. Important measures include an accelerated construction of industrial parks for overseas Chinese graduates and also accelerated expanded communication channels between China and its overseas students.

Approximately one fourth of undergraduate students in China receive training in engineering and science, while over fifty per cent of tertiary graduates in China receive degrees in these fields.[33] Simultaneously, there has been a gradual increase in the share of graduates in areas such as business, law and entrepreneurship. Furthermore, the educational authorities are currently revising curricula to improve the quality of higher education under the *University Teaching Quality and Teaching Reform Project*.[34] The OECD Report mentions that[35]

> The Chinese Ministry of Personnel has strongly promoted various policies to encourage highly skilled overseas Chinese to return to their home country, and their numbers rose on average by 13% a year in the 1990s. Related to this goal, the Ministry is supporting the construction of university laboratories and the development of science parks to provide venture opportunities for returning Chinese as well as employment opportunities for young graduates.

It has become increasingly obvious that China is not only attracting its overseas brain talent but also R&D activities of foreign affiliates. From a European perspective, the OECD notes that[36] foreign affiliates can generate benefits for their home countries, thereby compensating for some of the apparent disadvantages of outsourcing. Some of the current debate about outsourcing is fuelled by concerns that, unlike previous episodes in which outsourced work was low-wage or dangerous, the current phase is characterized by the transfer of professional work, such as software development, and of strategic activities like R&D. As recent analyses show, such outsourcing not only results in an outward movement of activity from the home country, but also generates benefits as knowledge from abroad flows into the home country.

The Report says that it is understood that such affiliates are located abroad to tap into knowledge created in foreign centres of excellence. Breaking links to such centres may undermine MNCs' innovation activities. A more effective strategy may be to build up local centres of excellence that can attract affiliates of foreign MNCs and encourage domestic firms to maintain a local R&D presence. The way to entice companies to invest in R&D in a particular region is to make the local area a strong supplier of knowledge or other critical resources. In this respect Europe may face growing competition from non-OECD countries.[37] This will force OECD countries to rely more on the creation, diffusion and exploitation of scientific and technological knowledge, as well as other intellectual assets, as a means of enhancing growth and productivity. The promotion of high-tech clusters and science-based regions has been one of the responses to strengthen the European Research Area (ERA).

At the same time Beijing and Shanghai have emerged as two important science regions in China, followed by second-tier science regions centred on Hangzhou and Nanjing in coastal areas, and Xian in the interior. High-technology clusters, for example in the automobile industry, are emerging in Guangzhou in South China, and in Tianjin near Beijing. Highly specialized technological clusters are being created and expanded for a number of products and technologies in many other cities that dot China's coastal areas. These developments are likely to challenge the EU ambition to become the world's most innovative region even if it reaches R&D expenditure of 3 per cent of GDP by 2010, with an announced expectation that the private sector will be the main actor.

A NEW KNOWLEDGE ENVIRONMENT

S&T globalization provides China with crucial opportunities to realize technology breakthroughs, as it encourages the inflow of S&T resources to developing countries in more efficient ways. In absolute terms transfer of technology from advanced countries to developing countries has expanded dramatically. R&D results and technology innovations held by investing countries would be diffused to developing countries sooner or later through international trade, technology exchanges or other channels, and developing countries could benefit from the diffusing process. It has given China more opportunities to participate in international S&T cooperation, exert its technology potential, narrow the technology gap with developed countries and also realize technology breakthroughs. Participating in global S&T cooperation includes many dimensions, among others, attracting persons from other countries is important.

Chinese entrance into the WTO has challenged R&D activities and the related strategies of domestic enterprises directly. At the same time, S&T globalization also has a deep effect on Chinese IPR institutions. Firstly, the Chinese IPR institutional system itself was facing more and more difficulties in the global background. Simultaneously, the system of Chinese applications for patents was not relevant in the new context of S&T globalization. In recent years, the number of applications for patents by foreigners has always been higher than those by Chinese residents. However, S&T globalization required substantial improvements in China's institutions with respect to IPR. Only with a rational and practical IPR institution could China ensure the inflow of advanced technologies. Such changes have to consider the following aspects. First, the regulations and laws of direct or indirect relevance to IPR must be improved to reflect the ongoing S&T globalization. Second, knowledge management in enterprises and other organizations must be improved. Third, people knowledgeable in IPR procedures and knowledge management must be trained. Fourth, China should establish close international relations and participate in the process of international regulations being established or modified. Finally, China should establish new or modify existing institutions to handle IPR.

China has realized that the unrestricted flow of S&T also has a direct bearing on national information security. S&T globalization requires unified development of language protocols and other instruments used in S&T activities. Standardization plays a crucial role in handling information, both in the infrastructure and the software. If this standardization is being monopolized by a limited number of organizations, it may create problems related to national security. Thus China has argued that it must attend to the following issues. First, national security must be one of the concerns in accepting an increasingly open S&T globalization process. Second, China has to pay special attention to highly-talented scientists and not lose domestic talents. Third, Chinese products related to information security should be researched and produced independently.

MNCs have extended their control not only in a geographic but also in a strategic sense. MNCs have been able to make better use of global R&D resources and cooperate more deeply with local universities, research organizations and enterprises. However, it is well known that advanced production sites in China depend heavily on FDI and foreign companies provide much of the needed technology. Both the sources and distribution of FDI in China have attracted much attention, and the different forms of utilization of FDI and the future trends in FDI are one of the crucial issues.

Thus, Chinese enterprises and R&D organizations face the challenges of MNCs directly in both labour and technology markets. Furthermore, Chinese enterprises participating in the international strategic technology alliance may occasionally make themselves more dependent on foreign technology. It may not be an altogether sensible policy for China to rely simply on FDI as the main pillar of its economic strategy. More and more negative results of FDI have come to light in recent years. These controversial issues are: the tendency for the imported technology to monopolize the domestic market; the threat of unemployment; the competition with national industries; the dependence on foreign technology; the outflow of domestic profits; the conflicts with the sovereignty of local governments; and issues of national security.

This expansion of foreign companies has raised considerable concern that China is losing control of its markets. However, the ongoing process has both positive and negative aspects which can be summarized in three important factors. First, the MNCs have focused their investment in capital and knowledge-intensive sectors and their impact on the economic structure has been quite different from domestic firms and from earlier FDI which came primarily from Hong Kong and Taiwan. Second, the MNCs have contributed towards higher technical efficiency and also higher effectiveness in the allocation of resources. Third, the entry of the MNC has forced the Chinese government to give up some autonomy over the economic structure, made the economy more vulnerable to changes in global markets, and changed the division of income among industries and within industrial sectors.[38]

There are many aspects to be considered and issues to be resolved, without any simple and immediate answers, but some concluding remarks are obvious. Firstly, it is reasonable to expect FDI and R&D investments in China to flourish during the next 15 years. Secondly, FDI is an important source for the creation of advanced technology in China. Thirdly, the diffusion of foreign technology brought into China through FDI is not efficient enough. Lastly, China should not depend solely on developed technology brought in by FDI. This is why some Chinese government agencies, academic institutions and national industries have sometimes expressed negative attitudes towards FDI in China. However, as FDI investments upgrade the level of technology in China and many other host countries, they undoubtedly provide many benefits to receiving countries.

However, there are also other important aspects to be considered. Chinese local enterprises cannot make dramatic technological progress by relying solely on FDI and technology transfer. In fact, technological diffusion of imported technology in China faces many obstacles, inherent both in the technological providers and the recipients. Investing countries try to prevent the outflow of their core technology, and adopt many measures to control its diffusion, such as enacting special laws or rules to limit the use of technology by enterprises in host countries, and often encouraging the use of less-developed technology in the host countries.

From the perspective of technology recipients, the biggest problem is the learning curve, and matching the acquired technology with their own, to make good use of the new technology. Therefore, it is necessary for the Chinese government to make some policy adjustments urgently. It has been maintaining an open attitude towards FDI for the past 20 years, in line with the mainstream pattern of technological globalization. However it should not try to attract FDI blindly, without any consideration of the impact on domestic industries.

NOTES

1. This chapter incorporates a contribution from Dr Jiang Jiang prepared for the project 'Emergence of New Knowledge Systems in China and their Global Interaction': *Technology Transfer in China*. The full contribution was presented at a conference, and a summary has appeared in Sigurdson (2004c).
2. A kombinat was an integrated production and social complex, with vertical processing plants, schools, kindergarten and hospitals.
3. Globalization and China's acceptance of participation in a global economy have generated a heated debate in China. Experts and researchers who were responsible for the Annual Report of Chinese S&T Development gave a comparatively complete definition of 'globalization'; they considered that globalization mainly points to the unrestricted flow and the reasonable distribution of knowledge, capital and labour resources in the global range. This process may penetrate different countries with different conditions in economy, culture, S&T, and so on. Some thought that the course of globalization has a close relationship with the development of Information and Communication Technology (ICT), and it also connects deeply with both the formation of market economies all over the world and the global enlargement of multinational enterprises.
4. The amount of FDI to China was by far superseded in 2003 by an inflow of capital amounting to US$160 billion, which indicates the expectation in the global financial community that the Chinese currency will appreciate. This financial inflow continued during the first half of 2004, although at a more modest level, while at the same time China reported a negative trade balance for the first time in recent years.
5. Moran (1998).
6. Ibid.
7. By 2006, Motorola aims to have made US$10 billion in accumulated purchases from China and to be producing $10 million a year in goods there. By the same year, Motorola also aims to have made investments totalling $10 billion inside China, including the construction of a global research and development centre and the hiring of 5000 researchers.
8. BCG (2004).
9. Hout (2003).
10. Source: website of Ministry of Commerce – Department of Science and Technology, http://www.mofcom.gov.cn.
11. Chen et al. (2003).
12. However, making adjustments for re-export markup, transport and insurance costs does not fundamentally alter the interpretation.
13. Wong and Chan (2003); however, an estimate in the UNCTAD World investment report 2004 (The Shift Towards Services) suggests that China's outward stock had reached

US$37 billion by 2003. Naturally, this figure does not include the substantial agreements that China entered into during 2004. Reaching accurate figures is complicated by the fact that a considerable amount of 'roundtripping' occurs. Four tax havens – the British Virgin Islands, Bermuda, Panama and the Cayman Islands accounted for 54 per cent of total Hong Kong outward FDI stock in 2002. Around 10–20 per cent of China's inward FDI is estimated to have its origin in outward FDI (*World Investment Report 2004*, p. 26).

14. UNCTAD, *World Investment Report*, as quoted in Wong and Chan (2003).
15. Wong and Chan (2003).
16. China National Bluestar is a state-owned conglomerate that is administered at present by Assets Supervision and Administration Commission (ASAC). *The Far Eastern Economic Review* reports that ASAC controls 'dozens of companies, one of them being Chonche Auto Service Corp. that needs the kind of technology that Ssangyong could provide' Murphy (2004). Chonche grew out of a decision in the 1990s to get the military out of business and consists of about three dozen former military companies with factories and car-service outlets.
17. Gong (2004).
18. 'China auto parts giant eyeing Japan – Wanxiang Group sets sights on buying local parts maker to help out with Japan sales', *Nikkei* 22 March 2004.
19. Yan (2001).
20. Another example is the acquisition of Akiyama Printing Machinery Manufacturing Corporation by Shanghai Electric Group Corporation after the Japanese firm filed for protection from creditors under the Civil Rehabilitation Law (*Nikkei* 22March 2004).
21. China's largest pharmaceutical company, Sanjiu Enterprise Group acquired Toa Seiyaku KK in 2003, a Japanese herbal medicine manufacturer, to exploit the Japanese brand name in overseas markets (*Nikkei* 22 March 2004).
22. Vatikiotis (2004).
23. Wang and Wang (n.d.).
24. SMIC website, http://www.smics.com: about Semiconductor Manufacturing International Corporation (PPT slide 4/42).
25. The figure for Shanghai was given by the Shanghai Foreign Economic Relations Trade Commission (interview on 11 January 2005). The figure for Beijing is based on a government list that includes all registered foreign R&D centres in the Capital (most likely including registrations until early 2004).
26. By courtesy of Anders Ohman (PPT Notes November 2004).
27. *The Novozymes Report 2003*.
28. For the following information on the Motorola R&D structure in China I want to express my thanks to professor Yun-Chung Chen of Hong Kong University of Science and Technology. The information has been extracted from a conference paper, 'The Localization of Multinational's Global Innovation Networks in China: Implication for other Developing Countries', HKUST Conference, 7–8 January 2005.
29. Ibid.
30. Courtesy of Dr Terrence Heng, Senior Vice President and General Manager, Motorola Global Software Group (PPT Rensselaer Polytechnic Institute conference, 5 September 2003).
31. The IC plant was transferred to SMIC in early 2004.
32. OECD (2004b), pp. 173–4.
33. OECD (2004b), p. 156.
34. Ibid.
35. Ibid.
36. OECD (2004b), pp. 173–4.
37. OECD (2004b), 'Executive Summary', p. 10.
38. Zhang and Zheng (1998).

4. Research and technological mastery in the corporate sector[1]

RENOVATING THE R&D STRUCTURES

The success of the national programmes should be judged by the extent to which the enterprise sector is benefiting and involved in R&D. China has initiated a number of national programmes in basic research and in technological development but they are still poorly linked to the country's economic development, and continued institutional changes are needed. The ongoing transformation of the state-owned enterprises poses a great challenge as the companies did not need to make any major investments in R&D to survive. They are now being forced to engage in R&D to secure their future since China joined the WTO in 2001. Thus the competitive climate has to be strengthened and the Ministry of Science and Technology (MOST) will need to modify policies to boost R&D spending such as tax incentives, and to expand programmes for small and medium-sized enterprises (SMEs). In the meantime advanced production sites in China depend heavily on foreign direct investment (FDI) where foreign companies provide much of the needed technology, and this is poorly integrated with the rest of the economy. This situation also explains the desire of Chinese companies to engage in major mergers with foreign companies, which was exemplified in late December 2004 by the agreement between Lenovo and IBM to take over the American PC division.

Today there are many indications that manufacturers in China are starting to realize clearly the possibilities of moving beyond cost advantage and copying capability and are beginning to display technological prowess. Many Japanese companies have seen their presence in China declining while they are transferring production there, which indicates strategic shortcomings or lack of understanding of the character of industrial progress in China. Some of the leading companies in Japan, such as Matsushita, Sanyo and Sony, once held large shares of the Chinese market for consumer electric and electronic goods, including television sets and refrigerators. In recent years they have seen their shares fall rapidly due to competition from local companies, and they survive now only in niche markets.

In the early 21st century, Chinese manufacturing industry has been attracting worldwide attention because of its rapid growth and its market dimension.

China has never paid such attention to corporate R&D as now. On the one hand, according to the statistics in the Chinese *S&T Statistics Yearbook* (National Bureau of Statistics and Ministry of Science and Technology 2003), the ratio of corporate R&D has increased over 50 per cent since the 1990s, which indicates that enterprises have overtaken the government in R&D spending. On the other hand, Chinese corporations have weak R&D capabilities and lack core technologies. This remains a serious bottleneck in industrial development. It seems like a paradox that substantial R&D resources are under-utilized and the Chinese government has introduced many measures to solve the problem. One of them has been the transformation of Chinese research institutes since the end of the 1990s. The data and analysis indicate that transformed institutes have become an important source for corporate R&D.

There was no real corporate research in the planned economy that came into being in 1949. The structure, location and management system were based on the Soviet Union model where the whole science and technology (S&T) system was allocated within industry and sectors of the economy. There were independent sections of R&D under separate ministries, the Chinese Academy of Sciences (CAS) and universities, the Ministry of Defence, and S&T organizations under local governments. The government was the principal source of S&T funding and allocated tasks by administrative orders. Before the reforms begun in 1978, many enterprises had their own S&T institutes or organizations. But most of these were very weak and were seldom able to carry out any real research. Their main objective was to meet temporary needs in production. Under the planned economy system, enterprises had no incentive to do R&D as all R&D results were considered to be a public good.

Reform of the R&D Institutes since 1978

China initiated changes starting with a national S&T conference in 1978. Assuming that S&T would become the leading force of productivity, China carried out many constructive measures. The most important ones included the science and research responsibility system[2] and the contract on charge system.[3] The reforms of this period introduced new methods for managing projects and allowed personnel to reap economic benefits from technology contracts.

In March 1985, China issued the Decision on the S&T System. This decision became the basis of the Chinese system of reform in science and technology, with a very profound impact. The Decision included three aspects directly influencing the transformation of corporate R&D. First, the government required that a fee should be paid when technology results transferred. Second, S&T institutes were encouraged to merge with production plants. The government encouraged institutes for technology exploitation to be combined with enterprises in different ways; one way was to make S&T institutes into enterprises. Third the import of technology was to encourage enterprises with a specific focus on the development of production technology and renovating existing equipment. Coastal cities and special economic zones (SEZ) such as Shenzhen should become leading locations to import advanced technologies. Following this reform Chinese S&T institutes were no longer economically independent with budgets guaranteed by the state. A number of enterprises came into being and employees became creators of technology-based companies such as Legend (now Lenovo) and Founder, just to mention a couple.

In 1992, the Chinese macroeconomic environment changed dramatically as a result of Deng's Tour of the South and Speech. In October 1992 a socialist market economy system was initiated and the S&T system underwent important structural adjustments. The S&T institutes were expected to implement organizational innovations and a series of laws and regulations were promulgated. These included the S&T Progress Law and the Climbing Programme. During the process, many institutes started to operate like real enterprises, though they did not change their legal status until the reform in 1999. State-owned institutes implemented a rent responsibility system. One hundred institutes were granted foreign trade operating rights and could independently import S&T equipment. In 1994, engineering design units were formally transformed into enterprises. This was the initial phase of 'Corporatization'.

In this way China began to create high and new technology industry zones and encourage private new technology enterprises. Supported by this policy, many state-owned organizations, universities and large and medium enterprises (LMEs) began to create state-owned but privately-operating new technology enterprises. Many employees became the creators and principal partners of these new technology enterprises.

In 1995, Chinese S&T reform had been proceeding for ten years. On the basis of this experience, China issued the 'Decision on Accelerating S&T Progress' and the 'Decision on Profound S&T System Reform'. In these documents the basic orientation, reform targets and main tasks were confirmed. The target of the reform was to establish a science and research system that was mainly composed of independent institutes and universities. The main tasks included making the institutes economically responsible, promoting technology exploitation through 'Corporatization', developing high technology, enhancing high technology industrialization, and improving the structure and geographical location of the basic science and research institutes.

The transformation of research institutes has applied to two categories of institutes since 1999 – technology exploitation institutes and public interest-related institutes (PIRI). The first category has received most attention.[4] Different entities were identified to undergo profound changes in terms of organization, statutes and systems.

In July 1999 a first group of 242 research institutes began their transformation. Among them 131 merged with corporations, usually industrial groups. Another 40 were transformed into S&T corporations and managed according to the local ownership principle, with an additional 18 institutes becoming agencies. The central government created 12 institutes by merging 29 related institutes and large S&T corporations owned and controlled by the central government. The remaining 24 institutes were merged into universities, transferred to other organizations or closed down outright. By the end of 2000, the formal transformation of the 242 institutes was completed and seen as a major accomplishment of the science and technology reform.

Then followed in July of 2000 the transformation of a second group of institutes, primarily belonging to the Ministry of Construction. Out of 134 research institutes, 46 merged with corporations, 57 were transformed into S&T corporations, 16 institutes were transformed into large S&T corporations owned by central government, 2 institutes merged with universities, and 13 institutes remained unchanged.

The main objective of the 'Corporatization' of technology exploitation research institutes was to promote a close interaction between science, technology and the economy in order to accelerate the use of S&T results in the economy. Following the transformation of ownership and control, each individual institute was to carry out its own institutional reform, which required changes for institute staff. By early 2002 a large majority of institutes had adjusted their mode of operation, and almost all of them had introduced an enterprise-oriented financial system. Three quarters of the institutes had adopted an employee contract system and almost all them had joined the local endowment insurance system, as well as the local unemployment insurance system.

Some ten transformed institutes went public such as the Central Iron and Steel Research Institute (CISRI), Ningxia Non-ferrous Metals Smelter (NNMS) and the General Research Institute for Nonferrous Metals (GRINM) (see below).

Assessing R&D Institute Reform

Naturally it is possible to use various indicators to analyse the results of the transformation and its effect on corporate R&D. One approach is to measure economic performance by indicators such as revenues, profits and so on.

Another method is to measure technology capabilities by using indicators such as S&T personnel, new products and so on.

The economic performance of transformed institutes improved very quickly. The total income of the transformed institutes was RMB19.91 billion[5] in 2001, which was 50 per cent bigger than the revenues of 1999, and employee incomes also increased substantially. Salary per person reached on average RMB21 400 in 2001, an increase of 42.6 per cent compared with 1999. The profits of transformed institutes reached RMB1510 million in 2001, having doubled since 1999. Similarly, taxes paid by transformed institutes in 2001 doubled, reaching RMB86 million. The number of employees had been decreasing year by year, mostly through retirement.

Technology innovation capabilities have been making progress in some aspects. Revenues from technology transfer, services, rent and consulting increased annually by 30 per cent between 1999 and 2001. Patents application from transformed institutes increased gradually, although patents granted decreased by 16.1 per cent in 2001.

The reform process has changed the technology landscape in China significantly and influenced R&D in two directions. First, R&D has moved into the enterprises and become a corporate responsibility. Second, the reform created mobility and former institute staff have become entrepreneurs and creators of new companies.

However, assessing changes in technology capabilities is much more complex and more challenging to measure. Technology capabilities, measured by number of S&T staff and expenditure on S&T activities have increased after transformation, although S&T activities carried out and patents achieved have been decreasing since 1999. Available data actually indicate that the technology innovation capabilities of many transformed institutes have decreased, with a subsequent significant reduction in S&T output. This suggests two different outcomes. One conclusion is that the transformed institutes may become ordinary enterprises without any special advantages in technology development. The more promising alternative may be to increase their efficiency and enable them to become successful fountains of industrial innovations in the future.

Thus, it has been difficult to interpret the results of the reform and a number of disturbing outcomes have been identified. Data indicate that S&T activities in enterprises increased slowly in the middle of the 1990s, followed by a decrease in the late 1990s, with a marked reduction during 1999–2001. The number of enterprises having their own R&D institutes has also decreased since 1991. The number of large and medium enterprises (LMEs) increased from 14 935 in 1991 to 22 904 in 2001, while the number having R&D units decreased from 7899 to 6000 in 2001. The same phenomenon appears to be general as 13 098 enterprises were reported to have R&D units in 1991 but this had dropped to 10 461 in 2001. Available information suggests that many LMEs do not have enough incentives to sustain R&D institutes and carry out science and technology activities as they pay more attention to surviving in the market and making profits.

Total funds obtained for S&T activities have increased substantially since 1991. From 1991 to 2001, total funds increased 4.42 times. Within this, the share of government grant and bank loans has decreased, while the share of self-finance has increased. The share of enterprises' self-financing in 1991 was 63 per cent, but after ten years, it had reached 84 per cent. The share of government grants had been reduced very sharply.

The available data and analysis clearly suggest that the technology capability and corporate R&D spending in China has expanded greatly since 1991, except for S&T activities in LMEs. Since the main target of the transformation of research institutes was to transfer their technology capabilities into corporate R&D, the result should become evident when the transformation process is completed.

The transformation described above has typically been driven in a top-down fashion where the government has been in complete control. In practice this is a reform of learning by doing; both institutes and government are learning during the reform and gradually adjusting their influence on actors and functions. During the process, many theoretical issues have been solved gradually, such as the nature of public interest related institutes, and the relationship between S&T and the economy. As an important institutional innovation, the transformation of research institutes should play a more important role in corporate R&D in China and become an opportunity to enhance corporate R&D in the future.

Public Interest Related Institutes (PIRI)

The Chinese science and technology system also includes public interest related institutes (PIRI) that have also undergone major changes, although the process is not yet completed. In the first stage experimental reforms were carried out in a select number of institutes. A second stage covered institutes belonging to nine different ministries, and in a third stage the transformation will be completed. Experimental reform started in 1996 but action did not start until November 2000.

The first group of PIRI institutes to be transformed included those belonging to four national ministries – the Ministry of Land and Resources, the Ministry of Water Resources, the State Forestry Administration and the State Meteorology Administration. These ministries controlled altogether 98 S&T organizations with 23 000 registered employees. The second stage of PIRI reform targeted 107 research institutes belonging to nine ministries, and implementation was completed by October 2002. The PIRI reform is a complex and ongoing project for which funding solutions are not yet completely sorted out. The Ministry of Science and Technology (MOST) provided funding for the first stage of the reform while the Ministry of Finance has identified special funds for completing the reform process.

ADVANCING ENTERPRISES, R&D EFFORTS AND SEARCHING FOR TECHNOLOGIES

Most international ICT companies are primarily concerned with expanding their markets and extending the lifecycle of their products. Chinese ICT firms are keenly aware of their need to gain access to technological know-how, and to be aware of advances in knowledge-related technologies. Furthermore, they are eager to reduce technology gaps through using foreign direct investment (FDI) and joint ventures (JVs) in developed regions. Many Chinese ICT companies have established a foreign presence as a window for business intelligence (BI) to follow the latest technology trends, occasionally combined with research activities that have a bearing on the future. The emphasis on learning means that gaining a long-term competitive advantage is more important than pursuing short-term financial returns.

Overseas branches focus on learning and developing technologies, tapping into technology sources and talent pools, while in China, there is still a large potential to learn from foreign partners through JVs, licensing and other means of cooperation. The benefits from technology transfers in formal collaboration may be limited and it may be more attractive to hire employees from foreign firms. Both the return of overseas Chinese, and the flow of employees from foreign firms to Chinese firms, are important trends. It should be noted that Western companies have set up operations in China primarily to gain market access, with limited intention of transferring technologies, although this is often required by Chinese agencies. Many domestic ICT companies are now making this negative assessment of foreign technology transfer.

Earnings in the home market are not sufficient for Chinese firms and there may be several explanations. First, as already mentioned, there is an obvious reluctance among foreign firms to transfer advanced technology. Second, Chinese firms have suffered from their inexperience in specifying the level of technology transfer. Third, many Chinese companies still lack the management techniques to handle efficiently the transfer of knowledge between individuals in different knowledge domains. A mechanism favouring technology transfer through JVs in China is still evolving, and firms like Putian with its involvement in several large JVs, expressed serious criticism of the reluctance of several of their partners to transfer technology.

Huaqiang Group

Huaqiang has recently been ranked by the Ministry of Information Industry as number 14 in China's top 100 ICT companies. Sales reached RMB8 billion in 2002 with 80 per cent being exported. Major customers for electronic goods include LG, Philips and Liteon in Taiwan. The Huaqiang Group consists of some 20 companies including six joint ventures with Sanyo in Japan. Huaqiang was one of the early company developers to participate in China's enterprise reform. It established its own brand, attracted foreign capital and introduced advanced foreign technology and management experience. Today the Group is a state-owned company directly under the Guangdong Government. It was established in 1979 by the personnel who were transferred to Shenzhen from three small munitions factories located in the mountainous region in Northern Guangdong. It has developed into a conglomerate with total assets of RMB4.4 billion from a small enterprise in which the province initially invested RMB6.42 million.

In recent years the company has changed its structure from an earlier focus on household electrical appliances into three main areas – electronic industry products, real estate business and the beer industry.[6] The latter industry, Zhaoqiang Blue, was acquired in 1997 and divested in early 2001. The Group has basically completed the transformation from a traditional household appliances industry to a high-tech conglomerate, and revenue from electronic parts and components accounted for more than 60 per cent in 2000.

The Group built the Huaqiang high-tech industrial park focusing on the production of laser pick-ups and its applications in an area which is a production base for similar products, providing the country's most advanced equipment and largest scale of operation. Other activities include the production of colour TV sets and floppy disk drives located in Dongguan City, where in the early 2000s more than 60 per cent of all floppy disk drives in China were manufactured. The Group has also built an industrial base for high-capacity batteries and DVD players, also in Shenzhen. Motorola and Nokia are the main customers for its batteries, which include lithium-ion, nickel-hydrogen and nickel-cadmium batteries.

The Group today has large-scale production lines for laser pick-ups supplying 25 per cent of global production, produces 15 per cent of all floppy disk drives, and delivers 90 per cent of all batteries used in Nokia handsets that are manufactured in China. Based on its own Intellectual Property Rights (IPR) the company has developed a number of high-tech IT products which include Global Positioning System applications and optical communication access products. To support its independent development the Group has established two research laboratories for broadband wide area information technology and for wireless multimedia technology. The company has given special attention to establishing high-level competence in software development by drawing on personnel returning from foreign countries and domestically recruited staff with Masters and Doctoral degrees. The Group has established a product design and development centre.

Household electrical appliances is a traditional industry for the Group that developed through cooperation with companies in Japan which have excellent teams for scientific research and design and equally strong ability in development and production.

In 1998 the Group acquired 70 per cent of equity in Shenzhen Yunwang Intelligent System Co., with a high level of competence in digital video and intelligent entertainment systems and related software development. In the same year the company established Shenzhen Huaqiang Communication Co. and Huaqiang Netcom Co., which are developing the communications industry by allocating highly qualified people to special projects. In 2001 the Group established Jingfeng Investment Co. to manage capital transactions. The Group was one of the initiators that built the busy commercial area of North Huaqiang in Shenzhen City.[7]

The company expects to develop into an inter-trade, trans-regional and trans-national shareholding company with a focus on high-tech industry and capital transactions. It considers itself to have a viable industrial structure, high quality of assets and several companies that are listed abroad and in China – with the ability to diversify and coordinate its future development. The company expects that by 2005 total assets, net assets and profits will reach RMB6.5 billion, RMB3.5 billion and RMB0.4 billion respectively, with an expected annual growth of more than 10 per cent.

The Huaqiang Group has established the following joint ventures with Sanyo:

1. Shenzhen Sanyo Huaqiang Optical Technology Co. Ltd that was established in 1993 and in 2003 employed 9500 employees. This company has several (Japanese) shareholders with a dominant position held by Sanyo. Revenue was about US$400 million in 2002.
2. Dongguan Huaqiang Sanyo Motor Co. Ltd was established in 1995 and is the next largest joint venture with 6500 employees, and is manufacturing various types of mini- and micro-motors. This company has several (Japanese) shareholders with a dominant position held by Sanyo.
3. Shenzhen Huaqiang Sanyo Technical Design Co. Ltd was established in 1995.
4. Shenzhen Sanyo Huaqiang Energy Co. Ltd was established in 1995.
5. Dongguan Huaqiang Sanyo Electronics Co. Ltd was established in 1998.
6. Guangdong Huaqiang Sanyo Holdings Co. Ltd.

Huaqiang Group has a total employment of 18 000 staff, which would indicate that the two Sanyo joint ventures for laser pick-ups and micro-motors constitute a major part of industrial activities. The total Japanese staff in the joint ventures is less than 50, mainly responsible for management and accounting. Sanyo is the most important source for technology transfer, although the company carries out some technology development. However, the joint ventures with Sanyo will remain the basis for further development, while Sanyo is one of the most important marketing channels for micro-motors and DVD laser pick-ups. Japan is also a key source for components like semiconductors and lenses.

The Group maintains an intelligence window in the US, where it collects information and analyses the market from offices in New Jersey and Las Vegas, with total employment of some 50 people.

Dongguan is located some 35 kilometres from Shenzhen and its choice for new industrial activities is justified as it offers land and labour at lower costs compared with Shenzhen, although the latter location has a number of advantages in terms of communications, skills and logistics. China is an excellent production base for mature products because of its good macro-environment in a number of places, of which Shenzhen is an excellent example. China has also become a preferred place for niche products, a good example being micro-motors. The explanation is that diligent workers are available at low cost in an environment that provides good communications and excellent logistics.

Putian Group

The China Putian Group has established over 90 joint venture companies with 47 companies, including long-term strategic partnerships with world renowned telecommunication manufacturers from 19 countries and regions, such as Motorola, Nokia, Ericsson, Alcatel, Panasonic, Toshiba, Sanyo, Lucent, Corning, Nortel and TI.

Putian was originally an industry bureau under the Ministry of Post and Telecommunications in charge of around 30 companies, and was changed into the China Posts and Telecommunications Industry Corporation in 1980. The Putian Group includes some 50 companies of which five are publicly listed, with a total of 54 000 employees.[8] Putian currently belongs to the State-owned Assets Management Commission. It has management control through several layers, and a complicated ownership and investment structure, which is typical in many state-owned companies.

The company's business range covers fixed and mobile communication. The communication equipment and terminal products it can provide include: mobile communication equipment, both the base station and handsets, network management and value-added service systems, internet protocol-related products, microwave communications equipment, optical communications equipment, optical fibre and cable, various power supplies, IC card payphones, multimedia terminals and fax machines, as well as project contracting, foreign trade, technology transfer, product import and export, international cooperation and so on.

Putian is committed to becoming globally competitive in IT products and services. The company's total import and export volume reached US$4.05 billion in 2002, being ranked fourth among the 500 largest import and export enterprises. Its export volume is US$1.72 billion and it is ranked fifth among the 200 largest export enterprises.

Datang and Putian have become competitors as both firms have their own research centres and production lines. Both firms were formerly managed by the Ministry of Post and Telecommunications. Datang was traditionally a research institute without manufacturing on its own, while Putian was traditionally in production, without its own research.

In recent years, the company has assumed the responsibility of developing new products of national importance, such as mobile communications systems and terminals. It has, in cooperation with key domestic and international companies, established COMMIT Incorporated and Beijing Putian Founder Communications, Inc. which participate separately in the R&D of TD-SCDMA and LAS-CDMA, and the company has also participated and co-funded the TD-SCDMA Industry Alliance. The Group has set up its central research institute – the China Putian Institute of Technology (CPIT) – research institutes in its major subsidiaries and four in Capitel, Eastcom, Runaway and Smartcomm. CPIT is responsible for the comprehensive management and planning of some 80 R&D units within the Group.

In interview discussions, Putian Group representatives argued about JVs and associated technology transfer in the following way:

> We still didn't learn much from the JV. JV is just a business mode, which was suited for the economic transition in China during the past 10 years. This mode has its own limits. We couldn't learn much from the foreign vendor through the JV. We were just used as cheaper labour presented in the production line. The foreign partner has always kept the control of the core technology. We were also used as a market channel for foreign vendors to get into the Chinese market.
>
> JV did bring both Chinese and foreign partners much profit. But the technology was not really being transferred. Now in the new JVs, we require the technology transfer. 'Market exchange technology' is our major goal in JVs.

Now, after China's WTO accession, there are less and less JVs. Foreign vendors can have fair competition in China and get direct access to the market. Foreign vendors have also accumulated much market knowledge about China.

IBM ThinkPad Becomes Lenovo

In early December 2004 the Lenovo Group and IBM announced a definite agreement under which Lenovo would acquire the Personal Computing Division from IBM and thereby becoming one of the world's largest PC businesses. The deal would bring IBM technology to Lenovo while the company would have a market reach beyond China and Asia. This is a logical strategy for Lenovo if it desires to remain a major and eventually global player in the personal computer business. In recent years much of China's growth has been driven by outsiders, many of them being Taiwan-based contract manufacturers. Furthermore, Dell being based in Xiamen was able to match the PC production from Lenovo before the merger. The size of China's market suggests that everlasting gains would accrue to companies that are able to control critical technologies. This is evident in the TV sector where China is already the world's single largest market, with a household penetration that is commensurable with developed countries.

The Legend Group originated in 1984 when 11 scientists from the Institute of Computer Technology of the Chinese Academy of Sciences (CAS) established a core unit, with funding from CAS, to deal in computers. In October 1994, the company was registered as the first civilian technology enterprise. Subsequently, Zhongguancun (ZGC) in Beijing Haidian District evolved as a chosen location for new forms of high-tech enterprises. The new companies were seldom dependent on government funding but were market-oriented in terms of product pricing and enterprise management, and with employees who could be hired and fired at will.

In the beginning of 1985, CAS imported 500 IBM computers for its more than 100 research institutes. Legend, with less than 20 employees at the time, took over the responsibility of receiving, checking and maintaining the computers, and training staff. Experts at the CAS Institute of Computer Technology developed technology for the 'Sinification' of foreign software. Legend invested all its profits from servicing the IBM computers into the design, production, sales and after-sales service of a sinification board called HanCard. The sales income from HanCard was RMB87 million in 1987.

Subsequently, Legend became the largest domestic supplier of personal computers. With an ongoing diversification, the company was reorganized into six groups under Legend Holdings Co. in 1997. A gradual change of logo and name took place during 2003–2004 and the company became clearly known as Lenovo before its alliance with IBM in late 2004.

Aside from industry projects and system integration, the company maintained a strong focus on Chinese markets and became the leading distributor for a range of PCs and related equipment. Legend established collaboration with a number of recognized US companies such as Motorola, Texas Instruments, National Semiconductors, and Computer Associates. This followed from a realization that the company would not be able to be in a leading position as far as computer and PC technology was concerned. Consequently, Legend never abandoned the strategy of acting as representatives for foreign makers.

In 1999 Legend employed more than 1000 people in R&D activities of which some 150 worked on projects that were not immediately related to ongoing products.[9] The R&D structure at the time reflected the company's origin from the Institute of Computer Technology of CAS. After 1996, when a new R&D organization came into existence, the Institute of Computer Technology remained an integral part of the Legend Group responsible for some 20 per cent of the 150 people involved in long-term research. Another 30 per cent worked in the Legend R&D Institute while the remaining 50 per cent were developing applications needed in business segments.

Legend maintained a special relationship with the Chinese University of Science and Technology, located in Hefei, that used to be an integral part of the Chinese Academy of Sciences, where it also established the Legend Computer Company on the campus while providing financial support in exchange for project research being done there. In addition Legend had a number of relationships with other universities. These included Tsinghua University, Shanghai Shuichan University, Shanghai Ligong (S&T) University and Xian Dianzi (Electronics) Jishu University. In a Silicon Valley centre Legend employed some 20 people under the direction of the Quantum Design Institute (QDI) that was responsible for Legend motherboards and also had the function of technology watch and collecting intelligence on changes in motherboard design and production.

This brief background lends support to a recent study that suggests that the critical contribution of CAS was not proprietary technology or significant start-up funding but seconded technical personnel and freedom for them to undertake commercial activities and thereby starting a stream of revenue. Furthermore, the Institute of Computer technology of CAS provided a critical amount of legitimacy in the early stage of company development.[10]

By mid-2003 Legend employed 12 600 people and spent 3 per cent of its revenues on R&D. Total R&D staff at the time were 2700, which corresponded to around 25 per cent of total staff. The company's main research centre was located in Beijing with branches in Shanghai and Shenzhen. In 2002 revenues from international sales amounted to a miniscule 5 per cent.[11]

This brief story of Legend has illustrated new opportunities to establish high-technology companies when the reform movement started, and also successful developments which continued for some 20 years. However, Legend was facing serious competition in the domestic market and had hardly been able to develop any technologies that would serve for successful and profitable entry into new technology areas. This happened at a time when IBM was shedding more and more of its less profitable hardware divisions, and earlier approached Toshiba for a possible deal.[12]

The new company, Lenovo, will become the world's third largest personal computer company with sales of approximately US$12 billion based on 2003 revenues. The agreement includes long-term strategic alliances between the two companies in PC sales, services and financing worldwide. Major operations will be located in Beijing and in Raleigh, North Carolina. The headquarters will be located in a suburb of New York and the CEO will be a former IBM manager. With a total employment of some 20 000, approximately 70 per cent of the workers will be inside China.

Lenovo has announced that the two companies will enter a broad-based strategic alliance. As part of the transaction IBM will be the preferred provider of services and customer finance to Lenovo, which will be the preferred suppler of PCs to IBM, enabling IBM to offer a full range of personal computing solutions to its enterprise and small and medium business clients.[13]

Huawei Technologies Goes Worldwide

Huawei Technologies started four years later than Legend and has pursued with great effort and apparent success a business strategy based on developing strong technology capabilities in telecommunications. This has been done in response to a rapidly developing domestic telecom market – similar to the expansion of the PC market that greatly benefited Legend.

Huawei was started in 1988 when an engineer in the military sector retired and together with five or six entrepreneurs also with engineering backgrounds, started a trading business. They began selling switching products supplied by a Hong Kong-based vendor – with only limited capital. At a later stage the Shenzhen government provided a loan. In the early stages Huawei was a trading company that imported PBX switches from Hong Kong and distributed the products in China. The company started R&D in 1990 and was able to develop the switch CMC08 in the early 1990s, from which followed optical communication, in which Huawei has established a strong position – even international markets. Huawei has become strong in the supply of optical networks (Optix) with a market share in Asia Pacific of 15 per cent in 2001, which in 2002 had increased to 18 per cent.

Huawei formulated plans in the early 2000s to internationalize and to grow into a much larger enterprise in order to capture more markets and increase its market share. Annual sales revenue in 2003 reached RMB31.7 million (US$3.82 billion) and the company Norson Telecom Consulting has indicated that sales in 2004 could reach more than US$5 billion; overseas sales in 2004 exceeded US$2 billion, which is double the preceding year.[14]

The company allocates about 10 per cent of sales to R&D. Long-term research, which constitutes 10 per cent of all R&D, is equally divided between in-house research and activities done outside the company – mainly in collaboration with universities. Allocation of funds to R&D has increased rapidly from US$87 million in 1999, US$179 million in 2000, US$342 million in 2001 and US$363 in 2002. Some 9000 of the company's staff work in R&D (45 per cent) and only 10 per cent in production. Its R&D staff, 45–50 per cent of whom hold masters degrees, work closely with major targeted universities noted for their achievements in the fields of technology in which the company is active. An important element involves developing links for research and joint ventures in project-based R&D. Through such collaborations, the company hopes to gain access to advanced research at universities and to be able to recruit bright students. Relations with universities include Beijing University and of Post and Telecommunications (BUPT), the Science and Technology University in Hefei, Tsinghua University in Beijing, Dalian Qigong University, Dongnan University, Nanjing, and Liking University. A research centre in Bangalore has a total staff of 1000 including 200 Chinese. The total staff in Bangalore is included in total staff figures. The centre has a high level of Computer Model Maturity (CMM) – from level 4 to level 5.

An initial and early focus on gaining a market foothold in rural China and then later gaining access to major big cities clearly resembles Mao's focus on winning the support of peasants before gaining control of the country. Internationally a similar strategy is pursued in first gaining market successes in developing countries before attempting to enter the market in developed countries.[15] Huawei has entered Russia, Ukraine and Pakistan and is also active in the Middle East and North Africa as well as Asia Pacific – Indonesia, Hong Kong, Singapore and India. The export share of Huawei's sales is presently around 15–20 per cent and is expected to reach 50 per cent within the next five years. Huawei has established a number of international headquarters: Latin America, North Africa, USA (Silicon Valley), Middle East and North Africa, Asia Pacific (Malaysia), Russia (Moscow), Pacific for Australia, Japan, Korea and Hong Kong.

Huawei started its R&D on W-CDMA in March 1995 and was able to make a first delivery to Shanghai in 2001, with successful handsets in March 2003. Huawei had delivered W-CDMA equipment for testing in 21 locations by January 2003. Almost all of the 2500 staff in Huawei R&D centre in Shenzhen have been working exclusively on W-CDMA for several years, which could possibly amount to 10 000 man-years. Huawei is considered to be strong in W-CDMA and also in GSM, which reflects its relationship with

China Mobile, while ZTE is strong in CDMA2000 through its relation with China Unicom.

Huawei has been engaged in R&D on overall solutions to communication networks, ranging from core layers to access layers. It also provides an open service platform in the form of standard middleware and moves toward broadband, packet-base and personalized network development. Many university professors were involved in Huawei's development of W-CDMA, both on hardware and software. Usually Huawei requested participation and appointed a team leader for projects which could last from a few months to 1.5 years. Huawei has tried to collaborate with foreign universities but with limited success because of costs and the language barrier. The setup in Bangalore basically works very well as Indian engineers are extremely good at programming, although they do not always realize the expectations of performance that are only implied at the beginning. Huawei has also been able to obtain a big contract in India.

In April 2003 Huawei entered into an agreement with NEC and Panasonic (Matsushita) to produce W-CDMA handsets jointly within a company named COSMOBIC, as the company aims to establish an integrated supply chain. The two Japanese companies will provide advanced manufacturing capability, while technology will primarily come from Huawei. NEC and Panasonic have long experience in the design and production of handsets while Huawei is strong in network technology.

All major products are based on self-designed application-specific integrated circuits (ASICs) that involved some 300 people in 2003. The development of ASIC design has partly been done under the Government Project 909, which involves a total investment of RMB10 billion, part of the national programme to significantly upgrade the country's chip making facilities.[16] A number of chip laboratories are included in the National 909 Project, including Huahong-NEC in Shanghai, with which Huawei collaborates. Huawei has established a number of R&D Centres which include Beijing for data communication technology, routers and so on; Shanghai for 3G mobile communication systems; Shenzhen for CDMA R&D; Nanjing for network management; Bangalore in India for software engineering, the Russian site for switching and transmission; Stockholm for Universal Mobile Terrestrial System (UMTS); and US sites for ICs, CDMA and data communication.

A Huawei–CISCO conflict on IPR in routers probably speeded up the negotiations between 3Com and Huawei that had been dragging on for some time, and were finalized in early 2003. The partnership with 3Com draws on Huawei resources in Hangzhou. Huawei has issued a statement that it has not infringed on any CISCO patents. However, the infringement claim appears to concern infringement of software copyrights and Huawei has agreed not to use the software in question in future projects. Huawei employs around 22 000 people with another 2000 if employment in the 3Com partnership is included.

Huawei is very aggressive in its pursuit of overseas markets and is able to offer high-quality solutions at reasonable prices. Thus the Dutch mobile operator Telfort BV signed an order for a 3G mobile telecom system in late December 2004 which may have surprised major competitors like Ericsson, Nokia, Siemens, Alcatel and Nortel Networks. In the same month Huawei and its North American subsidiary FutureWei, announced that US mobile operator NTCH Inc. had commercially launched Huawei's next-generation CDMA 1X system in California and Arizona.[17]

Less than two months later Huawei reached an understanding with Marconi in the UK. Marconi will distribute a certain number of Huawei's carrier-grade data products to its key customers while Huawei will distribute Marconi's next-generation access radio products to its key customers.[18] This is undoubtedly an indication that Huawei is entering the European market and a next step would be to develop new products jointly with Marconi and other European companies, and distribute them through mutual distribution channels. Huawei had already reached an agreement with Siemens in Germany in the autumn of 2004 that it would sell Huawei routers and switches to its corporate customers, a deal which is very similar to the one that Huawei signed with NEC in Japan in 2003.

Huawei has chosen to be engaged in a long-term partnership with IBM that is focused on integrated product development (IDP), now running for more than five years. Another area of cooperation is integrated supply management (ISM). A completely new Logistics Centre was completed by November 2001 with technology provided by Siemens-Dematic and with conceptual support from the Fraunhofer Gesellschaft in Germany. Before the completion of the Logistics Centre Huawei very often suffered from serious problems in the supply of components and materials. These days hardly any mistakes are made in the completely automatic centre. Huawei is considering outsourcing the centre as its capacity could possibly handle more than one customer. Many others are or will be outsourced, such as restaurants, cars, house management, cleaning, security and the production of chips, the last being taken care of by TI.

INFORMATION AND COMMUNICATION TECHNOLOGIES (ICT) IN CHINA – AN EMERGING GLOBAL FORCE[19]

Information technology (IT) has become one of China's pillar industries with an average growth rate of some 30 per cent throughout the last decade. In 2002, the sales revenue from the electronic and information products manufacturing industry reached RMB1.4 trillion, an increase of 20 per cent over the previous year, and an average annual growth of 25 per cent in the preceding five years, contributing 20 per cent to national industrial growth. The IT sector, the largest among domestic industrial sectors, was 2.5 times larger than in 1997.

Consumer electronics has become the major driver of the IT industry in China. In 2002, the investment-related products, consumer electronic products and electronic components ratio was 45:29:26. The investment-related products category was 3 percentage points less than in 2001; and consumer electronic products, 2 percentage points more. The output of investment-related products, mostly communications equipment and computers, increased by13.8 per cent over the corresponding period in the previous year; this accounted for 30 per cent of the growth of the industry, compared with 46 per cent in 2001.

The IT industry has been singled out for major development in the tenth Five-Year Plan ending in 2005. Following the development of the electronics manufacturing industry in China, many manufacturing lines located in Southeast Asian countries such as Malaysia and Singapore are moving to China, not only to exploit the labour advantage but also to gain access to the emerging clustering and abundant networks of parts suppliers. Furthermore, foreign companies are also moving their R&D centres to China to tap into the increasingly abundant academic talent. A very large share of high-tech goods in Chinese exports originates in joint ventures that are controlled by foreign multinational companies.

Foreign capital has been a major driving force for the development of China's electronic and information products manufacturing industry. For example the China Putian Group, that used to be the manufacturing arm of the former Ministry of Posts and Telecommunications, has established over 90 joint venture companies with 47 companies, including world renowned telecommunication manufacturers from 19 countries and regions. Some of the leading companies in Japan, such as Matsushita, Sanyo and Sony, once held large shares of the Chinese market for consumer electric and electronic goods, including television sets and refrigerators. In recent years they have seen their shares fall rapidly due to competition from local companies, and survive only in niche markets.

Today there are many indications that manufacturers in China are starting to clearly realize the possibilities for moving beyond cost advantage and copying capability and are beginning to display technological prowess. The experiences of recent years have taught Chinese managers that catching up with the developed economies will require effort over a long period of time, and it has become obvious that a narrow, product-oriented development strategy does not offer long-term competitive advantages. Historically,

Chinese firms have always been the followers of multinationals and tried to learn and catch up through foreign cooperation. However, the JV model has been a disappointment for many Chinese firms, especially in the manufacturing sector. An increasing entrepreneurial fervour, combined with quick learning, have made Chinese domestic manufacturers grow quickly and seriously compete with existing multinational vendors in the Chinese market, as illustrated by mobile handsets. Although Nokia and Motorola maintain strong market positions, their market shares have declined, and in 2003 a new sales ranking has seen a Chinese vendor, China Bird, at the top position.

Firms learned the lesson that dependence on technology transfer within China will leave China always a step behind. Going international, and looking beyond national boundaries, current product portfolios, and today's competition in order to create international competitiveness, are necessary for Chinese vendors. This is also reflected in government policy, where the Chinese government has formulated plans to transform national resources into a knowledge-based economy.

Internationalization of China's ICT Industry

The collapse of the global IT bubble only marginally affected the Chinese information and communications technology (ICT) industry as a major proportion of sales were inside China and the domestic market continued to boom. Chinese companies have realized the potential benefits of becoming global actors and many have entered the road towards internationalization. They have signed various cooperation agreements with foreign companies. Many Chinese ICT firms have stated internationalization as a current strategic priority, but they realize that implementation will require considerable resources and take a long time. There are a number of hurdles to overcome which require strategic decisions: how and when to enter foreign markets, different approaches for diverse activities, what kind of entry models to choose, and how to modify operations as successes are scored.

Chinese companies are, with a few exceptions, not yet directly active in global markets. However, a number of Chinese companies are contemplating early moves to establish their presence outside China. China has already accumulated production knowledge and grown into a formidable 'world manufacturing centre', and physical assets, product-specific knowledge and hard, factual information have become less attractive for Chinese firms which desire advanced technological know-how and high-value tacit knowledge. The forces that are prompting the Chinese ICT companies to become global actors include: searching for technologies, exploiting new markets, achieving economies of scale, establishing brand names, participating in setting technological standards, competing with global actors, and avoiding tariffs and border restrictions.

Exploiting Foreign Markets and Branding

Exploiting new opportunities in other markets is a natural next stage for many Chinese ICT firms which have consolidated and secured a home market. International markets can act as outlets for surplus domestic production, especially in consumer electronics, as the domestic market becomes saturated. The five year period ending in 2002 saw an annual average export increase for ICT products of 28 per cent. Analysts suggest that a major share of this increase emanates from foreign affiliates. However, the available information indicates that foreign affiliates contributed no more than 48 per cent of total exports in 2000–2001, which shows that more than half of the exports actually came from Chinese firms.

The intensive domestic competition forced many ICT vendors to search for a larger market. Overseas expansion for household electronics, mobile handsets and PC components seemed to be a logical next step, and Chinese firms have obvious advantages in price leadership and differentiation for low- and medium-end products.

Many Chinese firms have clearly realized that global brand recognition, or a global presence, gives firms an edge in reputation and credibility, and this has now become a driving force to go international. Some Chinese firms have acquired foreign firms and distribution channels to reach foreign markets. As a first cross-border acquisition case, in 1975 TCL acquired a colour TV producer in Hong Kong, and simultaneously acquired its technology. In 2002 the same company acquired Schneider, a bankrupt German TV maker, and will continue to use the German brand name. However, the global business community was surprised to learn on 3 November 2003 that TCL Corporation of China, based in Shenzhen, had signed a deal with Thomson SA of France to form a joint venture in the production of television sets with an annual production of 18 million units, compared to the 12 million units produced by Sony and the same number by Matsushita (Panasonic). Also Beijing Oriental Enterprise (BOE) has acquired the flat screen division from the ailing Hynix Electronics in Korea.

Huawei and other firms find plenty of buyers at home of course, but they are beginning to venture abroad as a matter of pride and being recognized as suppliers of advanced technology. Huawei displays its sales contracts from Hong Kong, Russia and the Baltic states as proof that its products are world-class. Konka Electronic Group is distributing its TV sets in the US market, and has set up a factory in Mexico. TCL has established TV production joint ventures in India and Vietnam. Skyworth Group, based in Shenzhen, has launched production bases in Mexico and Turkey. The Changhong Electronic Group has set up joint ventures in Russia, Mexico and Indonesia. Great Dragon is present in Cuba, Columbia and Bangladesh, and Lenovo, formerly Legend, is present in Russia.

A Chinese brand is traditionally associated with low-end, cheap products, and many Chinese firms have two strategies to overcome this negative image. One way is to allocate substantial resources to create a brand image. This is exemplified by Haier, which displays its name on luggage carts at John F. Kennedy International Airport in New York. A similar approach is exemplified by China Kejian, which sponsored the Chinese national football team, as well as the English Premier League soccer team Everton for two years, at a cost of US$3.2 million. Another way is to buy a recognized brand name. TCL bought Schneider in Germany, but whereas TCL uses its original brand name in Southeast Asia, it uses TCL-Schneider in Europe. The value of the Schneider brand name in international markets cannot yet be assessed, although it may immediately make a difference in the Chinese market as most Chinese consumers have high confidence in European brands.

Hurdles of Internationalization – Global Sourcing

Going global is a challenge for most Chinese ICT firms, and is also a recent phenomenon. Chinese firms need much country-specific knowledge, such as knowledge of the local market, business practices and institutional conditions, before getting into a new market. Many firms are still quite inexperienced at acquiring country-specific knowledge. They therefore behave in a reactive rather than proactive way when facing international business conflicts. It is obvious that locally-sourced personnel in foreign markets bring value to the acquisition of country-specific knowledge. Huawei has been very proud of its local sourcing of Indian software engineers at its R&D centre in Bangalore. BOE also highly values its Korean employees in the acquired TFT-LCD plant, Hydis, formerly part of Hyundai Electronics.

Much knowledge, such as foreign business culture and the sense of environmental variables, is not available in an explicit, codified form. Lack of industry-specific knowledge is one of the obstacles for Chinese ICT internationalization. This kind of knowledge is vital for strategic business decision making, requires time to build up, and calls for specific insights. Fortunately many overseas Chinese have offered great assistance to the Chinese firms, both in capital and human resources. In Silicon Valley, where many Chinese ICT firms have set up branches, overseas Chinese have become a natural contact between China and the US in business operations and technology development. There are, however, still gaps of knowledge and experience between the headquarters' management and the operation of the overseas branches.

When Chinese products are exported successfully, many local competitors, especially in the US, are quick to petition their governments for relief from the increased foreign competition by referring to anti-dumping regulations and so on. FDI may be one of the means to dodge tariff barriers and to avoid import harassment under the trade remedy laws of the host country. Chinese firms, especially home appliance makers, are setting up overseas plants, even though the production costs are higher. The Haier factory in South Carolina is one of the examples of being inside the tariff walls.

China is at the centre of the rapid technological change that characterizes the ICT sector. With increasingly short product life cycles, it is necessary for firms to keep up with the latest technology, which leads to the global sourcing of technology know-how. A substantial share of components, most importantly integrated circuits, must be imported. Thus, it has become important for Chinese ICT firms to have close relations with global suppliers, who can comply with the rapidly changing technological and quality demands, and shorten the time-to-market (TTM). Additionally, the international branches of Chinese ICT firms can act as platforms to access global talent pools and foreign capital resources, as we see from Huawei's recruitment of former Ericsson engineers in Stockholm and Lenovo's ambition to be publicly listed on the NASDAQ Stock Exchange.

SURFING ON R&D MANPOWER INTO FOREIGN MARKETS

The China of today is not generally perceived – in the traditional sense of fostering great innovations – to be an innovative country, in spite of its outstanding success in industrial production by capturing segment after segment in overseas markets. However, Chinese companies have in recent years shown great acumen in creativity and implementation which constitute critical pillars for the successful commercialization of innovations. Implementation itself requires the selection of an idea, its development and final commercialization in a receptive market. These processes can only be successful if they are carried out in a creative environment, which requires a substantial and ongoing learning process.

The traditional structure of large-scale industries in China involved a close coordination of actual production and the creation of a large cadre of engineering professionals under the direct control of line ministries. This structure often created confusion and heated debate on how to create linkages with other institutions in the economy responsible for breakthrough innovations. Despite its shortcomings the prevalent structure created a cultural understanding of the role of knowledge creation and transformation of engineering in itself, although relations between scientists and engineers remained bewildered.

The ongoing transformation of the Chinese industrial systems coincides with a number of far-reaching global organizational changes, of which the expanding network character of the company may be the most important, combined with new types of supply-chain management which have brought the concept of advanced logistics to the forefront. Simultaneously, industrial production has undergone a significant shift from manufacturing in favour of services, which has favoured new and expanding categories of entrepreneurs in new technologies.

There are few companies in China that would qualify in a list of multinational companies, although many of China's large companies are listed on the Hong Kong Stock Exchange. However the consumer electronics field has already seen the emergence of new players, some of which are aiming for global markets. One is Changhong in Sichuan, which has become one of China's leading manufacturers of TV sets in only a few years. The company has acquired other TV producers in Jilin and Jiangsu and the latter acquisition has been motivated by a desire to move into the overseas market. Changhong views both Southeast Asia and Australia as promising markets. The Haier Group, with headquarters in Qingdao, provides a similar example of successful foreign entry of home appliances. With rapid development over 15 years, the Haier Group has become one of the most well known Chinese brands, and has become one of the first Chinese companies with a brand name that is becoming recognized outside the country. After obtaining technology from Germany in the early 1980s, the company has in recent years expanded solely through its own efforts and vision, and has now set its goal of establishing production in both Southeast Asia and Eastern Europe. The group's major business is in electrical appliance manufacturing, including refrigerators, air conditioners, washing machines, microwave ovens and so on. The group has diversified into manufacturing TVs, fax machines, telephones and computers.

The experiences of the companies referred to above clearly indicate three major types of large companies in China. First, Huaqiang Group and Putian Group are large conglomerates which at an early stage entered into a number of joint ventures with famous foreign companies. These joint ventures started manufacturing products to meet demand in the domestic market as well as in overseas markets – usually under recognized brand names. However, the expected technology transfer to domestic partners and further transfer has only materialized in a very limited way. Second, Lenovo exemplifies a group of companies that has been able to establish a strong presence in the domestic market on its own, although with limited presence in foreign markets. This experience has until recently been shared with many other IT companies in the consumer segment. An important reason is that both types of companies have paid limited attention to R&D, when compared with foreign competitors. This explains why Lenovo and other such Chinese companies now attempt global recognition through mergers and acquisitions – to get access to technology and global brand names.

Third, Huawei Technologies illustrates a completely different corporate strategy, although originally starting by meeting domestic market demand with imported products. Huawei, at an early stage, like ZTE Technologies, recognized the importance of its own R&D activities. Both companies presently have striking employment profiles with almost 50 per cent of employees engaged in R&D activities. This has increasingly created a comparative advantage for knowledge-intensive products, as scientists and engineers are available at 25–35 per cent of the salaries of their counterparts in the US, Western Europe or Japan. This comparative advantage is now captured by an increasing number of Chinese companies, which also provides the reason why foreign R&D companies are congregating in China.

A rapidly shifting balance between manufacturing and services has a direct bearing on understanding the role of innovation that used to be closely linked to manufacturing. Subsequently, a number of activities in logistics, finance, planning and coordination need no longer be performed within the production plants but at distances that have both an organizational and geographical dimension. Thus, industrial production becomes increasingly dominated by services and by people who are trained in a large number of non-manufacturing activities.

The result in China is that the traditionally integrated manufacturing firms have been dismantled into brand name firms and contract manufacturers, with the former generally located outside China with almost complete control over product strategy, product R&D, functional design, marketing and form design. The combination of design and logistics capabilities has come to play a major role for a large number of IT products. The networking character in a large number of industrial sectors has increasingly blurred the understanding of where innovative processes occur and where new capabilities are created. The dominance of brand name firms and their access to market channels and early understanding of both gradual and major changes in the market place have in the past given them a dominant position. However, with the innovation processes in a networked economy being blurred, the intermediate contract manufacturer may become more important in the value-added chain in being able to attract and control a large share of consulting and coordinating activities. A rapid expansion of the talent pool provides China with a potential comparative advantage to capture more and more of such high-performance activities.

NOTES

1. This chapter incorporates a contribution from Dr Xinxin Kong prepared for the project 'Emergence of New Knowledge Systems in China and their Global Interaction': *Corporate R&D in China: The Role of Research Institutes*. The full contribution was presented at a conference, and a summary has appeared in Sigurdson (2004).
2. The science and research responsibility system means that institutes could receive projects from enterprises and other organizations and create revenues for themselves when they had finished the tasks allocated by the higher administration.
3. Contract on charge means that institutes signed contracts with users, including central and local governments, enterprises and other unit organizations. These contracts contained the regulations for technology, economy and projects.
4. One of the basic principles that Chinese reform abides by is 'first easy, then difficult', which is also the typical feature of progressive reform.
5. This figure and later data in this section all come from the questionnaire estimation of 319 transformed institutes held by MOST in May 2000. These 319 transformed institutes covered two groups of 'corporatizition' institutes started separately from July 1999 and July 2000. In total, 376 technology exploitation institutes were transformed. Among them, 319 institutes became corporations in the ways described in this section.
6. The company brochure of 2002 explains its acquisition in the following words: 'Participation in the beer industry was an important measure taken by the Company to conduct inter-trade business of famous-brand products by means of capital operation. Through the acquisition and control of the CBR Brewery company listed in America, the Company has made Blue Ribbon beer that has the true qualities of a world famous brand well received in Chinese beer market with high quality and strict management. The Company will strictly operate according to scientific standard and traditional technique and constantly put out high quality beer with light colour, pure taste and sufficient bubbles.'
7. The company brochure of 2002 has the following to say about real estate development: 'Facing the change of urban functions of Shenzhen Economic Special Zone, the Company has timely adjusted its industrial structure. Located at the "golden section" of Shenzhen City it has greatly developed the business of real estate and properties and opened up new channels for economic growth by making use of its resources of land and property. The Company established real estate company and property management company in succession and changed the functions of the original factory buildings and properties. It built Huaqiang Electronics Supermarket, the largest market of electronic parts and components and supporting equipment in China. The group obtains annual income of nearly RMB 0.1 billion from property leases through conducting leasing business. Meanwhile, the Company is building high-rise residential buildings including Huaqiang Building and Huaqiang Garden. Huaqiang Plaza with greater scale is being planned. After operation for several years, the business of real estate and properties has brought stable profits to the Company and become an effective supplement to the industries.'
8. In 2001–02, Putian ranked top in the MII list for sales revenue, and 10th in the overall listing of China's largest enterprise groups. In 2003 it ranked 2nd and Haier was number one.
9. Private interview notes August 1999.
10. Xie and White (2004).
11. Interview 12 August 2003.
12. Nishioka, Koichi, 'Focus. Lenovo IBM Deal to attest to China's leadership on PC mkt', *Nihon Keizai Shimbun*, 20 December 2004 (Nikkei Interactive).
13. As a consideration for the transaction IBM will receive at least US$650 million in cash and up to US$600 million in Lenovo Group common stock, subject to a lock-up period expiring periodically over three years. IBM will be Lenovo's second-largest shareholder, with an 18.9 per cent interest. Additionally, Lenovo will assume approximately US$500 million of net balance sheet liabilities from IBM. Lenovo will fund the cash portion of the considera-

tion through internal cash and debt. Source: Lenovo website, http://www.lenovo.com, 8 February 2004.
14. 'Huawei on a roll with 3G', *China Business Weekly*, 9 January 2005.
15. The founder, with his military background, has imposed a fairly strict management style throughout the company. Punch cards are used when starting and finishing work and people coming late are fined. The lunch period is between 12.00 and 13.30 and office workers are expected to have a short nap, on a pillow which they keep at their working place. R&D people are not expected to follow strict rules for working hours and dormitories close by provide a number of amenities. The company has issued a number of rules that are collected in a rather thick volume. The founder has also introduced Lei Feng as a role model, who during the earlier communist era was an icon of unselfish behaviour in serving the larger good of the society, and has also provided generous economic benefits. Huawei employees see the company expansion strategy as a parallel to Mao Zedong's strategy in winning control over China.
16. Hou Mingjuan (2001).
17. 'Huawei on a roll with 3G', *China Business Weekly*, 9 January 2005.
18. Li Weitao (2005).
19. This section is based on interviews carried out in Shenzhen, Shanghai, Ningbo and Beijing during 2003, jointly with Ms Vicky Long (now a PhD student at the Royal Institute of Technology in Stockholm Sweden). The project was sponsored by the Invest in Sweden Agency (ISA). Report: Sigurdson and Long (2003).

5. The information and communication technologies: example of institute reform

CHINA'S IT INDUSTRY AND ROLE OF ITS IC INDUSTRY

The importance of the IT industry was addressed in the sixteenth National Congress of the Communist Party, stating that the development of the IT industry should take precedence over other industrial sectors and that China should apply information technologies extensively in the economic and social fields. Following this national ambition the State Council in June 2000 announced a number of policies to encourage the development of the software industry and the integrated circuit industry in China.[1] The Deputy Minister of the Ministry of Information Industry (MII) stated at the end of 2002 that 'if the IT industry in China is to realize its potential and develop by leaps and bounds to make China the world power in this field, it is imperative that the bottleneck resulting from restrictions caused by the semiconductor industry and technology be removed'.[2]

China's IT industry (electronics information products) continued its rapid ascent in 2003 and registered a manufactured value of RMB1880 billion, which is an increase of 34 per cent over the preceding year. Added value of the industry increased at the same rate and reached RMB400 billion. The value of IT exports totalled US$141 billion, an increase of 53 per cent over the preceding year, considerably higher than China's overall increase in exports. The IT sector contributed 44 per cent of China's total exports in 2003. The combined value of the IT industry in the same year reached 4 per cent of GDP, one percentage point higher than in 2002. The IT sector employed 4.1 million people, with an increase of 325 000. The electronic manufacturing sector employed only 3.5 million, with another 0.6 million engaged in the software sector. However the export value of software totalled a miniscule US$2 billion, with RMB160 billion in domestic sales for software products and system integration.[3] Although having a huge market, competitive labour costs and highly skilled human resources, officials believe that the industry still has a long way to go and raise the concern whether the industry can become a real global player in the face of intense international competition.[4]

The semiconductor sector has developed through joint efforts from wholly Chinese enterprises, joint ventures and wholly foreign-owned enterprises. Although 2001 saw a downturn in the global market for semiconductors, the domestic industry increased to RMB120 million, which is about 30 per cent up over the preceding year. However, the Chinese semiconductor industry suffers from several shortcomings. First, domestic supply does not match the local demand and a large number of products have to be imported from abroad. Second, China's technological level still remains low and most core technologies are controlled by foreign companies. Third, China's semiconductor supply chain is far from complete and major pieces of processing equipment, instruments and materials have to be imported. These factors seriously constrain China's domestic development of its semiconductor manufacturing industry. In a medium-term perspective the situation will improve through policies such as industrial restructuring, the development of specialized science parks and support for the integrated circuit (IC) manufacturing industry.

Given these circumstances, the Ministry of Information Industry (MII) has decided that developments in semiconductor design promise more immediate progress in the short term, while laying a foundation for manufacture and its supporting industry. Here, emphasis will be given to products which are required in large quantities, great efforts will be made to train the engineering and scientific personnel needed to meet China's long-term ambitions in this area.

The computer sector continues to grow rapidly, although communication products, like switching equipment for mobile and fixed-line systems, continue to make the biggest contribution to the economic growth of the whole sector. Foreign IT companies, including those from Hong Kong, Taiwan and Macau, represent almost one fourth of more than 4000 IT companies. The China Centre of Information Industry Development (CCID) reports that 'In 2003, their sales revenue, profit tax, industry added value and export value all exceeded more than one half of the whole sector, at 67.5 per cent, 57.9 per cent, 52 per cent, and 82.7 per cent respectively'.[5] Given the importance of the IT industry not only in China's economy but particularly in its exports, it may not be surprising that the Government and its various development agencies felt challenged to shift the industrial balance in favour of domestic companies and their own development of key technologies. The integrated circuit industry plays a significant if not dominant role in the IT sector and is likely to become even more important. Thus, the great attention and major Chinese efforts within the domestic IC industry are understandable, as demonstrated by recent figures.

The demand for ICs in the Chinese market reached RB237 billion in 2003, which corresponds to 20.4 per cent of the worldwide market, an increase of two percentage points over the preceding year. In value terms the demand for products with a line width above 0.35 micron is less than 20 per cent while demand value for products with a line width of 0.18 or below amounts to 40 per cent. The main IC markets include computers, communications and consumer products. Other IC markets such as IC cards, ICs for meters and instruments and auto electronics have become more important in terms of market value.

China's domestic semiconductor industry, which consists of the IC industry, discrete devices and supporting sectors reported sales of RMB69 billion in 2003, which correspond to five per cent of the worldwide semiconductor market. The IC industry, which includes IC design, chip fabrication and IC packaging, reported total sales of RMB35 billion in 2003, which correspond to three per cent of the worldwide IC market.

China's IC design sector is developing rapidly and recorded a sales increase of more than 100 per cent in 2003, with a sales volume of RMB4.5 billion. Designers in China can now handle circuits with a line width down to 0.18 micron and more than five million gates, which indicates an ability to move into core technologies. Chip fabrication also expanded rapidly with an increase of 80 per cent, and reached a sales volume of RMB6 billion in 2003. With a number of new production lines and total investment of some US$10 billion, China is now entering into the global IC industry – strongly supported by major foreign investment such as Huahong-NEC, SMIC, Grace, ASMC, Hejian Technology and Tianjin Motorola. However in terms of value, it is packing that still dominates China's IC industry, and its production value of RMB24.6 billion in 2003 represents some 70 per cent of the whole IC industry in the country. A substantial number of large foreign IC manufacturing companies have established wholly-owned or JV enterprises for packing ICs in China – altogether more than 50 plants that have a total annual capacity of 17 billion units. CCID reports that many of these companies are approaching very advanced levels in terms of manufacturing equipment, process development and management.

China has a strong position in semiconductor discrete devices, with a 2003 sales amount of RMB34 billion, which corresponds to more than 15 per cent of global sales value. However, China's position in the semiconductor industry supporting sector is much weaker although everyone concerned has clearly realized that equipment in particular plays a very decisive role in the semiconductor industry chain. The sales value reached RMB600 million in 2003, covering a whole range of equipment, but in terms of industrialized manufacturing the capability is limited to 0.8 micron technology on 6 inch wafers, a level which has been phased out by major global manufacturers.

A total of 40 research organizations and equipment makers are involved in an attempt to upgrade rapidly China's capability for producing advanced semiconductor equipment. The national Programme 863 is supporting the development of lithography equipment for line widths below 0.25 micron, including etching and ion implantation equipment. Recently a number of Chinese companies have entered into the semiconductor and IC special equipment manufacturing field and overseas Chinese companies are considering establishing semiconductor equipment companies in China. The industry producing semiconductor material such as silicon is still not very advanced. Total production of silicon reached 300 tons in 2003 with a sales value of RMB1.4 billion. The production of LED frames totalled 10 billion pieces, with a sales value of RMB330 million, which meets 80 per cent of the demand for discrete devices and 20 per cent for ICs.

THE IC INDUSTRY IN CHINA AND ITS REGIONAL CHARACTERISTICS

In an overview of China's IC industry, an MII official has provided an illuminating perspective.[6] He suggests that the average annual rate of growth in the decade 2000–2010 will be more than 15 per cent and that annual sales to the world semiconductor industry[7] may amount to US$600–800 billion in 2010, which will support the electronic equipment market with a value in the region of US$4000–5000 billion.

The national semiconductor industry report of 2002[8] mentions that the Chinese industry consists of ten major chip-producing enterprises, dozens of packaging plants and more than 300 design units and R&D institutions, and also a number of enterprises that manufacture semiconductor equipment, materials and instruments. IC design, manufacture and packaging are closely related and are concentrated in three regions – Shanghai with neighbouring Jiangsu and Zhejiang provinces, Beijing and Tianjin, and Guangdong and Fujian provinces.

With an emphasis on design, it is still astonishing that China did not have more than 200 registered design companies in 2002 with a total employment of not more than 15 000 employees. Most of them are still small and in 2002 only four of them had an annual turnover of more than RMB100 million – Huada, Xike, Datang Microelectronics and SiLan. Mainstream design was at the time 0.35–0.80 micron with the highest level being 0.25–0.18 micron. Earlier bottlenecks have resulted from the lack of computer aided design (CAD) and of experience in using advanced software for IC design.

China's IC manufacturing industry is located in Shanghai, Beijing, Jiangsu and Zhejiang, with Shanghai Huahong NEC Electronics at present being the most advanced facility, able to reach 0.25–0.18 micron technology. The IC packaging industry has developed quite substantially and is oriented to markets both at home and abroad.

The Semiconductor Manufacturing International Corporation[9] (SMIC) is China's biggest supplier of made-to-order chips and will have a monthly production of around 150 000 wafers by the end of 2005, compared with a monthly production of 100 000 wafers at the end of 2004. Its major semiconductor foundries are located in Shanghai.[10] The domestic market in China is not affected by the lower growth rate for ICs in the global market and will continue to expand. Forecasts indicate that in 2005 it will overtake the US – the world's biggest IC market – after growing at 11 per cent to reach US$34 billion against a $33 billion capacity for the market in the US.[11]

The future prospects for China's IC industry looked quite promising in the early 2000s, because of the following factors. First, the demand was increasing rapidly with a significant demand for ICs to be used in communications devices, IC cards and specialized memories. Second, China was still enjoying a boom market in 2001 due to an expanding domestic market while the global market was suffering from a downturn. Third, domestic and international enterprises have made rapid technological progress – in terms of larger wafer sizes and finer line widths. Thus, the Chinese semiconductor industry is reducing the technological gap. Finally, some large IT manufacturing companies have given special attention to semiconductors and exploited their capability in IC design as an important entry for bridging IT and IC manufacturing. In the light of this favourable situation, the State Council Circular of June 2000 to promote the IC industry was followed by supplementary directives in September the following year. Although the original circular to promote China's IC industry suffered from serious shortcomings in the implementation stage, to be discussed later on, it had an important impact throughout China which has been clearly evident in several localities such as Shanghai and Beijing, where both technology and capacity levels have increased rapidly.

SMIC took a bold decision early on to establish China's first 12 inch wafer facility in Beijing, which will have clean room space of 18 000 square metres with a total floor space of 188 000 square metres. Construction started in autumn 2002 and pilot production started in the summer of 2004. The operation is done in collaboration with Infineon in Germany and Elpida in Japan for low line-width DRAM. SMIC will move into logic production and integrate with the Shanghai factories. The complete production chain at SMIC includes design services, mask making, wafer manufacturing, wafer probing, assembly and final testing. The facility will be able to handle chips with line widths down to 0.09 micron.

There are many reasons for selecting Beijing as the site of the new 12 inch foundry, although Shanghai might have been the preferred site for several reasons. Other locations that were considered include Suzhou, Hangzhou and Nanjing, as well as Shenzhen, although the latter two locations were not seriously considered. Beijing has a number of advantages that include: first, it is the location of a recognized Chinese Silicon Valley with activities in system development and design as well as hardware development. Second, it has most likely the best universities, such as Tsinghua University and Peking University, which are progressing in a way that in the future might make them comparable with Stanford University and Berkeley University in the US. Third, an exceptionally supportive government in many ways facilitated the SMIC investment in the new foundry.

SMIC has not experienced difficulties in obtaining semiconductor equipment for its foundry operations in China and is buying from suppliers in Japan, Europe and the US. The experience of buying equipment from the US suggests that the US system is heavily bureaucratic. Opinions expressed in an interview suggest that this is a reflection of the US being a world leader and undisputed superpower that wants to control everything. Thus, the US bureaucracy wants to evaluate all factors and issues carefully before approval is given, so it takes a longer time to obtain equipment, although approval is generally given. Observers from Japan have commented that there is a lot of equipment from Europe in Chinese IC factories, indicating that the semiconductor equipment industry in China is very weak and that it will take another ten years before products are commercially available – although great efforts are being made. However, around 2020 Chinese semiconductor equipment makers may have entered the global market.

Examples show, however, that local policies encouraged substantial investment in several locations. Motorola in Tianjin has established a large-scale plant for 8 inch wafers with a monthly production of 42 000 wafers, at a total cost of US$1500 million. Semiconductor Manufacturing International Corporation (SMIC) in Shanghai invested US$1400 million in an 8 inch wafer plant with a monthly capacity of 42 000 wafers. At the same time investment plans were also made for a number of projects, also for 8 inch wafers, and calculations have been made that China will have newly-added investment in its IC industry of more than RMB60 billion during the period 2001–2005.

Despite these developments China is still far behind the rest of the world in meeting its domestic demand for IC design and manufacturing. Its booming IT industry is serving not only an increasing export market but also its expanding domestic market. Given the existing concentration of the IC industry in the BoHai Rim Region and the Yangtze River Delta, there exists also a political concern that other parts of the country could be left behind as further IT development is of an encompassing nature. Following the strategy of 'Developing the Western Region' a number of attempts are being made to encourage the design and development of an IC packing capability in such regions wherever the demand and conditions permit. Against this background the following objectives were formulated.[12]

1. By 2005, the nationwide output of integrated circuits should reach 20 billion pieces and sales should reach 60–80 billion Yuan, which will account for 3–4 per cent of the world market and meet 30 per cent of domestic demand. Key special integrated circuits concerning important national defence projects and the security of the national economy should be provided by domestic manufacturers. The embedded CPU, DSP, radio frequency (RF) and IC cards for computer, communication, digital audio-video and informatization engineering should be designed and produced by Chinese firms; the 8 inch 0.25 micron technology should become the mainstream production technology of the industry; the packaging industry should enter into mass production; in the support industry there should be a breakthrough for equipment for the 8 inch 0.25 micron production lines; and materials needed in large quantities for widespread usage should enter mass production.
2. By 2010 the nationwide output of IC should reach 50 million pieces, and sales should reach 200 billion Yuan, which will account for 5–6 per cent of the worldwide market and meet 50 per cent of domestic demand. The ASIC needed for main IT products should be basically provided by domestic manufacturers. As for technological level, the technology of chip mass production should approach or reach the international mainstream level of that time, the integrated circuit products needed for main IT products should be designed and produced by local companies, special materials should be basically self-supplied, and there should be some innovations and breakthrough points in key equipment technology and in research on new processes and devices.

The immediate task is to establish a national-level research centre with the responsibility for developing mass production technology for ICs and systems on chip (SOC). This would include arrangements for R&D that would focus on equipment for lithography, plasma etching, ion implantation and silicon, all of which should be reaching 0.10–0.13 micron. This would be supported through funding from the 863 programme and other forms of government support for advanced technology development. Priority semiconductors are CPUs and a number of ICs which are required in large quantities in products like digital cameras, DVD/EVD (Enhanced Versatile Disc) and HDTV.

The role of advanced R&D will be given more attention and resources with due attention to the active participation of software and IC industry companies. The expectation is to have a couple of R&D centres that operate in 'open style' organized by local governments with support from the national government and active participation from the IC industry. They should be organized as modern joint stock companies and carry out research on IC manufacturing technology and advanced IC products of common interest to companies. The suggested approach in many respects resembles the SEMATEC project in the US.[13] Furthermore, the Government will also select ten universities which will have the responsibility of training high-level professionals in micro-electronics and related subjects, with 1000 graduates annually in IC design and another 200 specializing in process technology.

By June 2002 the semiconductor industry in China employed slightly more than 114 000 people, of whom 16 000 were involved in IC design. Jiangsu dominated in total numbers followed by Shanghai, Zhejiang and Tianjin, with Beijing maintaining a dominant position in IC design (see Table 5.1). Considering the role of China's three major regions, the figures indicate that 52 per cent of total semiconductor employees were active in the Yangtze River Delta, 24 per cent in the BoHai Rim Region and a miniscule 6 per cent in the Pearl River Delta, with the remaining 13 per cent in West China and 5 per cent in other parts of the country. The dynamic growth of production shows however, an increasing role for the BoHai Region (see Table 5.2).

Table 5.1 Employment in China's semiconductor industry (June 2002)

	Total employment	**Employment in IC design**	**Sales value (%) share of total RMB20 billion**
Jiangsu	24 191	2 370	20.2
Shanghai	20 464	3 874	22.6
Zhejiang	15 989	1 490	11.2
Tianjin	10 446	200	17.2
Beijing	8 500	4 800	9.1
Hebei	1 650	50	0.5
Liaoning	3 500	50	1.6
Shandong	3 166	100	1.3
Guangdong	6 523	1 800	8.6
Shanxi	7 269	940	0.8
Sichuan	6 808	240	3.7
Chongqing	857	80	0.2
Other	5 500	250	3.0
Total	114 863	16 244	

Source: Zheng Minzheng (2002), pp. 5–7.

Table 5.2 Total semiconductor production in China including exports (2003) (RMB billion)

	BHR	YRP	PRD	Other	Total
Value	10.705	22.569	1.001	0.865	35.14
Increase (%) over 2002	45.1	26.3	23.6	9.1	30.9
National share	30.5	64.2	2.8	2.5	100

Source: CCID Consulting.

China's consumption of semiconductor equipment has grown from RMB1100 million in 2000) to RMB27 300 million in 2003, which corresponds to 1.1 per cent and 13.4 per cent of the global market respectively.[14] Thus, a rapid expansion of the semiconductor market in China is simultaneously followed by a significant build-up of domestic production capability, with an assumed future capability of domestic suppliers to supply much of the necessary equipment for IC manufacturing.

REFORMING A RESEARCH INSTITUTE – THE GRITEK CASE STUDY[15]

GRINM Semiconductor Materials Co. Ltd (GriTek) is located in the Haidian Science-based Industrial Park in Beijing, where it employs more than 300 people. It was established by the Beijing General Research Institute for Nonferrous Metals (GRINM) in February 1999, by separating out the production, research and sales of silicon, germanium and compound semiconductors and related assets; it went public the following year. It expanded in 2001 with the construction of a 6 inch wafer plant in Beijing's Linhe Industrial Development Zone, with an investment of US$21 million. It has Class One clean rooms and is one of the leading units developing 12 inch silicon single crystals for IC production.[16] It has competition from the Shougang Group, a large Chinese steel company, which founded the Huaxia Semiconductor Manufacturing Company with three US companies, planning to build two 8 inch wafer lines with an investment of US$1.3 billion.

By 2002, GriTek was the largest producer of semiconductor materials in China, with an annual production of 2.5 million silicon polished wafers. The China Semiconductor Industry Association Report of 2002 states that its fabrication facilities are 'highly automated and feature the most advanced equipment and processes'. The case of GriTek shows how the science and technology system has been evolving in China.

Chinese nonferrous metals research and development activities started in 1952 when the Chinese Heavy Industry Ministry set up a nonferrous metals experimental laboratory. Before that R&D related to nonferrous metals had been carried out in a general industry experimental institute that was operated within the iron and steel and chemical industry.

The newly created laboratory made important contributions during 1952–57, mainly in the field of nonferrous metals developments, by selecting mines and supporting heavy nonferrous metals metallurgy. In 1955 it became the Nonferrous Metals Department under the Ministry of Metallurgy Industry, which appointed many technical personnel to join and strengthen the laboratory. The following year, in 1956, China announced the Outline of Chinese Science and Technology Development 1956–1967. A number of tasks were identified related directly to rare metal developments. Hence, the Ministry of Metallurgy Industry set up a second metallurgy laboratory that specialized in rare metals research, and in January 1958 the Nonferrous Metals Department was enlarged to become the Nonferrous Metals Institute, later to be known as GRINM. The mission of GRINM as an institute was to conduct basic and applied research that strengthens the competitive position of China's metal industry and materials science and engineering, and to explore new fields and opportunities in technology for the future.

By 1962, China had established the capability of manufacturing and refining many kinds of nonferrous metals. Early industrialization started in 1963, under the spirit of the draft of the Chinese 1963–1972 Science and Technology Programme, and the task of GRINM became focused on rare metals utilization. GRINM changed or enlarged the original experimental workshops or factories, and also set up several new factories.[17] During this period, another important work of GRINM was to research and manufacture new materials and new products for military use.

Up to 1978, the R&D activities of GRINM did not involve any economic transactions. Technology transfer was undertaken entirely through administrative measures. In those days many people felt as if they were entering a mystic world when coming into GRINM as no door-plates were to be seen, soldiers stood guard, and even the departments within the institute were all named by numbers.

In 1978, China started its earliest attempt to introduce the market concept and GRINM began to change from a pure science and research institute to a scientific R&D institute adopting economic principles. Thus it was able to take advantage of its technology capabilities to develop new products, set up experimental production lines and transfer technology results against economic compensation.

During the 1980s, GRINM began early attempts to consider market needs. In 1985, S&T reform was further promoted and GRINM was selected as one of the first experimental institutes. Although GRINM had already earlier started marketing attempts, it still belonged to that group of institutes that could get funding from the government budget. As the allocation system changed GRINM had to find means to continue its R&D activities and support its future development.[18] Similar to other institutes GRINM quickly established several enterprises.

The long-term orientation of GRINM dissipated as technical personnel no longer wanted to devote their efforts to topics that had previously been important, and paid less attention to basic research. Thus significant resources were distributed in different directions according to temporary economic interests. The initial reform period suffered from great confusion. Thus, GRINM was faced with the serious task of harmonizing its earlier research orientation with the needs and demands of the market, which had a direct bearing on the organizational structure of the Institute.

Subsequently, as a response to world market demands, GRINM was able to establish the fundamental importance of high technology.[19] This became the basis for a reconstructed GRINM. Science and research resources were divided into several parts. High-technology projects would receive more funding support, while some projects with limited or no market future were cancelled. Research topics of a similar nature or lacking resources were adjusted or combined. This process led to a concentration in five high-technology industry areas: semiconductor materials, rare and precious materials, rare earth metallurgy and materials, energy and related materials, and mineral processing and metallurgy.

Later developments vindicated the harsh decisions in the reorganization of GRINM. Some science and research advantages were lost, but others were strengthened, and the five main areas became a good basis for further institutional transformation. The institute has been able to deliver important results in high-technology fields. The rare and precious materials department has become the largest one in northern China. In collaboration with other institute partners, GRINM has further researched and developed a semiconductor materials centre.

Accompanying the organizational structure reform, a major change in the personnel promotion system of GRINM was also carried out. In 1992, ten young persons in their thirties were granted professorships. This was unimaginable before.[20] In 1995, president professor Wang Dianzuo recommended semiconductor expert doctor Tu Hailing who graduated from a university in the UK to be the president of GRINM. The changes in human resource management, especially of senior management and technical personnel contributed greatly to the subsequent success of GRINM.

The information and communication technologies 137

Early in the 1990s, GRINM allowed almost half of its administration personnel to leave the institute in order to set up various services companies.[21] This change did not take place without any dispute, but those who left were to realize higher salaries and improved welfare.

However, the new president Tu Hailing had to consider a number of tough issues for the future development of the institute. As a high-technology institute, most of GRINM activities were directly related to future commercialization, where it would face competition in the international market. At this stage it became clear that the institute should enter the capital market in order to have access to financial resources. Thus, in 1999, GRINM established GriTek, which became the first high-technology company – with an origin from a research institute – being listed in China.

The R&D institute transformation formally started on 1 July 1999, when the State Economic and Trade Committee took direct control of 246 science research institutes. GRINM transformed from an institute to a large state-owned science and technology corporation. In January 2000, GRINM became the first one to register to be a corporation under the State Industry and Business Administration Bureau.

GriTek can be regarded as the first attempt in China to combine the high technology and finance sectors. A spin-off company, Grikin Advanced Materials Co. Ltd, was set up on 18 October 2000, and became a hi-tech enterprise registered at the National Industry and Business Administration Bureau, with GRINM being its main charter member. Other members included four well recognized venture capitalists in Beijing and Shanghai.[22] Most of the company's products are its patented property and occupy a considerable market share thanks to their first-rate quality and steady performance guaranteed through internationally advanced processes.[23] Grikin Advanced Materials Co. Ltd has now developed into a hi-tech conglomerate of technology, industry and trade with extensive technological exchange and business cooperation with over 30 countries and districts around the world.

The earlier management system of the institute had to change. This involved the introduction of corporate costing methods and a fixed assets depreciation system, with a new operating strategy and control of cash flow. Financial management was concentrated within the institute, and including setting up an internal bank. Allowances for different positions were changed in order to support an efficient incentive system. Employment has been increasing, although S&T personnel in GRINM decreased initially. However, in 2001 S&T personnel started to increase. Total employment in 2002 was more than 2000.[24] Within this, there were about 950 workers and almost 500 engineers. An almost equal number were in the category of senior engineers. The managerial staff amounted to 11.5 per cent of total and 7.5 per cent of employees had been given professorial titles. The institute hosts an incubator for returned overseas Chinese professionals and two bases for postdoctoral studies with two specialities offered to PhD students and seven specialities offered to Masters students.

Between 1999 and 2002, the total revenues of GRINM increased from a little more than RMB6200 million to RMB11 000 billion. In 1999, R&D input in GRINM was RMB51 million, which doubled to RMB109 million in 2002. The transformation also improved living standards. An average salary of RMB14 658 in 1999 had almost doubled to RMB28 000 in 2000.

Since 2000, GRINM has been the leading researcher and manufacturer of semiconductor materials in China and hosts a National Engineering Research Centre for Semiconductor Materials.[25] It is a major R&D centre for nonferrous metal matrix composites and light metal structural materials in China. The National Engineering and Technology Research Centre for Nonferrous Metal Matrix Composites, located in GRINM, covers the following R&D: particle reinforced composites, laminated composites, bearing materials, materials for surface mount technology (SMT) and electronic packaging, ultra-high strength aluminium alloys, high strength and high toughness magnesium alloys, diamond synthesis catalysts, thermal spray materials, contact materials and functional gradient materials.

GRINM is also involved in advanced manufacturing technologies and equipment to produce new materials, which include rapidly solidified gas atomization technology, ultrasonic atomization technology, semi-solid casting technology, direct laser metal deposition technology, vacuum stirring casting technology, vacuum planar flow casting technology and rapid-cooling roll casting technology. GRINM also has extended experience in the R&D of superconducting materials, is a key researcher in power metallurgy and supplies pure metal powders and alloys and compound powders. Special expertise has been accumulated in the processing of nonferrous metals, including high-pressure deformation, spinning processing technology and equipment and hydrostatic extrusion technology and equipment.

Although achieving considerable success in its transformation, GRINM is still facing a number of challenges. One core problem is a low level of profitability, which GRINM shares with many other large state-owned enterprise. There has been a decreasing ratio since 1999, caused by two problems – international competition and the fact that GRINM is burdened with retired employees as a state-owned enterprise. The latter can only be solved through legislation on social security in China.

GRINM plans to enhance the development of core business and orient itself towards promising and competitive industrial sectors and products with an eye to long-term development and steady market shares.

IC POLICY AND PROGRESS

China's IC is constantly growing, from a demand of US$11 billion in 2000 to US$42.9 billion in 2004. The estimated demand for ICs in 2008 will be US$88.8 billion. At present China can meet less than 10 per cent of its demand; by 2008 the gap between domestic demand and domestic supply will have increased to US$82.3 billion. Foreign enterprises had established more than 600 R&D centres in China by mid-2004, of which some 170 are located in Beijing.[26] Infineon has established a new software centre in its headquarters in Shanghai and in nearby Hangzhou Samsung R&D activities include chip design and system solutions for LCD monitors. The foreign-operated R&D centres have so far involved an investment of US$4 billion.

The number of local IC design houses increased from 98 in 2000 to 483 in 2003. The revenue for the top ten IC design companies increased from US$86 million in 2001 to US$328 million in 2003. These include Datang in Beijing, Silan in Zhejiang, Ideabank in Jiangsu, Shaxing Silicon in Zhejiang, Wuxi Semico in Jiangsu, Vimicro in Beijing, Sigma-Jinghua in Beijing, SH Huahong in Shanghai, CEC Huada in Beijing and SSMEC in Guangdong.

A report on China's semiconductor R&D describes substantial efforts in five sectors where results have already been registered.[27]

1. IC chips developed and produced in China have been in trial production since early 2004 and have been designed by companies such as Tsinghua Tongfang and China Huada.
2. China's design companies have achieved breakthroughs in CPU products. This includes the Arca (Fangzhou) developed by the Cortech Corporation for 32 bit network computers, and the Longxin chip, which is a 32 bit general CPU developed within the Chinese Academy of Sciences.
3. Shanghai Jiaotong University has successfully developed a series of digital signal processors, of which Hanxin No 3 is close to world standards with a speed of 600 million calculations per second on chips with line width of 0.18 micron.
4. Stae (XinGuang) multimedia chips are already in mass production after having been developed by the Beijing Vimicro Company.
5. LED semiconductors have received considerable attention with the objective of replacing traditional illumination products. To this end the national Semiconductor Illumination Engineering Group has been set up under the joint leadership of MOST and MII.

The China Centre of Information Industry Development (CCID) argues in its 2004 report that sales in the country's semiconductor industry will increase by more than 20 per cent, and IC output in 2008 should reach RMB87 billion; they indicate that this development will primarily take place in YRP, PRD and BHR regions, taking into consideration their special characteristics. CCID indicates three areas which will have great importance:

1. With the existence of 370 million television sets and 100 million users of cable television, the future transition to digital television will create a huge demand not only for the terminals but also for semiconductors to be incorporated into the sets.
2. The rapid popularization of IC cards will create a demand for some 800 million IC cards over future years that together with IC readers will create a market in the region of RMB10 billion, aside from substantial markets in other areas such as radio frequency identification (RFID) cards and devices.
3. A third important area is auto electronics; CCID estimates that the cost of electronic systems in an ordinary car constitutes 20 per cent of the cost of the car.

Shanghai has the highest concentration of IC manufacturers, which includes the Semiconductor Manufacturing International Corporation (SMIC), Grace Semiconductor Manufacturing Corporation, Shanghai NEC Electronic Co. and Shanghai Belling Co. Two of the largest foundries in the world – Taiwan Semiconductor Manufacturing Co. and United Manufacturing Corporation are located close to Shanghai – in Suzhou and Songjiang.

The strength of Shenzhen in attracting IC manufacture lies in its concentration near the Pearl River Delta of final market product manufacturers such as TCL, Konka, Zhongxing and Huawei. These companies are being attracted to venture into design and manufacture of chips and also becoming involved in foundry development. Substantial research and educational institutions exist in Xi'an, where the IC industry has been rapidly growing and has also attracted a number of foreign projects into the region.

STATE COUNCIL CIRCULAR NUMBER 18

China applies a nominal VAT of 17 per cent on sales of imported and domestically-produced semiconductors. In June of 2000, in State Circular Number 18, China's State Council announced that all integrated circuits manufactured in China would receive a rebate of the VAT in excess of 6 per cent. The policy was amended in September 2001, with an announcement that integrated circuits both designed and built in China would be eligible for rebate of the VAT in excess of 3 per cent. This policy violates China's WTO commitments

– specifically, GATT Article III on National Treatment prohibits a WTO member country from engaging in activity that treats domestic producers and products more favourably than imported products. China should reduce or eliminate the VAT on all semiconductors, regardless of origin. Reduction or elimination of the VAT would not only help US chipmakers, it would further stimulate the development of China's high technology industries.[28] At that time, many foreign investors including SMIC, Taiwan Semiconductor Manufacturing Corp, Intel and LG in South Korea began to build semiconductor plants in China because of the fast growth of IC demand in China.

The tax rebate raised serious concern in the United States, Japan and the European Union (EU) for discriminating against imported products and setting import barriers, and was seen as contradictory to the rules of the WTO. It was a big occurrence for SMIC[29] – the biggest semiconductor manufacturing company in China – when the US and China reached an agreement on eliminating China's value-added tax (VAT) rebate for the semiconductor industry, which was signed on 14 July 14 2003. In this agreement China agrees to adjust the tax rebate policy and eliminate the 'rebate after collection' policy on semiconductor products from April 2005. As a result semiconductor products, designed in China, manufactured overseas and then imported, will no longer enjoy the tax rebate after 1 October 2005.

The China Centre for Information Industry Development (CCID) argues that 'the semiconductor VAT dispute is not only a trade war, although the United States said exports to China reached US$2.02 billion in 2003. The main purpose of the US side is to suppress China's production capabilities and ownership of technologies'.[30] The chief executive officer of SMIC estimated that the elimination of the tax rebate policy would cause US$1.7 million of losses for his company in 2004 – although 'The elimination of the VAT rebate will not materially impact SMIC's financial performance and its ability to compete in the global semiconductor market.'[31]

An important effect may be that China will give more support to semiconductor companies' research and development to help them upgrade their technologies. This has also prompted a concern that MII and MOST should set up a special fund for the semiconductor industry, as a reward to the companies who spend a lot on research and development. In Beijing, Liang Sheng, electronics and information development division chief of the Bureau of Industrial Development with the Beijing Municipal Government, said that China could increase electronic industry funding and suggested that the Ministry of Finance would allocate special funds.[32]

Liang Sheng has also provided additional comments on Circular 18 and its consequences. CCID provided advice on the contents, which were expected to be implemented from the end of 2000 and involve three national organizations: taxation authorities, the National Development and Reform Commission and the Ministry of Information Industry. However, Liang Sheng mentions that it actually required two years for those concerned to agree on how to implement the scheme and argues that it was never completely successful in defining the details of implementation, which raised three important questions. First, it never became clear which enterprises would be qualified to collect benefits. Second, if firms were qualified it was still not clear who should be responsible for administering the scheme. Third, the US government complained strongly against the scheme, based on the principles of WTO.

Although, the tax rebate scheme was abolished, Liang Sheng argues that the effect was still very substantial as it was a great 'promotion event'. However, he sees other possibilities for supporting semiconductor companies entering China, which include instruments such as support for local upgrading, support in the domestic supply chain and preferred procurement for domestic companies. Finally, he says that the size of the Chinese domestic market and its dynamic expansion provides the real influence in developing the Chinese semiconductor industry.

The Role of Beijing

Beijing City announced ambitious plans for its future IC industry in November 2000, encouraged by the fast-growing IC design and manufacturing industry in other cities like Shanghai. If all goes as planned, the Beijing area will ultimately embrace as many as 20 advanced semiconductor processing lines and dozens of chip design houses, as well as related materials and equipment producers.[33] This will be achieved with favourable tax conditions and other incentives for foreign investors coming into the area. The incentives should be seen in the context that the Capital wants to participate in a major way in the overall development of the China semiconductor market, which analysts project could become one of the world's largest by 2010. A proposed layout of the North China Microelectronic Industrial Base includes two areas – the North Microelectronic Technology Research and Development Base and the North Microelectronic Production Base. The R&D base will focus on state technical centres of excellence in the areas of microelectronic processing techniques, IC design, IC production equipment and materials.[34]

There can be no doubt that Beijing presently constitutes the intellectual and creative centre for China's future IC industry. Some of the best universities are located there with the cities having a total of some 400 000 students, of which 100 000 graduate every year. The universities in Beijing have a special advantage in creating a talent pool as they recruit not only from their own population base, which is little more than 1 per cent of the country's total, but also attract highly talented people from all over the country.

The attraction of Beijing also lies in its very large concentration of R&D institutions, which provide substantial support to the IC industry in the Capital. There are also many possibilities for risk sharing and for getting access to favourable loans from the banks. These factors may have contributed to SMIC investing another US$100 million in a second plant in Beijing. It is located in Beijing Economic Technology Development District, has the ability to process 12 inch wafers, and started production in September 2004. Discussions for this project started in November 2001 with a decision reached in May 2002. Construction started in September 2002 and completion was delayed until September 2004 for some six months because of SARS. This plant has basically procured most major equipment from European companies, with additional deliveries from Japan.

Aside from the recent SMIC foundry, Beijing has another six IC plants which include NEC-Shougang, Yandong, SPS, which produces Power ICs, and another three plants that are operated by Tsinghua University, Beijing University and the Chinese Academy of Sciences.

Beijing sees itself as a future national centre for IC design and for the development of IC manufacturing equipment, which today constitute close to 80 per cent of total investment in a 12 inch wafer plant. Various R&D projects already underway have a major focus on lithography and etching equipment. The national 863 programme is naturally one of the supporters for this development although it has been realized that China needs outside partners if Beijing is going to reach its ambitious goals.

In a comparison between Shenzhen, Shanghai and Beijing it is obvious that Shanghai and Shenzhen have more industrial engineering and production capabilities, with a focus on original equipment manufacturing in Shenzhen. However, the advantages for future development of the IC sector in Beijing, includes an ample supply of high-level engineering manpower, and many foreign companies that have decided to locate their headquarters there, including Microsoft and Intel. Silicon factories in Beijing including GRINM and Silicon Materials, which are already exporting 4 inch and 6 inch wafers while developing capability for 8 inch and 12 inch wafers.

The rapid advance of IC technology and the rapid expansion of the Chinese market also created a flux in the formulation of policies. As a recent example, the Beijing Government has decided to provide a grant of US$20 million to an R&D company with advanced system-on-chip (SOC) design capability deciding to set up shop in Beijing – whether domestic or foreign. IC circles in China have naturally realized that SOC technology will become prevalent in all IT segments such as HDTV, mobile communications, computers and networks, IC cards and recognition and car electronics. MOST is supporting two SOC projects under its 863 Programme – one of them focusing on costs while the other is focused on performance.

IC Design Capability

CCID reports that China in late 2004 has around 500 software design houses which employ some 16 000 people, being able to finalize the design of 500–600 ASICs in a single year. They are mainly providing ASIC designs for consumer products, computers and communication equipment. Line widths of 0.35 micron are most common although a few design houses are able to design circuits with line widths as low as 0.13 micron. Recently one of the design houses in Shanghai was able to deliver the design for an advanced mobile chip.

Some 100 of China's software design houses are located in Beijing with another 90 or so in Shanghai and around 60 in Shenzhen. The rest are located in several places although with a concentration in Jiangsu and Zhejiang provinces. The IC design industry in China in 2003 reported sales of RMB4.4 billion, which should be compared with sales of RMB50 billion generated by about 300 companies located in Taiwan. CCID estimates that employment in IC design will rise more than six-fold by 2010 with a reduction in the number of companies – possibly to around 300 – with estimated sales of RMB35 billion.

The Beijing LHWT Microelectronics Case[35]

LHWT provides an interesting example of a recent company that is focusing on large-scale integration (LSI) design and development. It was established in February 2000 by seven Chinese returnees from Japan where they had obtained PhD and Masters degrees.[36] They went to Japan in the 1980s, remained there for 15–20 years and accumulated rich experience and also financial resources. A scouting team from Beijing met them and encouraged a group of seven to return and establish an enterprise in Zhongguancun.[37] The company signed a contract with Sony in May 2001 for providing LSI IC design services for household appliances and wireless communication, and has established close partnerships with NTT DoCoMo, Oki and Maxell. The partner already in Japan had established close relations with Sony and its support for DoCoMo development. LHWT has a focus on IC design and many of its products relate to W-LAN, W-CDMA and PHS. The company's sales reached RMB15 million in 2003 with an export value of US$1 million.

The total staff in late 2003 reached 75, of whom 60 are involved in development. The ability to attract orders from Japan lies in the very large difference in costs for software engineers in Japan and China, which lies in the region of 4:1–3:1, although efficiency is lower in China, where many software engineers are lacking experience. The conclusion is that LHWT cannot primarily capture markets by having lower costs. LHWT states that it must capture markets by offering superior core technologies and from the very beginning paid great attention to applying for patents. The company has already developed considerable competence in designs for ICs used for mobile communications and has great proficiency in the FOMA 3G system used in Japan and in W-LAN. LHWT has been supported by the Beijing Software Association and was encouraged to apply for 863 funds to develop an IC circuit that would integrate the functions of both W-LAN and W-CDMA. The project involves several million RMB. A contract was granted in late 2003, with results to be demonstrated within a year and final findings by June 2005.

TELECOMMUNICATIONS, STANDARD-SETTING AND THE FUTURE OF TD-SCDMA

The ambition of China to enter the mobile telecommunications arena with its TD-SCDM system (the Chinese 3G system) – both domestically and internationally – has its roots in the 863 Programme. This was China's response in 1986 to the Star Wars Programme in the US and the Eureka Initiative in Europe. From the beginning the 863 Programme contained a number of projects that were directly related to the telecommunications sector and brought a number of universities and research institutes into action. This was the third stage in China's development of its telecom industry that has passed through the following phases:[38]

1. Self-sufficiency, 1949–1978, with very limited investment in the sector, limited technological change and a remaining focus on analogue technology; telecommunications was a political tool and there was limited ambition to develop commercial business.
2. Dependence on imports, 1979–1983, following from the ambitious reform to modernize the economy; using foreign technology was a very effective and easy way of catching-up.
3. Digestion and absorption, 1984–1993, included a four-step approach of import, digestion, absorption and generation to reduce China's heavy dependence on imports.
4. Creation from 1994 onwards: liberalization of the telecom sector, several structural reforms and entering the WTO in 2001 changed the industrial and technological landscape of the telecom sector dramatically. Researchers and entrepreneurs rapidly grasped market opportunities to develop competitive products.

The sources of success in the last stage can be found in the significant R&D efforts that were carried out by the Information Engineering College of the

People's Liberation Army (PLA IEC) and resulted in a system that is always referred to under the acronym HJD-04 in China. This was China's first successfully developed major stored-programme-computer (SPC) switching system. It can be assumed that the success of this project, completed in 1991, forced foreign manufacturers to reduce their prices while also transferring more advanced technologies to the Chinese market.[39]

Import statistics show that the government's R&D efforts were successful at an early stage. Only 54 per cent of installed SPC switches in 1992 were directly imported compared with 100 per cent ten years earlier. Out of the remainder, 36 per cent were provided by joint ventures in China while 10 per cent were provided by domestic manufacturers.[40] It is of interest to trace the changes in the structure of the telecom sector following the successful development of the HDJ-04, as PLA IEC could not shoulder any responsibility for its commercialization. Specifications and technologies were transferred to a substantial number of ambitious telecom companies that engaged in fierce competition for the domestic market, which was at the time rather small. Eventually Julong Telecom (Great Dragon Telecom) was given the task of integrating various manufacturing units, and in 1995 merged eight of them to undertake further development, manufacturing and marketing. Julong was initially successful and obtained a large contract for modernizing the telecom sector in Cuba. In 1996 China was home to five telecom companies – Huawei Technologies, Zhongxing Technologies, Datang, Julong and Jinpeng – with an expressed ambition to continue advanced development of the telecom system. In the mid-2000s the first three are active in launching TD-SCDMA.

Jinpeng belongs to the Guangzhou Jinpeng Group, controlled by Yue Xiu Enterprises in Hong Kong with backing from the Guagzhou Government, and was being restructured in early 2004.[41] Jinpeng was founded in 1998 and made an entry into mobile telecommunications through a partnership with Motorola. The Group is engaged in the manufacture of mobile telecom equipment and the development of value-added mobile telecom applications, for which China Unicom has been a major customer in the past. Jinpeng has no ambition to enter independently into 3G business and will focus on application features, based on its understanding of the domestic market. However, the company remains one of the major manufacturers of handsets for a personal access system (PAS) which in China is commonly known as Xiaolingtong.

Xiaolingtong, or Little Smart received a tremendous boost from the postponement of granting 3G licences in 2003, although the Ministry of Information Industry (MII) originally frowned upon the concept as it was not technologically advanced. Xiaolingtong is a wireless local loop (WLL) system that connects directly with fixed-line phone networks, and offers fees identical to those for standard fixed-line services and much lower fees than for standard mobile services. UTStarcom is the leading company for the PAS Xiaolingtong, and its President for its China operations argues that Xiaolingtong will not be quickly wiped out when 3G expansion starts in China. The system has limited roaming capability and is considered a backward technology as it cannot generally be upgraded to future generation wireless communications. However, Xiaolingtong can be used during the initial deployment of W-CDMA where the coverage is poor.[42] The late arrival of 3G services in China has provided a window of opportunity not only for Jinpeng but also for ZTE and UTStarcom, which have become the largest suppliers of Xiaolingtong services.

The PAS system has become one of the major telecommunication services in China with more than 65 million subscribers by the end of 2004. UTStarcom believes that the demand for affordable wireless telecommunications in China, especially with the urbanization of Western provinces over the next 10 years, will continue to drive the growth of Xiaolingtong.[43] The popularity of the system has increased since January 2005, when users have been able to send SMS messages via China Netcom, to China Mobile's all GSM mobile phone subscribers countrywide. The same services will follow for access to China Unicom mobile phone subscribers, the other major mobile operator in China.[44]

A TD-SCDMA Forum was established by the end of 2000, with eight members that included China Mobile, China Telecom, China Unicom, Datang, Huawei, Motorola, Nortel and Siemens. It is an open, international body for promoting the global uptake of TD-SCDMA technology.[45] It has become a communication bridge among government and enterprises, as well as among enterprises themselves. Simultaneously, the Forum is making efforts to attract more and more enterprises worldwide to be involved in the TD-SCDMA encampment, to complete a full industry supply chain, and to accelerate cooperation among enterprises. Internationally, the TD-SCDMA Forum has also become the representative partner in 3GPP, which extends the influence of TD-SCDMA technology among international organizations, thereby being one of the important international forum organizations to sustain the international development of 3G technologies.

A couple of years later, on 30 October 2002, the TD-SCDMA Industry Alliance was established, with ZTE[46] being one of the eight founders. Others included Soutec, Huali, Huawei, Legend (now Lenovo), China Electronics Corporation[47] and Putian. These companies are expected to cover the full range of TD-SCDMA products from chips, mobile phones and equipment. The alliance is maintaining contacts with some ten foreign providers of telecom equipment, including Siemens and Nokia, and STMicroelectronics of Italy will be the first foreign firm to join the alliance. This is interpreted as an indication that the alliance is becoming internationally accepted and will expand through participation of more international IT companies.[48]

To further its development, ZTE has established a number of partnerships which include DaTang Mobile, COMMIT Incorporated, Chongqing Chongyou Information Technology Co., Ltd (CYIT), Spreadtrum Communication and T3G. In early 2005 ZTE can offer end-to-end solutions for TD-SCDMA including service platform, Core Network (CN), Radio Network Controller (RNC), series of NodeB and mobile terminals.[49] While uncertainty remains about when 3G licences will be allocated and their choice of system, ZTE has identified a number of overseas market opportunities. There are two kinds of potential overseas market. In the areas where W-CDMA has developed rapidly, TD-SCDMA can be used as a supplement to W-CDMA data services. On the other hand, in the areas where mobile communication is less developed, TD-SCDMA can be used as a low-cost voice service solution. ZTE owns about 500 patents in TD-SCDMA and most of them are core patents or international items.

The TD-SCDMA Forum has more than 400 members, of which 15 are mentioned as director members, which means that they are recognized in the world and have the ability to make contributions in promoting the development and application of the TD-SCDMA technology.

The Chinese government has been making substantial efforts to promote the 'Chinese standard' with R&D and industrialization; this has been manifested through the signing of a supply contract for the field testing of 3G and the emergence of TD-SCDMA cellular phones.[50] Zhang Xinsheng, Vice Director of the Science and Technology Department, Ministry of Information Industry (MII) argues that the underlying reason for having drawn the world's attention lies in the technological advantages of the TD-SCDMA. The advantages are reported to include: TD-SCDMA can be connected with the existing telecom system, has features of intelligent antennas, and utilization.

A senior official of the Ministry of Information Industry (MII) in mid-2004 announced that a complete line-up of TD-SCDMA products, ranging from core chips, base stations and handsets to intelligent antennas, will be ready for commercial use by June 2005.[51] There is little doubt that China's interest in its own 3G system has benefited from the considerable delay in any major expansion of competing 3G systems anywhere in the world. This has apparently played into the hands of the Ministry of Information Industry, which stated in August 2003:[52] 'We should take advantage of the time delay to further boost the TD-SCDMA'.

Datang Mobile Communications Equipment and Siemens have been the main early developers of TD-SCDMA, which competes with the Europe-initiated W-CDMA and the US-backed CDMA 2000. It has become the conventional wisdom that China will not permit operators to build their 3G networks unless TD-SCDMA is ready to deploy. Nortel Networks signed a memorandum of understanding with China Putian Corporation in June 2004 to form a joint venture to research, develop and manufacture products based on TD-SCDMA and W-CDMA. Nortel earlier launched a laboratory with Datang to test TD-SCDMA. Siemens in Germany and Huawei Technologies also formed a joint venture to develop and market the TD-SCDMA system.[53] However, a number of observers suggest that it is very unlikely that TD-SCDMA will be used as a stand-alone technology to deploy 3G networks. Its main market possibilities would be as an add-on technology to some operators' networks, or as a stand-alone technology for minor operators to deploy networks.

The founder of the US Signal Research Group, Mr Michael Thelander, puts China's 3G strategy in a broader perspective in the following words:[54] 'We also believe TD-SCDMA is part of an overall ploy by the Chinese Government to receive more favourable IPR (intellectual property rights) terms from western companies, in particular Qualcomm'.

China has a positive outlook on the future introduction of 3G mobile telecommunications in China while maintaining a strong desire to support its domestically developed TD-SCDMA system, where the IPR are almost completely owned by Chinese companies. In late April 2005 two organizations – the TD-SCDMA Forum and the TD-SCDMA Industry Alliance – held their 2005 International TD-SCDMA Summit, endorsed by the Ministry of Information Industry, the National Development and Reform Commission, and the Ministry of Science and Technology.[55]

By then the future prospects for TD-SCDMA may have brightened, as the Forum in late 2004 reported success in developing and manufacturing chips to be used in handsets, following from a project that was launched in April 2004. An announcement stated that three companies – VeriSilicon Holding, Commit Incorporated and SMIC – jointly introduced TD-SCDMA phone chips.[56] The complete TD-SCDMA terminal chipset consists of five chips, Commit having taken a leading role, with the cell library from VeriSilicon to support the final manufacturing carried out by SMIC.

On the website of the TD-SCDMA Forum, the deputy director-general in the MII department of science and technology, Mr Zhang Xinsheng, indicates that several issues are critical for the success of China's system. These include a realization of 'smart antennas' as an integral element of future 3G systems, and the role of software radio and its possibilities for changing the direction of 3G technology development. The third, and possibly most important issue is whether TD-SCDMA will enhance the country's ability to capture intellectual property rights. He summarizes his views on development strategy in the following way:

> The Chinese government regards TD-SCDMA as our strategic industry to develop. On one side, we need to lead domestic enterprises to the whole

TD-SCDMA industry chain, on the other side, we also encourage international enterprises to involve themselves in the evolution of the TD-SCDMA industry. Currently, we are promoting the TD-SCDMA industrialization at full speed, we will establish a whole industry chain. The devotion of the Chinese government to TD-SCDMA clearly indicates that the Chinese government integrates TD-SCDMA into its whole strategy. In the meantime, the enterprises should continue to devote, to do some detailed work in order to perfect the development ... of the technology.[57]

A delay in allocating 3G licences in China has had the effect of bringing more supporters to the TD-SCDMA camp. This includes UTStarcom, which has announced that it is willing to launch its 3G business in China as long as it receives 3G licences. The company has stated that it is able to provide end-to-end solutions for both W-CDMA and TD-SCDMA. This US-based company was established in 1995 by a number of overseas Chinese students and is creating major R&D centres in Hangzhou. Its major product is Little Smart (Xiaolingtong), a handset based on the PAS system, which charges lower fees than GSM networks.[58] The number of Little Smart users reached 50 million in 2004.

Another important domestic actor to support TD-SCDMA is China Railcom. The company signed an agreement with Datang Mobile in 2002 to establish a field trial network in Chengdu, which has been done in collaboration with the Ministry of Information Industry. Early tests revealed shortcomings in the radio access network (RAN) and in terminals that needed to be improved. It has also become clear that full-scale practical applications are required to make TD-SCDMA into a mature system, and foreign carriers will not adopt it unless China uses its first. China Railcom has also advised the manufacturers to emphasize basic functions, stability and reliability rather than pursuing a variety of functions.[59]

Aside from Siemens, Nortel Networks is another company from outside China to support TD-SCDMA – being one of the first vendors to support the system and also one of the founders of the TD-SCDMA Forum. The company has indicated that it will pursue a three-step strategy.[60] First, Nortel Networks will support all access modes through a unified core network – by using the Nortel core network and collaborating closely with Datang in their joint laboratories. The next step will be to focus on radio network controllers (RNC), and Nortel has been working on bridging differences with W-CDMA, and analysing the probability of using dual-mode RNC to support the two standards. The third step is to expand research and development (node B) to follow market developments to the system. The Vice President of Nortel Networks in China, Mr Huang Jigong, says that he believes 'that the TD-SCDMA market is not only confined in China, but has more opportunities in the Asian market. TD-SCDMA will be very competitive if the mobile communications are less developed in some countries and regions. As such, TD-SCDMA will be able to replace other standards if market entrance is in time.'[61]

In the meantime the two major Chinese telecom equipment makers, ZTE and Huawei, are expanding their overseas markets for mobile telecom systems. After having received orders for four 3G systems in developing countries, Huawei Technologies announced in early December 2004 that it had received a 3G order from the mobile operator Telfort BV in the Netherlands. At the same time the company announced that its subsidiary in the US, FutureWei, that a mobile operator[62] had commercially launched CDMA 1X in California and Arizona – with deliveries from Huawei Technologies.[63]

The annual revenues of Huawei are expected to be more than US$5 billion in 2004, and the firm has reported that its overseas sales almost doubled during the year – exceeding US$2 billion. Norson Telecom Consulting forecasts that the company's overseas sales will continue to grow substantially and could reach US$6 million in 2008, thereby seriously challenging the dominant equipment vendors such as Ericsson and Nokia.[64] The private character of Huawei Technologies and its complex ownership are important reasons why the company has not yet been listed on any stock exchange. However, recent arrangements suggest that Huawei has secured its financial position to support its continued overseas expansion. In November 2004 the company secured a loan worth US$360 million from 29 overseas banks in a consortium that also included the Bank of China (Hong Kong). This was followed by a financing agreement with the China Development Bank (CDB) through which the CDB will support the international expansion of Huawei with a credit facility at the level of US$10 billion to finance both Huawei and its customers abroad during the next five years.[65]

CHINA'S ICT FUTURE

A strong economy and the 3G introduction will provide a substantial boost to the Chinese mobile telecom industry that already has 350 million subscribers. It was increasingly clear during Spring 2005 that the Chinese government will only allocate three 3G licences and that one of them will favour the domestically developed TD-SCDMA. In order to support such an outcome great attention is given to final development and testing and equally important restructuring of the telecom operator sector to secure a desired outcome. This will favour Chinese telecom makers and also strengthen market possibilities outside China. Equally important is the ambition of China to establish a strong presence in mobile radio technology which will find rapidly expanding applications beyond 3G, '4G' and '5G'.

Integrated circuits play an important role in this and all other ICT segments, and the Chinese government wants production and design competence to become highly elevated within its own borders. After the failure of the VAT rebate, abolished on 1 April 2005 to encourage relocation of IC manufacture to the Mainland, the State Council has moved in another policy direction. It was announced that government might provide funding of up to 50 per cent of the cost of chipmakers' research and development projects. Also, some chosen chipmakers might be exempted from paying income tax for up to 10 years. These announcements have been jointly made by the Ministry of Finance, Ministry of Information Industry and the National Development and Reform commission.[66] Subsequently, Intel has announced that it will set up a research and development centre in Shanghai Waigiao Free Trade Zone with an investment of US$39 million, already having three assembly and testing plants in Shanghai.[67]

For a number of years Chinese universities and colleges have expanded their training of software engineers, and soon China is to become as attractive for outsourcing as India has been in the past. More important is the domestic design capability that is created for advanced solutions in areas such as system on chip (SOC) and which may be able in a major way to increase the added-value of electronic products that are manufactured in China to meet the demand in the rest of the world. An eventual outcome of the indicated trends and changes is that within ten years China may become a serious competitor for certain highly advanced equipment to manufacture integrated circuits, a domain that is presently dominated by Japanese, American and European makers.

These long-term ambitions may temporarily be overshadowed by critical observations of daring movements as the TCL acquisition of Thomson Multimedia to become the global number one maker of CRT sets, and the subsequent acquisition of Alcatel mobile division. TCL may also face marketing and financial difficulties which may also occur when Lenovo consolidates its acquisition of IBM PC division.

CHINA IN THE GLOBAL R&D LANDSCAPE

The number of foreign R&D units in China has reached more than 700.[68] A preliminary estimate would indicate that they employ no less than 60,000 R&D personnel in China, if outsourcing is also included. This would correspond to some 7 per cent of the officially announced number of scientists and engineers employed in R&D in 2003. The still ongoing and not completed reform of the R&D system might indicate that foreign R&D personnel in China would constitute possibly 15 per cent of efficiently used domestic

R&D manpower. At the same time Chinese companies like Haier, Huawei and ZTE and a number of other companies have already established research laboratories, joint activities and listening posts in foreign countries. This tentative and speculative calculation suggests that China is already heavily drawn into a global innovation system and has already accepted a far reaching global division of labour.

NOTES

1. State Council Document No. 18 (2000).
2. Deputy Minister of MII, Gou Zhongwen, 10 December 2002, in Semiconductor Industry Association (2002).
3. The information in this paragraph is extracted from Semiconductor Industry Association and China Centre of Information Industry Development (2004).
4. 'Steps to become global player', *China Daily*, 4 August 2004.
5. Ibid.
6. Zheng Minzheng, 'China's IC industry under fast development', in Semiconductor Industry Association (2002).
7. Future technologies would include SOC, Micro-Electro-Mechanical Systems (MEMS), vacuum micro-electronics, Gallium-Arsenide (GaAs) IC and Silicon-Germanium (SiGe) IC, single-electron devices based on the quantum effect and quantum IC.
8. Semiconductor Industry Association (2002).
9. SMIC was co-founded by Richard Chang, a former Vice-President of Texas Instruments. SMIC is registered in the Cayman Islands and is listed on stock exchanges in Hong Kong and New York.
10. An earlier forecast from SMIC suggested that the production would have reached 185 000. This follows less growth than expected in the global market for electronics products. Thus SMIC plans to reduce equipment expenditure to US$1000 million against planned investment of US$1370 million, which is a reflection that overseas customers make up 90 per cent of the customer base. Source: SMIC cuts chip output target, *Shanghai Daily* 15 January 2005.
11. SMIC raised HK$14 billion in March 2004 to support expansion that would strengthen its position against its mains rivals, including Taiwan Semiconductor Manufacturing Co. and Chartered Manufacturing Ltd Co. in Singapore.
12. Zheng Minzheng, 'China's IC industry under fast development', in Semiconductor Industry Association (2002), pp. 4–5.
13. The SEMATECH concept originated in 1986 from an idea of industry-government cooperation and was expected to strengthen the US semiconductor industry. The initiative was prompted by several years of slipping US semiconductor market share. The SEMATECH (SEmiconductor MAnufacturing TECHnology) consortium was formed in 1987, when 14 US-based semiconductor manufacturers and the US government established the organization to solve common manufacturing problems. By 1994 both device makers and suppliers had regained strength and market share. Thus the SEMATECH Board of Directors voted to seek an end to matching federal funding after 1996, on the basis that the industry had returned to health and should no longer receive government support. However, SEMATECH continued to serve its membership, and the semiconductor industry at large, through advanced technology development in programme areas such as lithography, front end processes and interconnections, and through its interactions with an increasingly global supplier base on manufacturing challenges.

    The International 300 mm Initiative (I300I) was formed as a subsidiary of SEMATECH in 1995, with seven non-US companies and six US companies cooperating on 300 mm

tool standards and specifications; in 1998 five of those international companies opted to participate in more of the consortium's programmes through a subsidiary called International SEMATECH, and then ultimately made the decision to join SEMATECH as full members; the organization was subsequently renamed International SEMATECH.
14. CCID Consulting.
15. This section is taken from a case study by Dr Xinxin Kong and Dr Yuli Tang (both employed by the Ministry of Science and Technology) prepared for the project 'Emergence of New Knowledge Systems in China and their Global Interaction'. See also Tang, Yuli, 'Review of the Reform of Research Institutes', in Sigurdson (2004c).
16. Zheng Minzheng, China's IC industry under fast development, in *An Investigation Report of China's Semiconductor Industry*, China Semiconductor Industry Association, Beijing 2002.
17. These new factories were mainly distributed in Fushun (Liaoning province), Zuyi (Gansu province), Shanghai, Ningxia, Baotou (Neimenggu), Baoji (Shanxi) and Guangzhou (Guangdong).
18. To some extent, Chinese reform is a typical process of trials.
19. Concerning GRINM's orientation on high technology, there was a very interesting story. A previous president of GRINM, Professor Wang Dianzuo said, 'If we used our skills to bake the bread, undoubtedly, we can bake the best bread in the world. But can we do that? If we should be transformed into a corporation, our corporation should be different from the average ones. If we only use the high technology shipments to simply make a living, then we are nearly committing a crime.'
20. According to the Chinese personnel system, a person's length of time in a position mostly decided his promotion.
21. Actually, this led to many problems.
22. The company boasts of a highly intellectual staff with over 50 per cent of the 150 members holding an intermediate or advanced academic title, including 4 with a PhD degree and 30 with a master's degree. It has successfully completed nearly 100 key national high-tech projects and plucked scores of important awards, including 53 at the ministry level, 9 for national invention, 3 for national scientific and technological promotion, 2 from the national science congress and 1 for a special contribution to national scientific and technological promotion.
23. The company expects to develop into a modern enterprise directed by state-of-the-art management ideas. It is to go public for faster business growth to become a world centre for development and manufacturing of hi-tech shape memory alloy and precious metal products.
24. Employment includes seven nationally outstanding young and middle-aged experts, more than 100 experts awarded government special subsidies, and nearly 1000 professors, senior engineers and engineers as well as 300 with doctorates or master's degrees. Professor Dianzuo Wang, the Honorary President of GRINM is vice president of CAE and a member of the CAS. Professor Guocheng Zhang is also a member of the CAS.
25. GriTek supplies the following major products. IC grade 4–8" Si crystal and polished wafers, 3–5" FZ NTD Si crystal and polished wafers, 4–6" heavy doped Si polished wafers for Epi-substrates, GaAs, GaP, InP, GaSb crystals and wafers for photo-electronic applications, Ge crystals for infrared optical applications and $GeCl_4$ for optical fibres.
26. Ministry of Commerce.
27. Semiconductor Industry Association and China Centre of Information Industry Development (2004).
28. Semiconductor Industry Association website, http://www.sia-online.org (SIA issue backgrounders).
29. 'China's Semiconductor sector shake-up', *People's Daily*, 18 September 2004, http://english.people.com.cn/200409/08/eng20040908_156309.html.
30. Ibid.
31. Ibid.
32. Ibid.
33. Liu (2000).

34. Ibid.
35. Interview with LHWT staff member, 3 November 2003.
36. There are presently around 5000 Chinese with engineering background in Japan, many of whom might potentially return to take up positions in China.
37. The LHWT website (http://www.lhwt.com.cn) mentions that the company's President and CEO, Dr Guoliang Shou has a PhD degree from Tokyo University, and is a recognized expert in IC design, image identification systems, W-LAN, and RFID chips in read-write systems. He has acquired more than 100 patents in interrelated areas. Dr Shou used to work for the semiconductor development company Yozan Inc. in Tokyo before returning to Beijing.
38. Xu and Gong (2003), pp. 155–74.
39. Ibid.
40. Ibid.
41. Li Wenfang (2004).
42. Li Weitao (2003b).
43. Chen Zhiming (2004a).
44. Ibid.
45. http://www.tdscdma-forum.org/EN/index.asp.
46. ZTE Corporation, also known as Zhongxing Technologies, and Huawei Technologies are the two largest telecommunications equipment providers in China. ZTE specializes in offering customized network solutions for telecom carriers worldwide, fixed, mobile, data and optical networks, intelligent networks and next generation networks as well as mobile phones. Almost half of all ZTE personnel of more than 20 000 are involved in R&D. The company annually allocates over ten per cent of its revenues to R&D. ZTE has set up 13 wholly-owned R&D centres worldwide and has entered into research partnerships with major electronics companies like Texas Instruments, Motorola and Agere Systems. The company has been listed on the Shenzhen Stock Exchange since 1997 and recorded sales of RMB16 billion in 2003. ZTE currently owns some 700 patents, with more than 87 per cent of these being original innovations. (Website information available at http://www.zte.com.cn, accessed 27 January 2005).
47. China Electronics Corp. (CEC), one of China's electronics conglomerates, with many similarities to Putian, formed a shareholding company in August 2003 in preparation for overseas listing. The shareholding company, called CEC Engineering Co. Ltd, received RMB16 billion worth of assets out of a total of about RMB38 billion previously under CEC control. CEC used to control 26 affiliated subsidiaries and 35 shareholding companies, including Shanghai Huahong, one of China's largest IC manufacturing conglomerate, software maker ChinaSoft, Shanghai Belling and Amoisonic Electronics. Other activities included a wide range of business, including investments and financing, international trade, property and exporting labour services (Li Weitao, 2003a).
48. Chen Zhiming (2004b).
49. Private communication from Department of Public Relations, Marketing Centre, ZTE Corporation (January 2005).
50. '"Chinese standard" aims at international market', 14 June 2004, http://www.tdscdma-forum.org.
51. Li Weitao (2004).
52. Comment by Zhang Qi, director-general of the Department of the Electronics and Information Product Administration under the MII, as reported by Chen (2003).
53. Li Weitao (2004a).
54. Ibid.
55. The Summit is also supported by six operators: China Mobile Communications Corporation (China Mobile), China United Telecommunications Corporation (China Unicom), China Telecom Corporation (China Telecom), China Net Corporation Co., Ltd, China Railway Communication Co., Ltd, and China Satellite Communication Corporation.
56. 'China's locally designed and fabricated 3G TD-SCDMA Chip', http://www.tdscdma-forum.org/EN/news/see.asp?id=980.
57. http://www.tdscdma-forum.org.

58. 'UTStarcom ready for 3G business', *China Daily*, 14 June 2004.
59. Mr Jiang Chunsheng: the success of TD-SCDMA is the symbol of the success of Chinese 3G, Interview of the general manager of wireless communication department of China Railcom, by Liu Chunhui http://www.tdscdma-forum.org.
60. NortelNetworks wishes to be the leader of TD-SCDMA, Interview with Mr Huang Jigong, Vice President of NortelNetworks (China), TD-SCDMA website.
61. http://www.tdscdma-forum.org.
62. US mobile operator NTCH Inc., active in California and Arizona.
63. 'Huawei on a roll with 3G', *China Business Weekly*, 9 January 2005, http://www.chinadaily.com.cn/english/doc/2005-01/09/content_407215.htm.
64. Ibid.
65. Ibid.
66. 'Chipmakers get more tax breaks: after scrapping VAT rebate on April 1, mainland offers R&D compensation (Bloombergs in Shanghai)' (*South China Morning Post*, 7 April 2005).
67. Chen Qide (2005).
68. This figure was mentioned in Simon (2005).

6. Rising technological capability

CHINA BECOMING AN ENGINE OF GROWTH

China is now seen as the engine of economic growth in our global system. Let me start by referring to comments made by one of our leading development economists 25 years ago. Sir Arthur Lewis, the Nobel Economics Prize winner, in his lecture to the memory of Alfred Nobel on 8 December 1979 gave the following title to his speech: 'The Slowing Down of the Engine of Growth'. He started his talk with the following sentences:

> Let me begin by stating my problem. For the past hundred years the rate of growth of output in the developing world has depended on the rate of growth of output in the developed world. When the developed grow fast the developing grow fast, and when the developed slow down, the developing slow down. Is this linkage inevitable? More specifically, the world has just gone through two decades of unprecedented growth, with world trade growing twice as fast as ever before, at about eight per cent per annum in real terms, compared with 0.9 per cent between 1913 and 1939, and less than four per cent per annum between 1873 and 1913. During these prosperous decades the LDCs have demonstrated their capacity to increase their total output at six per cent per annum, and have indeed adopted six per cent as the minimum average target for LDCs as a whole. But what is to happen if the MDCs return to their former growth rates, and raise their trade at only four per cent per annum: is it inevitable that the growth of the LDCs will also fall significantly below their target?

In his speech Arthur Lewis on no occasion mentioned China, which would achieve a tremendous high rate of economic growth in the following couple of decades. China's experience is analogous to similar rates of growth in Japan, Korea and Taiwan. Like them, China's recent economic growth was initially export-led and overseas demand for industrial products continues to play an important role. However the export-led growth is now being complemented or even replaced by the growth of the domestic market, the size of which has no parallel in the earlier development in neighbouring countries. China has not only primed its own engine of growth after several earlier false starts in economic development; it has also created a great impact on almost every other country in the world that now has to react to challenges of competition and collaboration.

Furthermore, in its future industrial technological development China is basing itself on a combination of three important knowledge sources that will define its future technological capability. First, China is expanding its funding of R&D, which far exceeds its rate of economic growth, although much of those resources are used within a national innovation system that is suffering from a hangover from the earlier planned economy period. Second, the country has through its Open Door Policy attracted a huge amount of foreign direct investment in which manufacturing has played an important role, now increasingly followed by foreign direct investment in R&D. Triggering this development is the importance of the size and sophistication of the Chinese market combined with the talent pool that is being created, in which the universities will play an important role. Third, Chinese companies are beginning to invest critically in R&D, with the medium-term ambition to become leading global companies.

This chapter will exemplify China's ambitions, strategies and results in selected industrial sectors. Textiles are an example of a traditional industry where China has established a strong competitive advantage not only because of its low labour costs but also by significant technological upgrading. Technology plays an even more important role in sectors such as electronics. Here China started with an emphasis on consumer electronics at the lower end and is now following through with a strong entry into integrated circuits at the higher end. Aircraft and supercomputer industries are provided as illustrations of sectors where the future size of the domestic market – similar to electronics – could have a strong bearing on the outcome, and where China has already established a strong position as a component maker for Boeing and Airbus. Finally, biotechnology will be mentioned as an example where China, being less of a latecomer, has decided to join the world league nations in a research domain that has great future potential.

TECHNOLOGICAL BASE AND CHANGING TRADE PATTERNS

Textiles

The 1995 trade pact, Agreement on Textiles and Clothing, comes to end in 2004. The members of the World Trade Organization that signed that agreement have agreed to end the quotas that for the last ten years have strongly regulated global trade in textiles and clothing. The prediction is that American and European companies that currently buy from about 60 countries might source from as few as 20 by 2006 and less than 10 by 2010.[1] The expectation is that China might capture some 50 per cent of the trade compared with its present 16 per cent. To reach this level of dominance Chinese manufacturers might lower prices, as many of them already run very efficient operations.² Many Chinese companies are already offering improved supply chain management and value-added services in design, with the following illustration from Luen Thai, which is the largest apparel maker listed on the Hong Kong Stock Exchange.³

> At the Dongguan compound, Luen Thai provides major clients such as Polo Ralph Lauren and U.S. department-store chain Dillard's with dedicated teams of designers, merchandisers and sales staff – all housed in plush offices located on the factory site. Keeping designers close to production cuts development time and improves communication. Luen Thai teams are also able to learn about client expectations for materials, styles and costs. Eager to please, the company even bought a roomful of washing machines so it could carry out product testing previously farmed out to a laboratory in Hong Kong. There are top-loading Sanyo washing machines for clothes bound for Japan, Zanussis for Adidas-brand sportswear and outsize Kenmores for the US.

The looming drastic changes in global trade patterns for textiles and garments have triggered the US to protect its remaining garment industry.⁴ China agreed when it joined the WTO to let the US and other countries impose emergency restrictions on their textile imports until the end of 2008. Thus a WTO member state can continue to limit textile imports from China on the ground that they may cause 'market disruptions'. The US administration launched altogether seven investigations during late October and early November 2004 that could lead to import restrictions for clothing and textiles from China, including cotton pants, underwear and knitted shirts.

Electronics

Moving up the technological ladder – standards as competitive tools

China's news media on 22 October 2004 reported that 'the number of mobile phones supporting multimedia applications will reach 380 million and the size of the mobile phone industry will reach US$76 billion by 2007 globally.⁵ The same source mentioned that mobile communications multimedia represented by the 3G service are expected to enter into vigorous development: 'By 2008, the global market size of mobile game services, mobile photo services, mobile colour message services, mobile short message services and other existing services will reach US$17.5 billion, US$44 billion, US$60 billion, US$25.2 billion and US$8 billion respectively, which represents a great market potential for players in this field'.

To meet this challenge nine leading Chinese telecom institutions have formed the mobile multimedia technology alliance (MMTA) with the objective of making industrial alliances that could bring breakthroughs in the development of China's booming IT sector; backed by the government. Included in the MMTA alliance are telecom operators – China Mobile, China

Unicom, China Telecom and China Netcom – and equipment providers – Huawei Technologies, ZTE Corporation, Putian Corporation and Vimicro Corporation. MMTA also includes a research institution – the China Academy of Telecommunication Research of the Ministry of Information Industry (MII).

The alliance has been set up to boost technical innovation and development of standards and applications in a booming mobile multimedia industry.[6] 'The purpose of MMTA is to integrate the forces of various players in the value chain of the mobile multimedia industry to promote the innovation and standardization of networks, terminals and applications', according to MMTA president Jiang Lintao. A more general purpose is to boost the competitiveness of Chinese enterprises in the race to apply upcoming 3G (third generation) technologies. The alliance will include software developers and Internet content providers in addition to research institutions, operators, terminal and network equipment manufacturers and core chip technology providers. 'But Chinese firms need to co-operate if they want to grab a larger market share facing the stiff competition from international rivals.'[7] Foreign partners will be allowed to join, but rules will be quite different from those of domestic members,[8] and detailed rules will be rolled out.

The MMTA example mentioned above is only the tip of the iceberg of China's efforts to climb the technological ladder. China's industrial development that fuelled its success in exports has been based on abundant Chinese labour offered at low cost to foreign investors as well as domestic companies. Many and soon most of world advanced consumer electronics products will be manufactured in China but rarely do they carry a Chinese brand name or contain advanced technology of Chinese origin.

Standardization has in recent years become a very important element of China's technology strategy. There are basically two trends to consider, in the light of the tectonic shift of electronics to East Asia. One is the rising attempts to harmonize interests and development among three countries in East Asia – China, Japan and Korea. It has become obvious that the countries in East Asia have identified common interests and a more formal approach was established at a ministerial meeting in Seoul in September 2003.

The other and more important trend is that China wants to establish its own technological platform in as many areas as possible, in order to gain independence from foreign high-tech companies and drastically reduce the level of licence fees. Being able to establish or influence global standards has become vital for national technological efforts.

China is becoming the world center for mobile applications – together with Korea and Japan – with an installed customer base of around 360 million by the end of 2004. The three largest operators in the world include two from China – China Mobile and China Unicom – with the former being by far larger than Vodafone, which is the number two operator company in the world.

A large number of mobile applications will appear on the market over the next few years and a substantial share of them are likely to be developed by Chinese engineers. The large and expanding Chinese market will drive this development as it requires only limited creativity, the major requirements being engineering tasks of a fairly repetitive nature for which China would be able to provide ample manpower at low cost.

Semiconductor industry

The growth of China's semiconductor industry continues to be driven rapidly by an increasing domestic demand, and supported by government incentives. However, observers still suggest that throughout 2005 the foundries that have already been established in China will primarily rely on orders from overseas customers. The situation is likely to change during 2006 when domestic design companies will begin to deliver product designs for domestic manufacturing.

The global semiconductor industry has changed dramatically during the past few years in response to demand and incentives to locate new plants in China. In 2000, half of the plants under construction in the world were in Taiwan, with at the time confirmed plans to build other plants over the next eight years at a cost of $70 billion.[9] A number of plants are either under construction, operational or planned in China. One important factor has been the attractiveness of investment in China through its 17 per cent value-added tax (VAT) on imported semiconductors as opposed to a 3 per cent VAT on those made in China, although recent regulations have removed this incentive.

Other factors are China's industrial development strategy that has provided concessions in industrial parks and actively sought foreign investment. China has sought a step-by-step catch-up strategy rather than making large-scale investments in R&D projects as has been the case in Japan and Europe. However, the most important factor in the longer perspective is a rapidly expanding market for semiconductors, with the requirement of being close to customers to meet changing and more sophisticated needs. This is exemplified by German chip maker Infineon Technologies that is planning an assertive expansion in China, which it considers one of the world's most dynamic semiconductor markets.[10]

In September 2004 Infineon opened a memory chip assembly and testing operation in Suzhou, in the form of a joint venture China-Singapore Suzhou Industrial Park.[11] The venture will have a maximum capacity of 1 billion chips per year and employ 1000 people, eventually requiring a total investment of US$1000 million. An important factor in Infineon's strategy is that China should become a major car producer. Infineon will support China's automotive semiconductor operation by transferring know-how to meet higher emission and safety standards. The underlying factor is that China is the world's fastest-growing automotive semiconductor market, and Infineon wants to maintain a strong position. The research company iSuppli predicts that the size of China's car electronics market will grow from US$1000 million, in 2002 to US$5500 million in 2007 – with automotive semiconductor consumption increasing from US$416 million in 2002 to US$14 500 million, in 2007.[12] The annual demand for integrated circuits in China is expected to reach US$36 000 million by 2006.[13]

The start of the modern semiconductor industry in China can be traced to the eighth Five Year Plan[14] which covered the years 1991–95, during which time a number of multinational companies established initial operations. In 2000 China produced a low US$900 million worth of semiconductors compared to $11 000 million in Taiwan, and most of the production was rudimentary. The US government in the past restricted access to advanced semiconductor technology and still does for certain technologies; it is also influencing other countries to do the same under the Wassenaar agreement, which involves 33 countries. The objective was to hinder the transfer of technology that could be used in the military sector. An even more serious hindrance at the time was the lack of skilled technicians and managers, which partly reflected the turmoil of the Cultural Revolution. However, a report from the General Accounting Office in 2002 indicated that several of China's factories, using foreign capital and technology, were only one generation or less behind the world's leading semiconductor makers.[15]

The microelectronics industry in China, especially the integrated circuit segment, has always been regarded as strategically important, and was awarded priority when China formally announced the four modernization programmes in the late 1970s. However, in an interview at the Ministry of Information Industry in 2003 an official argued that China should rely on foreign direct investment to establish the costly and complex production facilities that are required for IC production.[16] This would be the quickest way to improve the IC industry, while China would concentrate its domestic resources on chip design. The China Center of Information Industry Development (CCID) argues that now is the right time for the multinational companies to enter China's IC market as it will be one of the important bases of global electronic and information products and there will be an expanding domestic market.[17] CCID argues that the development of a robust domestic IC is one of the government's main industrial priorities to be supported by favourable policies, with an advantage in low cost production and rich human resources.

Until recently, there hardly existed any wholly foreign-owned enterprises in any Chinese industry. However, the Chinese government has made it clear that such investments are now welcome and several 100 per cent foreign-owned semiconductor producing enterprises now exist or are being established, including the following:[18]

1. Motorola Tianjin Integrated Semiconductor Manufacturing Complex, operational in 2001 with a total investment in the range of US$1.5 billion.
2. Suzhou Matsushita Semiconductor.
3. Wuxi Huazhi Semiconductor Co. will be transformed into a wholly-owned Toshiba subsidiary.
4. Taiwan Semiconductor Manufacturing Corporation (TSMC) Shanghai, Songjiang HiTech Park.
5. Ultimate Semiconductor (a Malaysian enterprise that signed an agreement with the Shanghai government in 2003).

China has also attracted a number of multinational foundries. The report commissioned by the Semiconductor Industry Association (SIA) in the US says that the new semiconductor foundries being established in China are unique in the country's context not only because they separate the design function from production, but because the enterprises themselves much more closely resemble Western multinational corporations than any prior Chinese semiconductor enterprises, all of which have been at least partially government owned and controlled. They include the following:[19]

1. Semiconductor Manufacturing International Corporation (SMIC), Shanghai, with investment from Shanghai Industrial Holdings, Avant!, and others.
2. Grace Semiconductor Manufacturing International (GSMC), founded in 2000.
3. He Jian Technology Corporation, China-Singapore Suzhou Industrial Park.
4. Beijing Semiconductor Manufacturing Corporation (BJSMC), in collaboration with the Beijing Municipal Government, Beijing Economic and Technical Development Area (BDA) and Shougang Iron and Steel. (BJSMC will accommodate facilities for the partnership between SMIC and Infineon Technologies).
5. Wuxi CSMC-Huajing, which has been operating since 1997.

The SIA report also states that a long-standing source of weakness in the Chinese semiconductor industry has been the shortage of makers of semiconductor equipment and materials, assembly, testing, packaging and logistics firms. An earlier review stated that[20] 'China is still incapable of producing most of the equipment used in an 8 inch IC production line, though it can produce some supplementary machinery. As for 6 inch lines, Chinese firms are technically capable of producing almost all required equipment but no one firm is manufacturing enough to be considered a world-class producer.'

However, the SIA report argues that this problem is rapidly being addressed in both Shanghai and Suzhou as leading semiconductor enterprises, as well as materials, design and support firms are establishing operations in Zhangjiang Hi-Tech Park, with its Shanghai Zhangjiang Semiconductor Industry Base (ZSIB) and the Suzhou Industrial Park. ZSIB is well on its way to becoming a major centre for semiconductor production in China, while the Beijing Economic-Technical Development Area plans to establish a complete industry chain surrounding the new semiconductor foundries being located there.[21] China has in the past not been a major market for major equipment suppliers because of the low production volumes inside China as well as export control restrictions. However, by 2010 China could have become one of the biggest markets for semiconductor-related equipment, partly served by domestic companies. In the meantime China will have to import most of the necessary equipment.

SUPERCOMPUTERS AND AVIATION

Supercomputers in China – Usage and Production

In late 2004, China was in possession of 17 supercomputers out of a total of 500, which gave it fourth ranking in the world, equal with Germany and only behind the US, Japan and Great Britain. The Shanghai Supercomputer Centre assembled a machine in June 2004 that at the time became the world's tenth fastest computer, by using more than 2500 chips designed and manufactured by Advanced Micro Devices in the US.[22] A recent geographical trend is emerging more clearly. The number of systems in Asian countries other than Japan is rising quite steadily. Japan is listed as having 30 systems in 2004 and all other Asian countries accumulated an additional 57 systems. However, Europe is still ahead of Asia with 127 systems installed. Seventeen of the systems in Asia are installed in China – up from nine systems one year previously. The number of systems installed in the US has also increased to 267 – up from 247 one year previously.[23]

China has become the most rapidly advancing country in the world in terms of continuous thrust in high-performance computing (CHPC). However, China's progress in this high-technology field will not become fully recognized until a customer from Thailand or Malaysia is prepared to place its order for a supercomputer with a Chinese company. In 2003, China had four domestic companies that developed and marketed supercomputers:[24]

- Lenovo (formerly Legend), which was spun off from the Institute of Computing Technology (ICT) in 1981 and entered the HPC market in 2001. The ICT is a major research institute of the Chinese Academy of Sciences.
- Dawning, which was spun off from ICT in 1995 and the same year entered the HPC market. The fastest supercomputer in China – Dawning 4000A, which operates at a speed of 11 trillion calculations per second – was officially started at the Shanghai Supercomputer Centre in November 2004. It was jointly developed by the Institute of Computer Technology of the Chinese Academy of Sciences, Dawning Corporation and the Shanghai Supercomputer Centre.
- Langchao entered the HPC market in 2002.
- Galactic Computing became the most recent entrant in the computer industry in 2004, for which further details are given below.

High Performance Computing Centres have been established in a number of locations in China. Important ones include the Department of Computer Science and Technology at Tsinghua University, the Institute of Computing Technology, the Academy of Mathematics and System Science, Beijing Genomics Institute (BGI), and the Shanghai Supercomputer Centre.

Galactic Computing in Shenzhen could possibly be such a supplier in China's export market for supercomputers. The company primarily develops blade supercomputers, and was founded in 1999 by Shell Electric Manufacturing Holdings in Hong Kong.[25] A blade supercomputer launched in early 2004 is reported to be the world's fastest computer of its kind, with an average calculation speed of 1 teraflop, which can be scaled up to more than 50 teraflops at peak speed.[26]

In the past intelligence experts in the US were deeply concerned that supercomputing capabilities would aid China's weapons development. However, it has now become conventional wisdom that China's access to extreme computing speeds in itself does not represent a threat to the military security of the US. The attention has rather shifted to the threat that China would catch up more quickly with the United States in areas that have economic and scientific, rather than military, implications.

Galactic Computing will in the future focus on some ten sectors when developing industrial applications; this will be done in collaboration with universities and research institutions in China. Dr Chen, CEO of Galactic Computing, met with senior Chinese Government officials from two dozen ministries and bureaus in October 2004, including education, commerce, public security, information industry, health, science and technology, and the National Development and Reform Commission.[27] Galactic Computing has selected eight partners that will collaborate on areas which include healthcare, chip design, education, entertainment, national security, logistics, new drug discovery and clinical trials, bioscience and exploration for natural resources. These will include Tsinghua University for bioscience; China University of Geosciences for exploration of natural resources; Beijing Jiaotong University for transportation logistics and Nankai University in Tianjin for port logistics. When application systems are ready for marketing Galactic Computing will, with its partners, invite leading Chinese companies in each sector to promote relevant applications.

Dr Steve Chen, who used to be the chief architect of Cray supercomputers in the US is very positive about the future and states that 'With our technology transfer, I hope we can help China in leapfrogging in blade supercomputing technology and applications, and joining the global leaders in information'.[28] Galactic Computing expects to receive favourable responses within China from universities, research institutions and companies.

Some major computer vendors in the US have taken an interest in blade supercomputers, and the US Government has invested US$90 million in an ongoing project which has been designed to support energy-related research. By October 2004 Shell Electric had invested US$20 million in Galactic Computing, and the company clearly suggests that its development costs will be considerably lower as a result.[29] The company also suggests that the users of blade supercomputers need not discard old computers to buy new ones as they can recycle the old ones and add new ones for greater performance or new applications. Many customers may appreciate blade computers as a revolutionary design concept as it significantly boosts the sustainability of computer capabilities.

The situation has changed dramatically since the 1980s, when it was difficult to import supercomputers into China. Today advanced components, for building high-performance computers have become mature and commercially available. The open availability of high-speed computer chips has shifted the expertise needed for building supercomputers to the software domain. The challenge lies in developing software that can efficiently link hundreds or thousands of processors together. Despite this change in the technology environment the US State Department in October 2004 renewed calls for maintaining the arms sales embargo against China, which also extends to restrictions on the fastest computers.[30]

Another Chinese company has also made an entry into the supercomputer field. Dawning Information industry delivered one of the most powerful supercomputers to the Chinese Academy of Sciences in 2004. However, the processors forming the computer cluster were obtained from Advanced Micro Devices (AMD) in the US, although the cluster will run a Chinese-designed Linux operating system.

Aircraft Industry

The aircraft industry is one of the sectors where China presently perceives a most conspicuous trade imbalance. China is expecting a rapid and substantial expansion of air transportation over the next 15 years, primarily to meet domestic need. The country is now in the process of consolidating and combining its various aircraft-related plants into new and viable structures. It is following a twofold strategy with a rejuvenation of existing plants and entering into agreements with foreign partners, as described below.

China Aviation Industry Corporation I (CAICI) and China Aviation Industry Corporation II (CAICII) originated in the late 1990s from the former Aviation Industries of China, and are still state-owned holding corporations controlled by the Government. The latter includes 54 industrial enterprises and three institutes which are involved in helicopter aircraft, engine and airborne equipment and so on, and another 22 enterprises and institutes that are directly controlled by CAICII, which also holds 50 per cent of shares in the China National Aero-Technology Import and Export Corporation (CATIC), and the China Aviation Industry Supply and Marketing Corporation.[31]

CAICII was the first company to be listed among the country's ten most sensitive and largest military business conglomerates under the direct control of the State Commission of Science Technology and Industry for National Defence (COSTIND), the top governmental body in charge of the country's national defence industry.[32] With a focus on the car industry, it is different from CAICI as it has fewer military businesses.

CAICI received approval from the State Council in 2003 to utilize RMB5000 million for a preliminary study of the ARJ21 regional jet aircraft and launched manufacturing bases in four industrial cities.[33] In parallel CAIC II had entered into a joint venture with Embraer in Brazil, with an equity investment of US$25 million, of which 51 per cent was taken by Embraer – with a planned annual production capacity of 24 ERJ regional jets. Mauricio Botelho, president and CEO of Embraer, said the successful maiden flight suggests that this advanced regional jet will soon enter the Chinese market.[34] China's domestic development of the ARJ21 and its joint venture with Embraer to produce the EMB170 in China are clear indications of China's long-term ambitions to join the league of the two global makers of large transport and passenger planes – Airbus and Boeing.

The Aviation Industry Development Research Centre of China estimated that the country's civil aviation fleet would have to add 1400 large jet liners by the year 2022, which if all are imported as in the past, would cost in the region of US$100 billion. The Ministry of Science and Technology (MOST) will together with other agencies be involved in a sizeable science and technology development programme to support the aviation industry over the coming 20 years. A senior aviation industry executive declared in March 2004 that he anticipates that China's first large aircraft will fly by 2018.[35] This means an aircraft with a load capacity exceeding 100 tons and a passenger capacity of more than 200 passengers.

Today decision makers within the industry consider that China's aviation sector will be incomplete without developing its own civil aircraft industry. They gain support for national defence purposes as well as the size of China's market – although facing the fact that Boeing and Airbus combined are *de facto* sole suppliers of large aircraft in China. Liu Gaozhu, President of AVIC I realizes that China has sought to develop civil aircraft since the 1970s, first by itself, then through subcontracting and cooperation, and that the efforts have virtually ended in failure, either because the planes made could not match clients' needs, or because China did not own independent intellectual property rights. However, he and others consider that the time is ripe for China to initiate a large aircraft programme that would be successful. China should first develop large cargo aircraft before developing large passenger aircraft and the involvement of private business would be important to speed up the reform of needed research and development.

Boeing and Airbus are increasing their production in China, thus making the country a major world producer of aircraft parts and components.[36] Their increased interest in sourcing from China reflects the technological achievement of China's aircraft industry. Boeing plans to increase the production rates of model 737 aeroplane assemblies built at plants in Xi'an, Shanghai and Shenyang. Furthermore, Boeing is also planning a new major aeroplane maintenance and repair facility in Shanghai. The ARJ21 jet, a new regional jet model to be built by China Aviation Industry Corporation I, is supported by Boeing technical contracts. A joint venture, BHA Aero Composites Co. Ltd, will have additional opportunities as aeroplane production rates and support contracts with Boeing will increase.

Boeing and China's industrial cooperation started in the mid-1970s, and there are now more than 3200 Boeing aircraft in service that include major parts and assemblies built by China, corresponding to 25 per cent of Boeing aircraft presently in service. Boeing has procured about US$500 million in aviation hardware from China. It might double by 2009 and reach US$1.3 billion by 2010.[37]

Airbus and its parent company European Aeronautic Defence and Space Company (EADS), the world's second largest aerospace and defence company, has also entered into major collaboration with China's aviation manufacturing industry. Airbus plans to considerably increase its procurement from China, which would be worth US$60 million annually by 2007 from the 2004 level of about US$10 million annually.[38] Industrial cooperation between Airbus and the Chinese aviation industry began in 1985, when the General Administration of Civil Aviation of Shanghai, now China Eastern Airlines, became the first carrier in China to operate the European consortium's aircraft. Contracts for Chinese companies to build sections of Airbus aircraft followed, as did further orders from Chinese airlines.

Aerospatiale, which is now Airbus France, signed the first product subcontracting agreement in 1985 with Xi'an Aircraft Company for manufacturing and assembling access doors for the Airbus A300/A310 wide body aircraft. Since then, the total value of projects subcontracted by Airbus to Chinese manufacturers has exceeded US$500 million. In 2002, Chinese manufacturers delivered more than US$12 million worth of aircraft components to Airbus.

In November 2003, Eurocopter, another subsidiary under EADS, signed an agreement with Harbin Aircraft Industry Group Corporation, China National Aero-Technical Import and Export Corporation and Singapore-based Technologies Aerospace to produce jointly the five-seat EC120 helicopters in China. In October the same year, EADS subscribed to five per cent of the issued shares of the Chinese aviation automobile manufacturer AviChina Industry and Technology Co. Ltd in the Chinese company's initial public offering on the Hong Kong Stock Exchange. EADS and AviChina will cooperate in the development, manufacturing and upgrade of aviation products, including helicopters and trainers.

Global Integration

In early November 2004 the president of Brazil took the decision to recognize China as a market economy. This move was part of a deepening economic, industrial and technological relationship between the two countries.[39] In the preceding month the countries agreed to work jointly on a third satellite for resource observation, scheduled for placement in orbit in 2006. The satellites will observe and plan for urban expansion and resource conservation, as well as detecting river and ocean pollution and deforestation. However, the Federation of Industries criticized the move and argued that it would hurt Brazil's industry, although China had become Brazil's third-largest trading partner after the US and Argentina.

Aside from joint interests in space technology, the two countries have also found the aircraft industry to be a complementary economic sector, which is exemplified by Empresa Brasileira de Aeronáutica (Embraer). The company has become one of the largest aircraft manufacturers in the world by focusing on specific market segments with high growth potential in regional, military and corporate aviation.

The international market for regional jets is strongly dominated by Embraer of Brazil and Bombardier of Canada. China has been approaching its future aircraft industry for smaller jets by a combination of strategies. While CAIC I is building the ARJ21, a sister company China Aviation Industry Corporation II (CAIC II) signed an agreement with Embraer in 2003 to produce jointly regional jets with 30 to 50 seats in Harbin.

China will need hundreds of small regional jets of up to 100 seats to serve the burgeoning domestic air travel market over the next two decades. Around 50 new airports are scheduled for construction over the next five years and feeder traffic from regional jets is forecast to grow 12 per cent annually in China over the next decade.[40] Shanghai Aviation Industrial (Group) Corporation planned to begin production of the first batch of components for the country's first self-designed regional aircraft in 2003, expected to enter service at the end of 2006 or early 2007. The aircraft is known as ARJ21, which stands for 'advanced regional jet for the 21st century'. With this and other initiatives China aims to capture a portion of the world's fledgling regional jet market.

Shanghai was chosen as the location for assembly production based on existing equipment and rich experience and know-how from building MD passenger planes in the 1980s and 1990s. The factory will join with domestic partners in Chengdu, Shenyang and Xi'an, where many components will be manufactured for the ARJ21 jets. Approved by the State Council, China Aviation Industry Corporation I (CAIC I) has set aside RMB5000 million for a preliminary study of the jet and launched manufacturing bases in the four cities.[41]

A Boeing report suggests that China's airlines will add some 1500 new planes by 2020 to serve domestic markets, and regional jets and single-aisle aeroplanes will account for more than 78 per cent of the planes serving the domestic market.[42] Shanghai Aviation Industry (Group) Corporation predicts that orders for the ARJ21 will total 500 over the coming two decades, which include exports to Asia, Africa and South America.

China Aviation Industry Corporation II (AVIC II), also referred to as CAICII, controls a newly established Harbin Aviation Industry (Group) Co. Ltd which is based on the Harbin Aircraft Industry Group and the Harbin Dong'an Engine Group Co. Ltd. AVIC II is the parent company of the Hong Kong-listed AviChina[43] Industry and Technology Co. Ltd. It reported sales of RMB23 000 million in 2003. Its major business is the manufacture of minicars. The company is the largest maker of such cars, with the production of 321 000 units in 2003 that provided RMB14 700 million with an additional RMB2000 million from sales of auto parts.[44] AVIC also produces helicopters[45] within companies that belong to the group, and has entered into a regional joint venture with Embraer in Brazil to assemble regional jets in Harbin. Embraer is expecting its EMB170 jet to win a considerable share of the domestic market in spite of growing competition.[46] There are several factors that explain the Embraer joint venture in China. A very important reason is the imbalance between the capacity of China's fleet of aircraft and the low passenger demand on most flight routes. Embraer has estimated that China will demand 230 regional jets with 30 to 60 seats over the next two decades. Furthermore, demand for 60 to 90-seat aircraft as well as 90 to 110-seat aircraft will both exceed 200 units. Embraer has entered the Chinese market to gain a strong foothold in China's highly promising regional jet market over the coming five years. The EMB170 is able to accommodate between 70 and 78 passengers, and Embraer has reported 145 confirmed orders and 150 orders of intent for the plane from airlines worldwide.

However, the EMB170 may come into direct competition with the ARJ21, which can accommodate 70-plus passengers – now being developed by China Aviation Industry Corporation (AVIC) I. Commercial Aircraft Co. Ltd (ACAC), which is marketing the ARJ21, has announced 35 orders with more orders to come during 2004 – although main deliveries are not scheduled until 2007.

PERSPECTIVES ON BIOTECHNOLOGY IN CHINA

China has started to create a durable foundation, based on a number of significant measures that could develop into a globally competitive research-based pharmaceutical industry. The pharmaceutical industry in 2002 had a market size of US$6.8 billion and ranked as the seventh largest in the world, and is expected to be the fifth largest in 2010.[47] Aside from China's rapidly expanding gross domestic product (GDP) and the increased access of the population, in general, to healthcare products, there are a number of key factors that could propel an industry based on research to the front rank by 2020. They include attention given to the sector in government priorities, the expansion of public research institutions, increasing acceptance of intellectual property rights (IPR) and their integration into research strategies, an already existing industry and the attention given to the development of human resources. Thus, the International Federation of Pharmaceutical Manufacturers Associations (IFPMA) envisages the appearance on the global scene of novel Chinese-originated medicines.[48]

Biotechnology is a science-based industry and innovations drive the competition within the sector. The industry took root in the late 1970s and the 1980s in the US, which still dominates the sector with its specific culture and institutions. Researchers within universities have played an important role, often in collaboration with venture capitalists. This has prompted attempts to establish new integrated companies that specialize in biotechnology products or services. However, these companies are faced with a scarcity of long-term funding and often lack complementary assets that are needed for marketing, clinical tests and product development. These factors have enabled the large pharmaceutical companies to incorporate biotechnology gradually into their activities, initially as research tools, as they have needed financial resources and marketing channels.

Collaborative agreements have become very common within the biotechnology industry in order to capture innovative activities in various fields. The networks provide both coordination of innovative activities as well as a basis for division of labour, and networks have become more numerous and extensive. It has been observed that the large pharmaceutical companies have a comparative advantage in drug development, which always requires clinical tests, although drug discovery as such does not offer any economies of scale. This suggests that the advantages enjoyed by large pharmaceutical companies in market power and access to resources enable them to embark on more risky ventures than small biotechnology firms, which can seldom manage the rising costs of R&D and marketing.

Molecular biology has created a fundamentally new knowledge structure, which has impelled changes in the way that academic researchers, new biotechnology companies and large pharmaceutical companies organize themselves. One important outcome is the formation and successive evolution of R&D networks, as the knowledge base has continued to expand. It is far from clear if this phenomenon is a transitional or a permanent feature of the biotechnology knowledge landscape in which a distinct pattern of division of labour in R&D will emerge. However, it is evident in the US and to a lesser extent in Europe that new actors have been driving the expansion of networks, with established R&D-intensive pharmaceutical companies, and they have simultaneously absorbed new knowledge within such networks. This would indicate that there are essential competencies within time-honoured corporations with R&D intensity in multi-technology domains, to take in new knowledge and techniques generated outside firm boundaries. China's biotechnology policies and advances in its fledgling biotechnology industry should be interpreted in the light of the broad panorama which has been outlined in this section.

Since 1991, China has emphasized the development of its biotechnology sector, which has been included in its Five-year plans. Technological priorities are not very different from those of other large countries and include functional genomics, proteomics, biochips, animal and plant bioreactors, medication and vaccines based on genetic engineering, gene diagnosis and therapy, transgenic technologies for animals and plants, bio-pesticides, bio-fertilizers and biosafety.

The aim of biotechnology development in China, as in several other high-priority sectors, is to catch up or come close to levels of advanced research and production as practiced outside the country. This will not only require raising the overall research competencies, but also becoming a leader in a number of important fields, which can be summarized in the following objectives. First, China should make breakthroughs in frontier research fields which are of a strategic character to the further development of the biotechnology sector. Second, China needs to develop and support innovation structures that need to be driven by the combined forces of entrepreneurial spirit and high-talented research manpower. Third, biotechnology companies have to move to the forefront of international competitiveness, which would involve partnership with foreign companies and international collaboration.

There are a number of favourable factors that would indicate that China is likely to become an important player in global biotechnology within the coming decade. The most important factor and an absolute prerequisite is the already existing talent pool which is rapidly increasing through graduate and post-doctoral training. China also has the potential to draw on the extensive number of highly trained Chinese scientists now residing overseas, primarily in the US. Estimates indicate that China at home has a total of 200 000 R&D staff trained in biotechnology (life sciences) although most with relatively recent graduate degrees.[49]

The second important factor is the priority that the government and its agencies have given to biotechnology development – through funding, new institutions and policies that are formulated and implemented not only at national level but also at lower levels, where Shanghai and Beijing have decided to undertake major efforts.

Third, with limited innovative biopharmaceutical research undertaken in the industrial sector, a number of recently created public research institutions have been given a major responsibility for the sector. Basic research outside China has traditionally been the responsibility of universities in specialized departments and research institutes. China has in recent years been extending support to such institutions active in pharmaceuticals, biotechnology, traditional Chinese medicine (TCM), and other selected areas.

Fourth, given the importance of intellectual property rights (IPR) in biopharmaceuticals, the sector is now benefiting from China's move to protect IPR strongly. This includes the reform of the patent law in 1993 and becoming a member of the WTO in 2002, which also includes a commitment to the Trade-Related Intellectual Property Scheme (TRIPS).

Finally, although the industrial sector is small, 'China has a possibly unique combination of attributes in its existing industry. China has a major generics industry with good manufacturing capabilities, a small biotech sector, an important TCM sector, and an engaged foreign-owned/invested sector.'[50]

Government Policies and Initiatives

The factors mentioned above provide China with a platform to reach its objective of establishing a globally-competitive research-based industry. Although China has been working rapidly towards creating a market economy since 1978, many economic activities are still framed in a planned-economy framework with frequent references to Five-Year Plans, and the expected future of the pharmaceutical sector is no exception. In June 2001 the State Economic and Trade Commission (SETC) issued the tenth Five-Year Plan for the Chinese pharmaceutical industry, which covered the period 2001–2005. Aside from including economic goals such as annual growth in output value of 12 per cent, it also included goals and strategies in the following three areas.[51]

1. Product structure of the industry. On this, SETC stated:

   > Significant emphasis will be given to the development of biopharmaceutical technologies. Between 10 and 15 new biopharmaceutical products independently developed by China should be launched, and some of them will enter international markets during the period.
   >
   > To foster development of the sector, the government will control the establishment of new biopharmaceutical enterprises in order to concentrate resources in existing companies, develop a mechanism of venture capital involvement in the sector, establish a pharmaceutical industry fund to support the development of new biopharmaceutical products, and introduce various incentive schemes and subsidies.

2. Technology structure of the industry. On this, SETC stated:

   > Large companies should increase their investments in new drug R&D to above 5 per cent of revenues, while smaller companies should also increase their investments appropriately.

3. Organizational structure of the industry. On this, SETC expected that:

   > The modern enterprise system will be adopted, separating state ownership and management. Reorganisation of small state owned pharmaceutical companies through a variety of means including merger and acquisition, leasing and contractual management is encouraged. Pharmaceutical companies are encouraged to develop international markets.
   >
   > Ten large pharmaceutical groups with over RMB 5 billion [$600 million] in annual sales that can compete with multinationals will be created during the period. These companies will have a 30 per cent market share in the entire pharmaceutical market and will be created through various means including mergers, acquisitions, reorganisation or stock listing.

The following year biotechnology was reconfirmed as one of the nation's key high-technology areas, in the State Policy on Industry and Technology 2002. This included an emphasis in the following three areas:[52]

- Carrying out R&D in genetic engineering, cellular engineering, enzyme engineering, biochemical engineering and bio-medicine.
- Following closely developments in the Human Genome Project, gene therapy and transgenic plants and animals.
- Promoting applications of biotechnology in agriculture, medicine, energy and environmental protection.

The Shanghai example
Shanghai became one of the most important life science research bases of the CAS after the Academy established a number of research institutes during the 1950s and 1960s, covering cellular biology, biochemistry, plant physiology, neuroscience, entomology, pharmacology, the chemistry of natural products, the chemistry of synthetic drugs and drug design. Shanghai is seen as the cradle for China's development in biotechnology as the total organic synthesis of the world's first biologically active protein, crystalline bovine insulin, was successfully completed there in 1965. There now exist three tiers of R&D organization – universities, national or municipal research centres and research institutes/centres under the Chinese Academy of Sciences (CAS). The city is the location of seven biotech-related national research centres. They include the Chinese National Human Genome Centre at Shanghai and the National Centre for Drug Screening. As national centres they focus primarily on genomics, new drug discovery and modernization of TCM.

In 1994 the Shanghai Science and Technology Commission restructured a number of local biotechnology research facilities into some 20 municipal research centres. They included the Shanghai New Drug Research and Development Centre, the Shanghai Centre for Bioinformation Technology, the Shanghai Research Centre for Biomodel Organism and the Shanghai Biological Chip Engineering Research Centre.

In 1999, the CAS consolidated its eight biotech-related institutes and two research centres in Shanghai into a comprehensive research park – the Shanghai Institute for Biological Sciences of the CAS. The objective was to provide a more ideal environment for innovative research in the life sciences. Activities focus on fundamental issues in human health and promote basic and original research, while at the same time encouraging the application and expansion of biotechnology, particularly in the fields of new drug discovery and modern agricultural biotechnology. The Institute has also become an important training base for advanced biotechnology scientists.

The Institute of Biochemistry and Cell Biology was formed through the merger of the former Shanghai Institute of Biochemistry and the Shanghai Institute of Cell Biology (SICB) in April 2000. The Institute is one of the leading research institutes in China with scientists of high academic level. Among SIBC's more than 700 researchers, there are 11 members of the Chinese Academy of Sciences. The institute has three laboratories which belong to the category of State Key Laboratories:

1. The State Key Laboratory of Molecular Biology, officially opened in 1987, was the first state key laboratory in life sciences. Its main research programmes are 1) RNA and protein interaction, 2) gene expression and regulation and 3) chromosomal and functional genomics.
2. The Key Laboratory of Cellular Biology was established in 1988. The research areas are 1) signal transduction, 2) genetic and molecular mechanisms of cell growth, development, differentiation and apoptosis.
3. The Key Laboratory of Proteomics is a new addition with research interests in the proteomics of basic biology and diseases.

Shanghai is also home to a large number of universities, many of which offer undergraduate or postgraduate programmes that are biotechnology related. These include Fudan University, Shanghai Jiaotong University, Shanghai Second Medical University, Shanghai University of Traditional Chinese Medicine, the East China University of Science and Technology and the East China Normal University. These universities annually award thousands of PhD and MSc degrees and create the talent basis for Shanghai's future expansion in biotechnology.

China's Pharmaceutical Market and Industry

China's pharmaceutical industry is highly fragmented, with around 7000 companies, with a market of slightly more than US$7000 million in 1971, at ex-factory prices. The IPMRC report[53] mentions that 5000 manufacture medicines (small molecules and biotech products), while the remainder are engaged in pharma-related activities such as packaging and equipment supply. The same report mentions that leading companies have market shares that barely exceed 2 per cent with the top ten companies having a combined market share of less than 20 per cent. The same source mentions that 'analysts estimate domestic companies have 65–75% market share, and only one foreign invested pharmaceutical is in the top ten in terms of sales value'. It is estimated that the market profile includes 15 per cent over the counter (OTC) medicines, 62 per cent generic drugs, 14 per cent branded generic drugs and only 9 per cent patented drugs. According to the same source the market will have changed dramatically by 2010 with 23 per cent OTC, 37 per cent generic, 19 per cent branded generic and 21 per cent patented drugs.

Some of the large producers of generic drugs, such as the 999 Company, Hua Bei and Dong Bei, are expected to become truly innovative R&D companies focusing on small molecules. Another company, North China Pharmaceutical Group, is focusing on areas which include biotechnology products, small molecule compound chemistry technology, traditional Chinese medicines based on natural products screening, and formulation technology. Several of the pharmaceutical companies in China are very substantial entities such as the Shanghai Pharmaceutical Group Corporation (SPGC), which has around 45 000 employees.

SPGC is one of the largest pharmaceutical enterprise in China with total sales of RMB14 billion in 2001, which gives it a dominant position in Shanghai, constituting 80 per cent of the city's drug output and 8.5 per cent of China's total gross drug output. More than 10 per cent of the company's production is exported. SPGC has more than 40 subsidiaries, of which ten are state-owned enterprises and another 20 are joint ventures, including Shanghai Bristol-Myers Squibb and Shanghai Roche, as well as joint ventures with Johnson and Johnson, Abott, Siemens and Boehringer Ingelheim. The Group manufactures more than 1500 products which cover almost every clinical application, with antibiotics being one major product category. SPGC has been keen to form strategic alliances with multinational pharmaceutical enterprises with the assumed objective of getting access to technology, at least in the production stage. It has established a number of R&D organizations and its R&D expenditure is in the region of several hundred million RMB to support new technology and new product development.[54]

Similar to the Putian Group in the telecommunications industry, SPCG is highly diversified, and provides a good illustration of the complexity of the Chinese industrial landscape. It has two subsidiaries in the medical instrument field, with almost identical names. One is the Shanghai Medical Instrument Group Company, which is a group of companies by itself. The other is The Shanghai Medical Instrument Co., Ltd. Under its management there are more than 30 medical device manufacturers producing a large variety of medical equipment.

SPCG owns ten TCM companies including Shanghai Lei Yun Shang, and has been very active in the development of TCM products. Similar to its strategy in chemical drugs, it has entered into a number of foreign joint ventures which include Tsumura and Co. in Shanghai, Shanghai Hutchinson Pharmaceuticals and Shanghai TCM Compound Purification Centre. Furthermore, SPCG also includes China's largest listed drug distribution company – Shanghai Pharmaceutical Co., Ltd. This company has more than 60 subsidiary companies, 30 distribution outlets and approximately 1500 drug stores located all over the country.

Traditional Chinese medicines

In China as in several neighbouring countries, traditional medicines have played a very important role in healthcare. There are possibly close to 5000 different TCM drug preparations, although with different dosage forms the total number would be much higher. In 1999 the annual value of TCM sales in China was RMB33 billion, which corresponds to 27 per cent of total drug sales in the same year. The total number of specialized TCM manufacturers was reported to be more than 1000 in 2001, although many pharmaceutical companies include a few TCM products in their portfolios.

An Assessment[55]

Biosciences in China have major centres in Beijing, Shanghai and Shenzhen, the latter location being important as it has easy access to the capital. China is not yet strong in the field but has become very 'hot' as it offers a huge future market. China offers interesting opportunities in a number of pharmaceutical areas. Patents are expiring for a number of large-volume products like insulin, hormones and interferon. Traditional Chinese medicines remain important as they are now produced to a high quality and with good laboratory practices. Furthermore, a number of new products are coming on to the market. Thus China has become an attractive place for returning scientists who have been trained in the US and Europe. They return as they perceive and are able to realize exciting opportunities as China is becoming a technological powerhouse. One interesting example of a returning scientist is Dr Wang Xiaodong, who was trained at Texas University, was subsequently appointed director of Beijing Life Science Institute and has recently become one of the youngest members of the National Academy of Sciences in China.

Furthermore, a substantial number of promising students in life sciences are graduating every year at Tsinghua University and Beijing University. These two universities and a handful of other top universities in China can select the very best university students from all over China. The number of new graduates in life sciences from only one university such as Tsinghua University may be as many as 60–80 every year. Many of them are now being given very good training and opportunities. However, it will take more than ten years for promising candidates to pass through doctoral programmes and post-doctoral positions, with subsequent intensive experience, to become senior highly competent researchers by the age of 35–40. China will have a great pool of highly talented scientists in the life sciences within ten years.

Simultaneously, the creation of new institutions is impressive. Together with returning scientists and returning entrepreneurs, China is creating a technological powerhouse in life sciences and could capture Nobel prizes in medicine within a period of less than ten years.

However, the Chinese research system is still suffering from a number of problems. First, many professors are still too much involved in technology transfer, which may cause a neglect of teaching duties. Second, the system is still not sufficiently market-oriented to capture new market opportunities and regulations are still not strong enough to stop fake products being marketed. Finally, as shown by the SARS epidemic, organizational barriers are still a great hindrance to the exchange of information and knowledge.

However, China has in recent years been able to surprise the world with outstanding scientific successes, which is exemplified by genome sequencing for the chicken, shrimp and rice at Beijing Genome Institute. The last achievement was highlighted in a special issue of *Science* on rice genome research, published in 2003. China's progress in genome sequencing can be partially attributed to two scientists, including Dr Yu Jun, who have returned from Washington University in Seattle, which has been at the forefront of sequencing research. The returnees were able to achieve rice genome research ahead of colleagues in Japan. Another important researcher is Chen Jing, now active at Tsinghua University and active in nano-gene technology.

ECONOMIC INTEGRATION WITH NEIGHBOURING COUNTRIES

FDI in Chinese industrial sectors has forced a rapid acceptance of international division of labour with its constantly shifting comparative advantage, and contributed to the expansion of bilateral trade between China and the nearby countries from which FDI originated. However, the challenge for these countries is to develop new areas of comparative advantage if they wish to maintain economic growth, as China will itself become a source for increasingly sophisticated products and services. The expansion of internal trade in Northeast Asia clearly indicates that industrial sectors and the economies of the region are entering a phase of close integration. This process involves not only the Chinese Economic Area (CEA) but also economic relations with South Korea and Japan. One may assume that close relations with neighbouring countries was a significant factor in developing the three major regions that are discussed in this section.

The Chinese Economic Area

Since the late 1990s, or rather since the end of the Cold War, East Asia has experienced a growing economic, albeit not political, integration. This change has been influenced by a growing worldwide integration as exemplified by the EU, the North American Free Trade Area (NAFTA) and the ten-member Association of Southeast Asian Nations (ASEAN). The formation of an increasing economic interdependence in East Asia has led to the emergence of common regional interests. However, critical observers note that there have been no signs that China and Japan, being the two major actors in the region, can be involved in deeper integration before they reach a resolution on their relations between 1895 and 1945.

Although the growing closeness of economic relationships with Japan, Korea and Taiwan has been remarkable, China's integration with the rest of the world is equally noteworthy. It is not limited only to trade flows but has also become very significant in China's outward flow of foreign direct investment. This reached US$2.9 billion in 2003, with an accumulated total of 33 billion in the same year.[56] China is becoming the world's fifth largest outward foreign direct investor after the US, Germany, Britain and France, thus displacing Japan. By the end of 2003 some 3400 Chinese enterprises had established 7470 companies in 139 countries.[57] Companies in manufacturing, wholesale and retail businesses are the largest investors.

Since the early 1990s, the CEA has become a concept widely used by economic commentators, the IMF and the World Bank. The concept refers to the economic integration of a geographic area which encompasses China Mainland, Hong Kong, Macao and Taiwan. The economic exchanges between the two sides of the Taiwan Strait have been growing rapidly and have become a dynamic core of the CEA. This agglomeration of economies has become a new agglomeration of industry, trade and finance since the mid-1980s, although lacking intergovernmental coordination. The CEA has substantial technology resources and an enormous manufacturing capability, marketing and service skills and extensive business networks. It also has very large foreign exchange reserves, an abundant supply of natural resources and an unparalleled market potential. This influential network is based on a shared culture and language and in many cases the extensions of family and other ties.

Despite the lack of formal political relations between China and Taiwan, the latter has in a short period of time become the third largest external investor on the Mainland, next to Hong Kong and the US. There are several conditions which determine continued development of Taiwan–China economic relations. First, high-tech industry is reaching a critical mass with high-level sophistication in many sectors, which is attracting new investment from high-tech companies in Taiwan. Second, the investment from Taiwan has shifted from neighbouring provinces like Fujian and Guangdong to Shanghai, which has become China's economic centre, and the surrounding Yangtze River Delta. Third, investors from Taiwan are now focusing on a combination of labour-intensive and capital-intensive industries. Fourth, an increasing number of large companies from Taiwan are also investing in China. Fifth, the business community in Taiwan is exerting pressure on the government to change its policy of restricting investment into China and also lift its ban on the 'three direct links'.[58]

The integration of the CEA differs significantly from other regions such as the EU and NAFTA, where integration has in a major way been shaped by political initiatives and intergovernmental coordination. The economic integration of Taiwan and China is an informal process with the following three characteristics. First, it has primarily been driven by the entrepreneurship and self-interest of businessmen and other actors, and politicians and grand designs have hardly played any role. The integration has been driven from below as the process has lacked intergovernmental coordination and has had to deal with prolonged mistrust and political hostility between the two sides. Second, there is no necessity for mutual recognition as a prior condition for cooperation. This has implied a tacit understanding on both sides that the other side has a *de-facto* administrative power over the territory under its jurisdiction. Third, the economic integration of the CEA has been greatly facilitated by mutually reinforcing comparative advantages, and shared culture and language. In sum, the informal integration between the two sides seems to have gained its own momentum despite the lack of intergovernmental cooperation and coordination. However, such an informal integration is likely to encounter difficulties, which need to be resolved through government-to-government contacts through institutionalized channels of dialogue.

For Beijing, enhanced institutionalized dialogue and cooperation could facilitate an eventual peaceful reunification with Taiwan. An integrated economy within a well-institutionalized structure might be the best guarantee of a successful reunification. Intertwined economic and non-economic cooperation is more likely to create a vested interest in Beijing to peacefully resolve disputes in order to ensure continued prosperity, and could lessen the nationalistic outlook on both sides.

Taiwan–China Technological Integration

A network of global industrial and technological relations is the conduit for an integration of human resources across the Taiwan Strait and US relationships also play a very important role. Networks of IT industries connect high-tech centres in Taiwan, China and the US which constitute integral parts of global production networks (GPN). They control cooperation and manage the division of labour in R&D, design, production and logistics. The division of labour in such networks is constantly undergoing changes as conditions in terms of costs and/or sophistication of production become more favourable in one or another location. The networks are borderless and an essential ingredient in this process is skilled manpower, which by its very nature has become highly mobile.

Globalization of the electronics industry over the past couple of decades has facilitated the interaction and cooperation between Taiwan and China despite a lack of political understanding, and the result is a high-tech network in which capital, technology and human resources are flooding on to the Mainland. However, both Taiwan and China are facing a serious challenge from the attraction of high-calibre universities in the US and subsequent job opportunities in American companies. The result is that an outflow of some of the brightest students constitutes a potentially serious brain drain, although this could at some time in the future turn into a brain gain. In the 1999–2000 academic year China's students made up more than 10 per cent of all international enrolment in US universities.[59]

With manufacturing moving to Mainland China, Taiwan is increasingly being transformed from a production base into a knowledge centre. This meets the expectations of the global IT companies because they believe they can design and develop products faster and at lower cost by setting up R&D bases in Taiwan and partnering with local companies with experience in production technologies.

Taiwan, based on its strength in design, development and distribution, has become the IT hub for the network of foreign production bases in China.[60] Taiwanese IT firms are relatively unknown because their core business is producing notebook personal computers on an OEM (original equipment manufacturing) basis. But in terms of production volume they hold 65 per cent of the world market. In 2003, they made two-thirds of their computers in China, increasing the ratio of production there by a factor of 13 over the 2001 level. While their Chinese operations produce goods at low cost, the Taiwanese parent firms handle design, development, parts purchasing and distribution. The main beneficiaries of this division of labour are IT companies in Japan and the US. Taiwan worried about deindustrialization when its plants first moved to China, but soon found its niche as a design and IT distribution hub connecting Chinese production bases. Quanta in Taiwan is setting up a 7000-strong R&D team in northern Taiwan, scheduled to start work in 2005. Taiwan will probably continue to produce core electronic parts and be the IT hub.

Drastic changes have taken place in Taiwan's trade structure. Exports to China have grown bigger than exports to the US and are still increasing, above all for goods like liquid-crystal display panels and other electronic components.[61] Simultaneously China has become a major exporter of goods to the US, but it depends on Taiwan and other nations in the region for parts and materials. A horizontal industry structure has been created in the IT sector, and affects not only Taiwan but the entire East Asia region, including South Korea and Japan. Increasing demand from the US for finished goods is the engine to growth for China's assembly industries, and they in turn have increased their demand for imported parts and materials.

Since the early 2000s the IT industry in China has developed by leaps and bounds, and covers basically all sectors and stages of the industry although still largely dominated by foreign companies.[62] In the process a strong symbiotic relationship has developed between the IT industries in Taiwan and China. Although the rapport was initially focused on assembly operations, it now involves Taiwan companies establishing advanced production facilities for semiconductors.

BOX 6.1   THE SEMICONDUCTOR INDUSTRY

To understand the evolving relations between Taiwan and China in the IT sector it is essential to understand the character of the global semiconductor industry, which consists of three distinct sectors. First, central processor units (CPUs) make up the largest segment with more than 40 per cent of a global production value of around US$200 billion. This segment is highly design-intensive as well as capital-intensive. Intel is the absolute leader in this segment. Second, the next largest sector is made up of memory, where DRAM dominates. This sector constitutes roughly 30 per cent of global production of semiconductors. Samsung is the absolute leader in this segment, which is highly capital-intensive but less design-intensive.

Third, application-specific integrated circuits (ASICs) constitute a little less than 30 per cent of semiconductor global production. This segment is highly design-intensive. The production of ASICS can be separated into two distinct stages – the design and the manufacture – which need only be loosely coupled. Thus it is actually possible to see a parallel in the publication of journals where the editorial offices are responsible for all design and can freely choose the printing companies. Similarly the design of ASICs is done by software companies (like editorial offices) which can choose freely from different IC fabricators, where they can have their designs printed into silicon.

China has only a limited interest in and opportunity to enter into the first two segments as they require huge investments, since the Chinese banking system may have difficulty in bankrolling such extensive projects. Furthermore, there exists hardly any industrial group in China that has the acumen of Intel and Samsung. There was at one time a Chinese industrial group which wanted to acquire the ailing Hynix Electronics Company in Korea, but only a minor deal was accomplished with the acquisition of the Hynix flat screen division by a Beijing-based company.

China is actively pursuing a sophisticated strategy in its development of ASICs, where Taiwan has a highly developed manufacturing competence in a number of companies that are now setting up IC production plants on the Mainland. The dynamic development of the electronics industry in China has fuelled the major changes in its software industry and IC design capability. Thus, the superiority of Taiwan in supporting China's electronic industry with advanced software and IC design is now being gradually eroded – at a time when ASIC manufacturing capability is being transferred to the Mainland.

Furthermore, a substantial number of Taiwan-based IT firms have already given R&D mandates to their subsidiaries in China. A recent survey[63] suggests that these relations include five different ways of performing R&D in China. First, engineering and manufacturing-related R&D is undertaken in China for products where production lines are concentrated on the Mainland, with major product development maintained in Taiwan. Second, many IT companies outsource their software development to China. Third, Taiwanese companies have started collaborations with universities and research institutes to perform basic research. Fourth, there is also a shift towards China of R&D activities for mature products while core development for new products is maintained in Taiwan. Fifth, some firms have relocated system R&D to the Mainland for products where the final product consists of several modules.

Taiwan's IT companies have concentrated their investment in the Yangtze River Delta region in recent years, with a strong agglomeration in Shanghai. Tentative conclusions suggest that this flow of investment is becoming both more capital-intensive and more technology-intensive, with 40 per cent of Taiwan's FDI in China going into the electronics and electrical appliances industry.[64] Locating R&D in China has been prompted by an increasing demand for R&D personnel with their higher costs in Taiwan, and the increasing availability of R&D personnel on the Mainland at considerably lower costs.

Taiwan ceded its low-technology and labour-intensive IT industries to China during the 1990s while keeping a comparative advantage in high-technology manufacturing. A recent study suggests that this advantage may not be lasting, and Barry Naughton provides the following comments:[65]

The comparative advantage can be transferred to the Mainland, and Taiwan companies have in fact developed a high stake in the emerging mainland industry, currently concentrated in Guangdong and the Shanghai region. This means that as expertise develops in Shanghai and Beijing, Taiwan's share of the business may not necessarily decline. Instead, Taiwan could be pulled into increasingly close involvement with mainland industries. This will have significant implications for economics, politics, and the development of new information technologies.

Korea and Japan – Influence on China's Regional Development

Taiwan saw its exports to China surpass its exports to the US in 2003. Similarly, South Korea, another of the world's main production bases for IT components, also experienced a change in its trade pattern. Since the 1980s Korea has made major investments in China, which has become Korea's third largest trade partner.[66] As a result of China's rapid transformation, its exports have in many cases replaced Korea's exports to the US and Japan, which is partly a reflection of an expanding movement in intra-industry trade between the two countries. It has now become apparent that Korea must succeed in gaining a comparative advantage in technologically more advanced industrial sectors to maintain its economic growth. The changes in China's export structure are concurrent with its mounting ability to manufacture technologically more advanced products. Since the turn of the century the structure of China's exports has changed dramatically with an amazing increase in medium and high-technology products. This is particularly evident in engineering and electronics products, for which the combined share of exports increased from less than 2 per cent in 1986 to more than 35 per cent in 2001.

China is for the time being only able to achieve this level of exports in high-technology products with the support of foreign investors. The results depend on foreign technology, capital and management but may increasingly originate from technological improvements that are based on genuine domestic capabilities, which could replicate similar developments in Korea and Taiwan at an earlier stage. When looking at the revealed comparative advantage (RCA), it is very clear that China in the period 1986–2001 quickly dropped its RCA in primary products, keeping it in textiles while also gaining in other labour-intensive products. At the same time China's increase in the revealed comparative advantage in electronics and engineering products was very remarkable. The effects have been extraordinary in major export markets like Japan and the US, where a comparison between exports from Korea and Japan shows an enlightening picture.

China's share in Japan's import of telecommunications, sound recording and reproducing equipment increased from 1.5 per cent in 1987 to 28.8 per cent in 2001, while Korea's share fell from 27.4 to 7.5 per cent. China's exports achieved an almost equal success in the US market during the same time while Korea's export share for these categories remained at basically the same level. These developments indicate that the engineering and electronics industries in China are catching up with their counterparts in Korea. A Korea Development Institute (KDI) investigation shows that China's export intensity with Korea increased significantly during the period, which indicates that bilateral trade between the two countries expanded quicker than trade with the rest of the world.[67] However, intra-industry trade with the world as a whole is more important than that with Korea. An intra-industry trade is generally understood as a consequence of product differentiation and economies of scale and has increasingly become related to an escalating international fragmentation of many manufacturing processes.

A substantial share of exports from China's high-technology industries comes from companies that are foreign-owned or are joint ventures with Chinese partners, where assembly operations are of a significant magnitude, in order to take advantage of low labour costs. Between Korea and China there is an increase in the trade of parts, which clearly indicates a growing division of production processes and growth of production networks. Thus, it is quite possible that a loss of market in Japan or the US for companies from Korea and other nations is at least partly a displacement of exports that are now coming directly from affiliates in China.

Foreign investment in production facilities in China is made in order to gain access to the huge market, benefit from low-cost labour or a combination of both. For example, textiles and footwear industries had been major export industries in Korea until they lost comparative advantage in the 1980s because of high wage increases. With established international marketing networks, many Korean firms were able to shift production to affiliates in China. Companies from Korea have in the past concentrated their investments in Shandong, Hebei and Liaoning, all of which are close to China. The importance of closeness is further supported by the development of airlinks between Seoul and nearby Chinese cities.

However, low labour costs in China have also been a significant incentive for firms to invest in capital-intensive industries. This is explained by the progressively widespread practice of manufacturing based on intra-firm, inter-process structures which have become prevalent in the era of international fragmentation. The reason is that heavy industries contain a large number of sub-processes which can be separated and have very different requirements for labour, capital and technology. Thus, China as a developing country has become a natural site for manufacturing parts for advanced and heavy machinery that would previously have been totally integrated, thereby acquiring skills and technology by participating in the production process.

WINNING AND LOSING SECTORS

When, after ten years' absence the World Table Tennis Championships were coming back to China by the end of April 2005 millions of fans were expecting to see Chinese players sweep the board in Shanghai. Moreover, China also wants to see the science and technology scoreboards filled by names of Chinese scientists and Chinese companies. This ambition may take longer to accomplish and will require a leadership that not only accepts that China needs to fully integrate into the global economy but also requires the nation to invest more intellectual and financial capital to make breakthroughs in key technologies. In early April 2005 Mme Chen Zhili of the State Council urged Chinese scientists and enterprises to become more innovative in building up the nation's research strength.[68] She said that the country will invest more capital to make desired breakthroughs in key technologies and will offer preferential taxation and financial support to help enterprises to become involved in this important national endeavour.

International partnerships will play an important role in this process. The Minister of Science and Technlogogy, Mr Xu Guanhua, emphasized deeper collaboration when speaking at the China–EU High-level Forum on S&T Strategy, held in Beijing on 11–12 May 2005. He said that foreign research institutes should be encouraged to operate in China by linking up with local partners for new technology and project developments. He suggested that China and the EU should further expand cooperation in the fields of hydrogen energy and nuclear fusion resources, space and aviation, environment, biological and information sciences.[69] At the end of the Forum the two sides issued a joint declaration stating that information technology, biotechnology and nanotechnology would be at the forefront of their future scientific cooperation.[70]

China is using a number of policy instruments to support its future development in science and technology. The Ministry of Foreign Trade and Commerce (MOFCOM) in its annual report in May 2005 stated that export of high-tech products continues to grow at a higher rate than overall exports. The same report also referred to the worldwide industrial restructuring that, since the late 1980s, has promoted technological progress and facilitated the introduction of advanced international technology into domestic enterprises.[71] In its portfolio of policy instruments China will encourage trade through science and technology promotion and in its major efforts will rapidly urbanize large parts of China.

However, there are multiple dilemmas. The auto sector provides one serious illustration of the dilemma faced by Chinese policy makers and domestic enterprises in establishing an indigenous technological capability. It is generally agreed that Chinese car makers have wasted huge amounts of capital on technology transfer while failing to enhance their own technological capabilities. In all likelihood Chinese car makers would be discarded as partners by large foreign auto companies if the latter were allowed to establish wholly-owned manufacturing plants in China.[72] This was stated even more bluntly at the CEO roundtable discussion – Auto China – at the time of the Shanghai Car Exhibition in April 2005: 'Unless (Chinese manufacturers) can figure out the third revolution in the auto industry ... they will never be able to catch up. ... They don't have the required capacity to play the same game'.[73]

The reason is that most of China's state-owned car manufacturers are busy assembling foreign cars and appear to have little time to upgrade their own technological capability and develop their own brands of vehicles, although there are many Chinese firms outside the industry eager to enter into the sector. However, another factor may favour the Chinese car industry, as foreign car makers are increasingly forced to use local components in order to cut costs. The strength of the euro has propelled FAW-VW, the joint venture between Volkswagen and First Automotive Works (FAW), to rapidly increase the amount of locally produced components in its cars.[74] The joint venture is now encouraging its suppliers in Europe to set up wholly-owned units or joint ventures with partners in China. What happened is that cost-controlling has become one of the most important tasks for car makers operating in China. Although much technology is controlled by foreign car component makers such as Delphi and Bosch there are better opportunities for Chinese companies to compete in the market for car components where cost and quality are more important than brand names and marketing channels.

NOTES

1. Gida (2004).
2. A *Time* report refers to data compiled in an International Monetary Fund working paper that provides the following information. 'The average Chinese garment worker was paid $1,600 in 2001, more than double his Indian counterpart's salary and four times what he'd make in Bangladesh. Despite the Chinese worker's higher pay, the study found his productivity was significantly higher: he adds $5,000 a year in value to the garments he processes, compared with $2,600 by his Indian equivalent and $900 by a Bangladeshi worker. The difference reflects China's greater investment in modern manufacturing equipment and in infrastructure such as transportation.' (Ibid.)
3. Ibid.
4. 'US curbs loom for Chinese textile industry – seventh probe under way, on fears of job losses when cheap imports flood the US market', *The Straits Times*, 20 November 2004.
5. 'Industrial alliance boosts IT sector', http://www.chinamet.com.cn/english/MEspecial/invest/invest-envo-detail.jsp?id=1890.
6. Xiao Huo (2004).
7. Ibid.
8. Ibid.
9. 'How China is quickly capturing the world's semiconductor industry', *Manufacturing and Technology News*, Vol. 10, No. 15 (August 2003).
10. Li (2004b).

11. Infineon already has a production facility for logic devices in Wuxi, and a subsidiary in Xi'an focused on IC design, and another subsidiary in Shanghai for applications. The new facilities in Suzhou will include an IT Development Centre that will support manufacturing processes.
12. Li Weitao (2004b).
13. CCID presentation at Asian Semiconductor Industry Conference 2004, organized by Sangyo Semiconductor News Agency, 1 January.
14. This included Project 909 with ambitious goals for building IC plants and developing necessary technical expertise.
15. Iritani (2002).
16. Private communication from MII official, November 2003.
17. CCID presentation at Asian Semiconductor Industry Conference 2004, organized by Sangyo Semiconductor News Agency, 1 January 2004.
18. Howell et al. (2003).
19. Ibid.
20. Simon (2001).
21. Howell et al. (2003).
22. Markoff (2004).
23. '24th Edition of TOP500 List of World's Fastest Supercomputers Released: DOE/IBM BlueGene/L and NASA/SGI's Columbia gain Top Positions', 9 November 2004, http://www.top500.org/news/articles/article_51.php.
24. Stickel (2003).
25. Shell Electric Manufacturing Co. is a Group company that has been a manufacturer of consumer electronic goods since 1952. The Group in 2004 employed more than 55 000 employees worldwide and is involved in three business areas – manufacturing, investments and technology. The last one includes enterprise software and the development of blade computers.
26. 'China to lead supercomputing sector', 26 October 2004, http://www.excitecity.com/china/chat/military/messages/39494.html.
27. Ibid.
28. Zhu Boru (2004).
29. 'China to lead supercomputing sector', 26 October 2004, http://www.excitecity.com/china/chat/military/messages/39494.html.
30. Markoff (2004).
31. 'China's aviation industry has broad investment portfolio', 11 January 2002, http://english.people.com.cn/200201/10/eng20020110_88412.shtml.
32. 'Flagship aviation company takes up historic IPO mission', *People's Daily*, 25 June 2002, http://english.peopledaily.com.cn/200206/25/print20020625_98520.htm.
33. Zhang Yong (2003a).
34. 'First China-built regional aircraft takes to skies', *China Daily*, 17 December 2003, http://www.chinadaily.com.cn/en/doc/2003-12/17/content_290965.htm.
35. 'China's first large aircraft to fly by 2018', *China Daily*, 17 March 2004.
36. Xu Dashan (2004b)
37. Ibid.
38. Ibid.
39. 'China's President in Brazil for space talks and more', *Space Daily*, 14 November 2004.
40. Zhang Yong (2003a).
41. Ibid.
42. Ibid.
43. AviChina raised HK$1.9 billion from an initial public offering on the Hong Kong Stock Exchange in 2003.
44. Xu Dashan (2004a).
45. At present, China has less than 80 helicopters for general use, meaning there are only 0.06 helicopters for each million people in China. This is far less than the world average of 3.9 helicopters per million people.
46. Liang Yu (2004a).

47. Lee Zhong (2004).
48. International Federation of Pharmaceutical Manufacturers Associations (2003).
49. International Federation of Pharmaceutical Manufacturers Associations (2003).
50. Ibid.
51. Ibid.
52. 'Government Policies on Biotechnology industry', *Asia-Pacific Biotechnology News (APBN)*, Vol. 7, No. 5 (2003), Special issue on Shanghai.
53. International Federation of Pharmaceutical Manufacturers Associations (2003).
54. *Asia-Pacific Biotechnology News (APBN)*, Vol. 7, No. 5 (2003), Speical Issue on Shanghai, p. 213.
55. These comments are partly based on an interview with Professor Hew, Director of the Department of Biosciences at the National University of Singapore.
56. 'UNCTAD World Investment Report 2003', *China Daily*, 22 October 2004.
57. 'China pours more money overseas', *China Daily*, 22 October 2004.
58. Embedded into the WTO is the assumption of free trade in goods and services and direct trade between all WTO members. After joining the WTO China and Taiwan will have to establish direct trade relations, direct postal services and direct transport services by sea and air.
59. Leng (2002), pp. 230–50.
60. 'Analysis: Taiwan's IT hub status may help improve China relations', *The Nikkei Business Daily*, 20 May 2004 (Nikkei Interactive).
61. Murayama, Hiroshi, 'Taiwan atop East Asia trade triangle' (Nikkei Interactive), 15 March 2004.
62. An insightful presentation can be found in Naughton (2004).
63. Chen, Shin-Horng (2004), pp. 227–349.
64. Ibid.
65. Naughton (2004).
66. Kim et al. (2004).
67. Ibid.
68. Fu Jing (2005a).
69. Cui Ning (2005).
70. Fu Jing (2005).
71. MOFCOM (2005b).
72. Gong Zhengzheng (2005b).
73. Comment by Paul Gao, partner Mckinsey China (Luo and Wan, 2005).
74. Gong Zhengzheng (2005a).

7. Space and defence technologies

AMBITIONS IN SPACE

Planners and think tanks in the US are obviously of the opinion that national security and the future economic health of the country both depend on maintaining a comparative advantage at the frontier of science and technology. The threat of Japan to the technological hegemony of the US some twenty years ago, has mainly vanished.[1] Japan was not agile enough to adjust quickly to the twin challenges of rapid globalization and rapid scientific and technological (S&T) progress in almost all technological and scientific fields – and developments during recent years suggest that the country has become a limping technological superpower.

The European Union, with 25 countries, has not been perceived as a threat to the US hegemony, although the EU at its General Assembly in Lisbon in 2000 declared that it would become the most innovative region in the world. It is only of late that China's advances in technology and science have been perceived to provide a foundation for a changing world balance not only in trade but eventually also in the military arena. Space technology constitutes an integral and very important arena where the United States enjoys an obvious and dominant military advantage. A continued deployment without regard for China might compel the country to respond, resulting in a space competition with unavoidable military repercussions.

There is already an early indication that such a development is gathering force. In August 2003 China launched its first manned satellite. China also indicated that it was embarking on an ambitious space research programme and earlier indicated its desire to collaborate in space. However, China's ambitions to move forward with its space programme prompted President Bush, five months later, to announce that the US would formulate a new national space programme that would include sending astronauts to the planet Mars.

China has launched 80 space flights using its Long March rocket carriers, with its first home-made satellite put in orbit in April 1970. By late 2004 China had sent about 70 satellites into orbit,[2] including those manufactured by foreign companies. China's manned space flight in 2004 has its roots in an ambitious project that was formulated in early 1992 and initially known under the code name 921.[3] The China National Space Administration may have benefited from the collapse of the USSR and obtained important technologies used in Soyuz, the standard Russian spacecraft used for transporting people. However, the launch of Chinese rockets failed on a number of occasions in 1991, 1992 and 1995, and twice in 1996. Hughes Space and Communications at the time provided assistance to solve some of the remaining problems in the launching systems, and in October 1996 China formally announced plans to send its astronauts into orbit.

Shenzhou V (Divine Vessel), with its Chinese astronaut, sent two messages to the technology community. First, it made it clear to the Chinese people that the country is able to handle successfully any challenging task. Second, it also told the world that China has entered the league of space nations, with ambitions to surpass Europe and Russia and impinge on the US dominance.

The China National Space Agency (CNSA) received the instructions in 1999 to follow though with a ten-year space plan, and was given the responsibility for building up an integrated military and civilian Earth observation system using meteorological, earth resource, oceanic and disaster monitoring satellites, all coordinated for receiving, processing and distributing data to both civilian and military users. The effort began with the launch of the Ziyuan-2 satellite in 2000.[4] In the early 1980s, China started to utilize the American Navstar and Russian Glonass systems to develop the application technology of satellite navigation and positioning.[5]

Shortly after the successful manned flight in 2003, China announced plans to build a space laboratory and station within the next few decades. China has been working on a new generation of more powerful launch vehicles with high reliability, low cost and low pollution to meet the needs of the next three decades, and more rockets for launching heavy satellites and lunar probes.[6] In 2004 the CNSA declared that China was scheduled to launch its second manned spacecraft in 2005. China has also publicized that it is planning to launch a satellite to orbit the Moon by 2007 and examine its surface and geography as part of a three-stage lunar project, the Shenzhou Circumlunar Mission. The second phase will be to launch an unmanned device to land on the moon by 2010 and the third is to send a robot to the moon to collect samples of lunar soil by 2020.

Another important programme is a constellation of small environmental monitoring and disaster forecast satellites which the Chinese Academy of Space Technology has announced will be established by 2010.[7]

SPACE INDUSTRY

The CNSA 20-year plan will focus more on maturing the industrial foundations necessary for supporting government space operations. This would include modernizing the aerospace industrialization process, marketing of space technology and applications, and establishing an integrated space infrastructure and a satellite ground application system integrating spacecraft and ground equipment.[8]

China has developed a new generation of small satellite launch vehicles, Explorer I, which will use solid fuel. It has been designed to take small and micro satellites into space, and complement the Long March group, the country's large-scale liquid-fuel space launchers. Explorer I will be able to carry loads weighing less than 100 kilograms. The low costs and high thrust of solid-fuel rockets make them an important factor in the commercialization of the space industry.[9]

The change from a planned economy to a market economy has had a profound effect on the development of the aerospace industry in China and has passed through several stages – mandatory planned regulation, planned economy in the main with market rules as a complement, and finally planned guidance based on the market.[10] The aerospace industry consists of three major components, which are military production, civilian production and tertiary industry. Two organizations play important roles – the China Aerospace Science and Industry Corporation (CASIC), and the China Aerospace Science and Technology Corporation (CASC).

The military production system is the most important part in the aerospace industry and is managed directly by the government because of its significance for national security. Civilian production results from transfer of aerospace high technology to civilian production. There it is guided by market forces, although also regulated by the government through market mechanisms. The tertiary industry system serves both the military production system and the civilian production system. Today both the military and civilian aspects determine the character of technology management in the aerospace industry. Technology management within the civilian sector can be implemented in the same way as technology organization in manufacturing enterprises, while technology management of military production must fully reflect the national interest.

An earlier focus on specific tasks resulted in scattered resources with duplicated investment and core technologies being developed separately. Integration of technologies was difficult and serious wastages of resources occurred because of the management system indicated in Figure 7.1. Each research institute was responsible not only for scientific research in the military sector but also for the development of the civilian industry. Managing such diverse technology interests in separate places hindered the efficient use of resources. Thus it was urgently needed to find a new organizational model.[11]

The management and operation of China's space activities are carried out by the China National Space Administration (CNSA). It was established in June

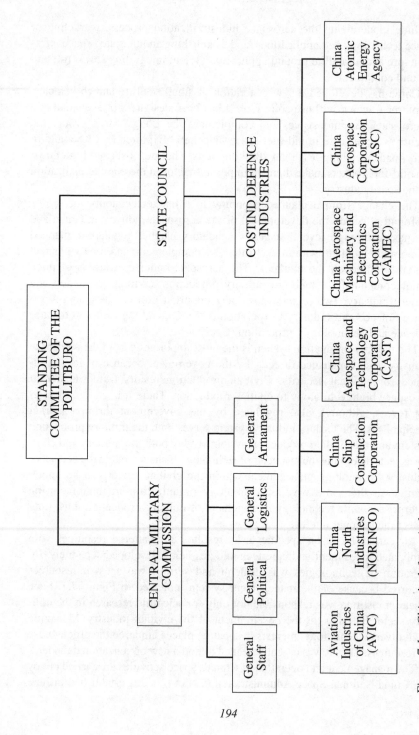

Figure 7.1 China's defence industry within the political system

1993, being roughly the equivalent of the National Aeronautics and Space Administration (NASA) in the US. The organizational structure and evolution of China's space programme is complicated and constantly undergoing change, as China's military and civilian programmes are highly intertwined.

A separate group of organizations is charged with supplying the hardware, software and research and development for the CNSA under the patronage of the China Aerospace Corporation (CASC).[12] CASC and its subordinate companies, research academies, and factories develop and produce strategic and tactical ballistic missiles, space launch vehicles, surface-to-air missiles, cruise missiles, and military (reconnaissance, communications or other) and civilian satellites. CASC was previously known as the Ministry of Aerospace Industry, also known as the Seventh Ministry of Machine Building. Since April 1998, China's military, the People's Liberation Army (PLA), has exercised control over PRC satellite launches under the new General Equipment Department.

CASC is divided into two large organizations: China Aerospace Machinery and Electronics Corporation (CAMEC), and China Aerospace Science and Technology Corporation (CASTC). The administration and management of facilities, payroll and other similar functions fall to the State Commission of Science, Technology, and Industry for National Defence (COSTIND). CASC headquarters is in Beijing, and the organization oversees the activities of 250 000 employees.[13] The majority of space-related industrial activity is conducted by CASTC, and projects are mostly divided among the three key organizations under CASTC.

The China Academy of Space Technology (CAST) oversees institutes and factories related to research, development, and production of communications, space-based military systems, navigation satellites data relay satellites, and weather satellites. It also conducts research and development on space station and rocket launch vehicle (RLV) technology. The China Academy of Launch Technology (CALT) conducts research, development and production of launch vehicles, liquid-fuelled surface-to-surface missiles, and solid-fuel surface-to-surface and submarine-launched missiles, among which is the Long March family of space launch vehicles. Shanghai Academy of Spaceflight Technology (SAST) is responsible for tactical air defence missiles and carrier rockets.[14]

A third grouping of organizations also falls under CASC. Commercial launch campaigns are contracted through the China Great Wall Industry Corporation (CGWIC). This is managed under a conglomerate called the China Great Wall Industry Group (CGWIG), which also oversees the China Precision Machinery Import and Export Corporation (CPMEIC).

CGWIC[15] has been China's commercial space launch company since 1986. It markets the use of rockets developed by the China Academy of Launch Vehicle Technology (CALT) and other aerospace academies.

MILITARY INDUSTRY

This century, China's defence industry has become far more productive than in previous decades.[16] The defence industrial reforms implemented in the late 1990s, unlike the ones adopted in previous years, were substantial and have positively influenced the quality of China's defence industrial output. Gone are the days of widespread inefficiency and a paucity of innovation in defence production. Chinese defence firms have improved their R&D techniques and production processes, and as a result the quality of their output. Improvements have been gradual and incremental, but can be expected to continue to accumulate in the future, assuming that the Chinese economy continues to grow. China's defence firms produce a wide range of increasingly advanced weapons that, in the short-term, are relevant to a possible conflict over Taiwan as well as China's long-term military presence in Asia.

The Rand Testimony of February 2004 reports that 11 state-owned enterprises are the main industrial actors in military areas that include nuclear, aerospace, aviation, ordnance and electronics. These are:[17]

1. China National Nuclear Group Corporation (www.cnnc.com.cn)
2. China Nuclear Engineering and Construction Group Corporation (www.cnecc.com)
3. China Aerospace Science and Technology Group Corporation (www.cascgroup.com.cn)
4. China Aerospace Science and Industry Group Corporation (www.casic.com.cn)
5. China Aviation Industry Group Corporation I (www.avic1.com.cn)
6. China Aviation Industry Group Corporation II (www.avic2.com.cn)
7. China State Shipbuilding Group Corporation (www.cssc.net.cn)
8. China Shipbuilding Industry Corporation (www.csic.com.cn)
9. China North Industries Group Corporation (www.norincogroup.com.cn)
10. China South Industries Group Corporation (www.chinasouth.com.cn)
11. China Electronics Technology Group Corporation (www.cetc.com.cn).

China's defence industry has in the past revealed a number of shortcomings in both procurement policies and factory production. The RAND report suggests that the defence manufacturers had limited interest in improving the efficiency with which they were produced or designed and had little direct influence on the overall quality of the weapons systems.[18] The hierarchical system, the strictly planned economy inherited from the Soviet Union, deterred the horizontal knowledge flows that are critical to technological progress in any sector, civilian or military.

The defence industry landscape changed quite dramatically in the late 1990s when the Chinese government started to increase weapons funding. The RAND report cites the following information:[19]

> From 1990 to 2002, the official defence budget allocation for weapons procurement grew from RMB 5 billion to RMB 57.3 billion. These increases are twice the rate of growth of the official defence budget. Also the share of the budget devoted to weapons procurement increased from 16.3% to 33.8% in this time period. From the period 1997–2002, according to official Chinese budget figures the amount of funding for equipment grew 124%, more than the other two categories in the official defence budget.

The same period saw two important organizational changes. First, in March 1998 the Chinese government closed down the Commission on Science and Technology and Industry for Defence (COSTIND). The Commission, having been created in 1982 and dominated by the military circles, was replaced by a rigorously civilian agency that maintained the same name but under direct State Council control. The new COSTIND no longer has direct control over the management of defence manufacturing enterprises and can be seen as a regulatory and administrative agency. Second, during the preceding year a new general department – General Armaments Department (GAD), with direct relations to the People's Liberation Army – had been created. GAD took over many of the former responsibilities of the old COSTIND, including life cycle management of weapons systems. This included R&D activities from the early stages to final rolling out of systems, as well as testing, evaluation and training.

RAND suggests that the 'civilization' of the Commission of Science, Technology and Industry for National Defence (COSTIND) and creation of GAD not only led to a centralization of procurement but also to a separation of buyers from suppliers. These changes were accompanied by reforms at the enterprise level with the aim of encouraging innovation and efficiency by making them more market-oriented. Defence enterprises have benefited from new partnerships with civilian universities and research institutes, which have become most evident in the IT sector. Like state-owned enterprises in the civilian sector, the defence industry enterprises have also been forced to reduce employment that was estimated to be around three million before the reforms in the late 1990s.

Increasing access to foreign weapons technology has also contributed to changes in China's defence industry, of which co-production with aircraft manufacturers in Russia provides a good example. The RAND report mentions that Israel has provided assistance with avionics and air-to-air missiles and the French have assisted with the development of air-to-air and surface-to-air missiles.

As a general comment RAND suggests that

> in the last two years alone, Chinese defence factories have produced a variety of new weapons systems based on novel Chinese designs. Many are highly capable weapons platforms. The development of these weapons importantly reflects improvements in R&D techniques, design methods and production processes, especially compared to the 1980s and 1990s. Not only are the new systems more advanced, but China's production of them is faster and possibly more efficient.

Jiangnan Shipyard in Shanghai[20] provides one illustration with its serial production of 7000 ton destroyers – Luhu – based on stealth design and with enhanced capabilities for air defence and anti-submarine attacks. The Luhu-class missile destroyers, the first Chinese naval system using indigenous blueprints, were designed by the China Warship Design Institute, formerly the Seventh Academy of the Ministry of National Defence. The ships – the first one ordered in 1985 – make use of a number of advanced foreign technologies.[21] Much of the foreign technology was obtained before the Tiananmen incident in 1989 when Western countries were still assisting China in its military modernization. Later orders were delayed because of problems obtaining key systems from the West, including more gas turbines from the US. The Chinese Navy is being provided with a variety of auxiliary ships from other shipyards as well as recently designed conventional and nuclear submarines.

Considerable technological and manufacturing capability is also evident in the aerospace sector, with a capability for serial production of short-range ballistic missiles.[22] Accuracy is improving and China is moving towards satellite-aided navigation. China is making progress in the development of land attack cruise missiles as well as anti-ship cruise missiles. China's mastery of satellites is further evidence of its capability in the aerospace industry. RAND states the present situation in the following words:[23]

> The modern capabilities of China's defence electronics and IT sectors has facilitated the modernization of PLA's command, control, communications, computers and intelligence (C4I) systems. The Chinese military is in the midst of a C4I revolution, characterized by the wholesale shift over the last twenty years from relatively insecure analogue communications to digital, secure communications via fibre optic cable, satellite, microwave, and enhanced high-frequency radio.

The origin of apparently successful development of military technologies can be traced to two important factors. First, China's full-scale acceptance of globalization has created a booming and advanced IT industry within China, which is fuelled by knowledge transfers in logistics, management and new production technologies. Second, the Chinese Government has substantially increased R&D funding to government programmes, research institutes and universities, while at the same time undertaking major institutional reforms to make national R&D spending more efficient. These changes have greatly benefited the development of defence technologies and the procurement of new systems by the People's Liberation Army. However, many of China's aviation platforms still make use of foreign imports for a number of fundamental sub-systems such as propulsion and avionics, and system integration remain a serious weakness in many areas.

Civilian–Military Integration (CMI)

China's defence budget was reported to have been RMB212 billion in 2004 and has expanded a little more rapidly than the rate of economic growth – from RMB171 billion in 2002 and RB191 billion in 2003. Total defence expenditure is divided into roughly three equal parts – personnel including pensions, maintenance expenses, and procurement of equipment.[24] The Chinese People's Liberation Army is still a people army with a planned overall strength of 2.3 million by 2005, after a reduction of some 700 000 troops since 1996.[25] The reduction in number of troops, which is likely to continue, reflects a rapid development of military science and technology, global competition in military affairs, and an overriding importance of information technology and information systems. The Chinese defence white paper of 2004 lists five areas which will require increased budget allocations. They include costs for higher salaries for military personnel, a social security system for servicemen, structural reforms in the defence sector, training and skill upgrading, and new and upgraded equipment. This may enable the reduction of military commercial activities that in the past have made up for inadequate budget resources, thereby often negatively affecting the modernization of the defence system.

China has become intensely aware of the potential benefits of a civilian-military integration (CMI) that offers the possibility of reducing the costs and risk of weapons development and production, thereby speeding up the process of military modernization. China had already in principle accepted the urgent need to develop a high-technology national defence in the 1980s. However, the real impetus for change did not occur until after the 1991 Gulf War with a further accentuation after the Iraq War in 2003. This has, among other things, caused added attention to be given to the development of an advanced air force, as reflected in an increased number of advanced aircraft bought from Russia.

Chinese attempts to exploit CMI started in the 1980s when a number of military factories were converted to civilian production. One illustration of such a conversion is the present Beijing Oriental Enterprise (BOE), established in 1955 as military Factory 774 to produce vacuum tubes and related devices. The conversion into a predecessor of BOE – Beijing Electronics –

met with financial disaster and the company had to be reconstructed in 1993, after which followed listing on the Shenzhen Stock Exchange in 1997. BOE is today a major player among China's makers of advanced display devices and in 2002 acquired the flat display division, Hydis, from the ailing Hyundai Electronics in Korea. BOE had close to 10 000 employees in early 2005.

This is only one example of a broad and deep transformation that has affected the IT industry in China. With a defence budget of RMB72 billion for equipment procurement in 2004 there is little doubt that this goes much further when many components and sub-systems are easily available in a domestic dual-use technology market. Such benefits not only accrue in the IT sector but also in shipbuilding where Chinese yards are moving into more sophisticated production, as well as in the aircraft sector where China has become a not insignificant subcontractor for Boeing and Airbus. Furthermore, the 863 high technology programme initiated in March 1986 has been organized to promote China's overall competitiveness in both the economy and national defence fields that include aeronautics and astronautics, information technology, materials science and bioengineering.

TECHNOLOGY COLLABORATION

In early 1950 China decided to develop an aircraft, mainly for defence, and relied on the Soviet Union for technology imports as the West had imposed an embargo. China faced a completely new situation in the early 1960s after ideological differences led the USSR to an almost complete withdrawal of industrial and technological collaboration with China. By then China had just started to manufacture advanced military aircraft like the MIG-19 fighter and the TU-16 bomber under licence. Thus independence was forced on China as imports of important raw materials and components ceased. Based on available Soviet designs the country continued its manufacture and also developed transport aircraft, a development path that was seriously disturbed by the Cultural Revolution. With its Open Door Policy from 1978 onwards China was again able to rely on advanced technology imports as the West, in particularly the US, recognized China as an important partner in its continued struggle against the Soviet Union. By 1990 this situation had completely changed again after the Tiananmen incident and the collapse of the Soviet Union. These two events and China's emerging economic strength created a new geopolitical situation. First, the Tiananmen events prompted Western countries to impose far-reaching economic sanctions while the US increasingly identified China as a potential foe in military affairs.

From 1979 Chinese companies became important sub-contractors for the large aircraft companies. China has been assembling aircraft and components for years. Both Boeing and Airbus, world leading makers of passenger aircraft, contract the manufacture of parts like tail fins, nose cones and aircraft doors to factories in China. Boeing was first to consign the production of aircraft doors to a company in Xian in 1982, and a Shanghai factory started assembling McDonnell-Douglas aircraft in 1985. More recently Empresa Brasiliera de Aeronautica (Embraer) in Brazil produced its first Chinese-made 50-seater aircraft in Harbin in late 2003. Building on this proven capability China at the end of 2003 embarked on the ambitious aviation project to build a small aircraft for less than 100 passengers. The company, AVIC-1 Commercial Aircraft Co. (ACAC), will rely on foreign suppliers for advanced aviation and aerospace technology. The booming Chinese market has attracted General Electric for supplying the engines, Parker Hannifin for providing the fuel and hydraulic systems, Rockwell Collins will supply avionics and a Honeywell-Parker joint venture has preliminarily agreed to provide the primary flight control system.[26]

China today relies on three sources for its continued development of the national aircraft industry, of which military aeroplanes pose the most critical challenges. First, China is directly benefiting from its international partnerships which provide limited technology transfer but substantial learning in management techniques. Second, the country allocates substantial R&D funds for its aerospace industry, which has in recent years undergone a fundamental restructuring which is likely to have made it more efficient and innovative. Thirdly, China is relying at least for the time being on significant and substantial imports and technology transfer from Russia, and also from the Ukraine, for which examples will be provided below.

In a brief from the Jamestown Foundation it is clearly stated that Russia-China aerospace industry cooperation has been gradually improving the knowledge and skills of Chinese engineers and technicians, turning them into potential competitors in the world markets.[27] It has yet to reach a situation where it can compete with Russia and Western Europe, although continued improvement in the design and development of Chinese military aircraft raises the possibility of competition with other manufacturers in the not so distant future.

Russia expects that China will remain a long-term partner, although national ambitions for independent design and manufacture of aircraft and weapons systems might limit the partnership to specific projects and technologies. The quoted reports suggest that China can achieve the following objectives on its own within its aerospace industry:

1. Increased investment in the aerospace industry infrastructure, which was sharply curtailed from the early 1990s when China started to divert substantial funds towards space programmes.

2. Enhancing the Russian-built air fleet inventory using Western/South African avionics and Israeli components.
3. Procurement of airborne early warning and control systems (AWACS), after China failed to obtain Israeli-built Phalcon. China is trying to develop AWACS, but will require outside assistance which could come from Russia.
4. Enhancing domestic development and manufacture of advanced unmanned aerial vehicles (UAVs) and unmanned combat aerial vehicles (UCAVs). Here Russia could offer possibilities for joint R&D and potential joint manufacture.
5. Continued investment in space with an emphasis on military aspects.

A sixth Chinese objective is to design and manufacture a new military transport aircraft.[28] Furthermore, the experience of advanced avionics as evidenced from the First Gulf War has provided a catalyst to increase funding for China's aerospace projects, including modernization of the Chinese Su-27 fleet which was obtained in the early 1980s. The change in China's military priorities occurs together with decreased cooperation with Russian counterparts. This reverses a trend of substantial increases in cooperation between both countries' aerospace industries that began in 1992.[29] Chinese aerospace engineers and technicians have learned a great deal from their Russian colleagues but, after their successful space launch, China has been less willing to cooperate with Russia on space issues. For instance, China declined to take part in improving the Russian Glonass system of navigation satellites.

In redefining its relationship with Russia, China has increasingly become interested in technology transfers rather than in direct defence purchases. The technology transfer share may still be around 30 per cent, while China would increase the share to some 70 per cent in pursuit of self-sufficiency. Russian defence industry sources have disclosed that China's engine makers are close to mastering the complex skills needed to build the Su-27's AL-31 engine – something that most observers had thought would be controlled by Russia. If this proves to be the case, then China will have used the Su-27/J-11 project to establish a total systems capability for advanced combat aircraft in little more than ten years.[30]

The PLA has steadily procured approximately 1.5–2 billion dollars worth of weapons from Russia every year. The Russian share of Chinese arms imports has reached 90 per cent.[31] China is eager to buy new-generation military technologies but in early 2004 indicated 'the practice of deliveries of finished military hardware from Russia has exhausted itself, and the time has come to seek new forms of interaction'.[32] Chinese cloning of advanced military aircraft has in the past not been very successful.[33] The same shortcomings have appeared with aircraft carriers.[34] However, Russia acknowledges that

China has even surpassed it in military prowess. The military region NOAK is armed with 280 fourth-generation fighters, principally the Su-27 and Su-30. The number of these aircraft will increase to almost 400 in two years. By the year 2007, the Chinese Navy will have almost twenty low-noise Kilo class Russian-made submarines and a minimum of two squadrons of powerful combat vessels, chiefly Sovremennyy class destroyers made in Russia. The PVO (Russian missile defence) system of China, which consists of a dozen S-300 battalions and 30 Tor-M1 systems, is the best in Asia. At the same time India has become the biggest purchaser of Russian weapons, with 42 per cent of total deliveries in 2003, while China had only 38 per cent. The same commentator offers the following judgement:

> Of course, if China were to pay us ten billion dollars for new technologies, that would make it possible to finance some promising projects in the areas of aviation and weaponry, but far from all of them. Once we sell modern technologies to China, we will be closing off the Chinese market to ourselves in the future. So would it not be better to sell only those weapons that are in service with the Russian army? China has no way to get away from us, since they have no place else to buy military hardware, after all.[35]

The Analysis of Strategies and Technologies Centre (AST Director Konstantin Makiyenko), has offered some calculations on the transfer of the latest arms and technologies to the PRC and indicated that it would bring Russia no less than ten billion dollars. 'The technologies for creating missiles could be transferred in exchange for an order of a hundred and fifty Su-30MKK fighters for a total of seven or eight billion dollars', says Mr Makiyenko. 'If Russia assists the PRC in creating its own aircraft carrier, the question has to be raised of China's procurement of no fewer than four destroyers, at a cost of three and a half to four billion dollars.'[36] The income from the sale of licences for engines and radars could bring roughly another two billion dollars to the treasury.

The NOAK is now procuring from Russia the latest modification of the Su-30MKK fourth-generation fighter. Those aircraft are considered some of the best in the world. They are not even in service with the Russian army. The Russian Air Force, owing to a shortage of funds, has been forced to content itself with the modernization of the less advanced Su-27. China is nevertheless demanding that the Bars phased-array radar and air-to-air missiles with an active seeker head be installed on the Su-30MKK. The Chinese are tying the acquisition of new aircraft to the transfer of the technologies for the manufacture of both the radars and the missiles to them, as well as selling them a licence for the production of the AL-31F engines with which the Su-30 is fitted.

The interests of the Chinese are even broader in the area of naval arms. Before the break-up of the USSR, it designed the type 956U destroyer with a gas-turbine power plant. That ship was expected to become the main strike force of the USSR Navy in the medium ocean zone. But the project was never put into series production due to a lack of money. The NOAK is now offering to finance the building of the pilot destroyer, but only provided the ship itself and all the documentation for it are transferred to China. The PRC wants to become a co-investor in the project to create floating nuclear power plants with Russian KLT-40S reactors, and the Chinese COSTIND Defence Industries have a plan to adapt those reactors for their own production of nuclear submarines. Finally, the PRC is coming out with plans to bring Russia in as the design engineer for a heavy aircraft carrier that is expected to be laid down in the next two or three years. The Chinese COSTIND Defence Industries are thus not in a position to reproduce foreign technologies at the present time. So the sole chance for the Chinese defence officials to fulfil the new party task is to acquire the latest technologies from Russia, and bring in our specialists to develop modern weapons systems.

What price are the Chinese prepared to pay to see that their army is mightier than the Russian army? Judging by everything, the Chinese consider even a total of ten billion dollars, which is the figure calculated by the Director of the AST Centre referred to earlier, to be highly exaggerated. The leaders of the PRC prefer not to cite specific figures now, hoping to achieve the revocation of the embargo on arms sales to that country that was imposed by EU authorities in 1989, after the student demonstrators in Tiananmen Square were fired on. The Chinese calculations are simple. If the embargo is lifted, they will not have to pay Russia the maximum.

EUROPEAN TECHNOLOGY COLLABORATION

Cooperation in science and technology between China and Europe started in 1981,[37] actually following the Sino-US Agreement on S&T cooperation that was signed by Deng Xiaoping and President Carter in 1979. The Ministry of Science and Technology (MOST) in a newsletter mentions a number of important fields in its European collaboration such as telecommunications, energy, biology, aeronautics and space, as well as the China-German Joint Laboratory on Molecular Medical Sciences that was established in 1998 and early collaboration with Italy on cosmic ray detection. The cooperative research programme between China and France has led to the establishment of three China–France cooperative research institutes in the fields of information, automation and applied mathematics. In addition to collaboration between China and European nations there are also a number of projects that involve the EU and examples include Global Digital Telecommunication, Digital Audio Broadcasting and Biomass Energy.[38]

> **BOX 7.1   THE EUROPEAN UNION COUNTRY STRATEGY PAPER**
>
> The European Commission on 1 March 2002 approved the Country Strategy Paper[39] (CSP) for China, which set out the framework for EU co-operation with China during the period 2002–2006. The cooperation strategy outlined in the CSP is designed to support the implementation of the EU's broad policy objectives regarding China. These include further integrating China into the world economy and world trading system; supporting China's transition to an open society based on rule of law and respect for human rights; making better use of European resources by improving coordination between EU assistance and bilateral spending by Member States; and raising the EU's profile in China.
>
> A budget of €250 million is planned for the period 2002–2006 to support EU programmes in China. The Country Strategy Paper will provide the instrument for guiding, monitoring and reviewing EU assistance.

An S&T cooperation agreement agreed in December 1998 set the stage for a new development stage between China and the EU, both in terms of research level and scope. Subsequently, the EU Fifth Framework Programme for Science and Technology was made open to China (now followed by the Sixth Framework Programme) and conversely China has offered participation in China's 'basic studies and high-tech research plan' to foreign research institutions and scientists.[40]

Chinese participants have been encouraged to join the Information Science and Technology (IST) programme which requires that it should conform to EU interests, provide substantial added value in implementing science policy and involve the required participation of entities from the EU and associated states. China is presently participating in 11 IST projects through its major universities with a total investment of €30 million, of which the Chinese side has contributed €5 million. The indicated activities are an outflow of China's stated policy to achieve participation of overseas research institutions and scientists in the country's basic and high-technology programmes.

Although the direct funding and involvement of the European Commission is important in supporting and guiding China–Europe relations in science and technology, the most important factor will be the rapidly expanding trade between the two regions. Furthermore, direct bilateral contacts between scientists and researchers will also expand rapidly. China will over the next

10–20 years provide the rest of the world with extraordinary possibilities for commercial and intellectual exchange – on the assumption that China's experience in smooth political transformation and rapid economic development over the past two decades continues. China as an emerging technological superpower will have far-reaching consequences for Europe (EU) and the rest of the world.

These possibilities will not fully materialize unless there exists a deep understanding of the process and initiatives are formulated to bring mutual benefits. The developments within science and technology briefly indicated in this section present challenges to three different EU communities – the business community, the science and technology community and the political community.

The Galileo Project

In order to capture possibilities there is a need for deep knowledge of regional changes and understanding of the underlying factors. Without such recognition it will be difficult to be aware of ongoing processes and impossible to anticipate future trends and discontinuities. This requires a two-pronged approach of building competencies inside the EU and establishing long-term relations with institutions in China. The Galileo project gives a good illustration of emerging China–EU relationships in science and technology.

Galileo is Europe's own global navigation satellite system and will work alongside the US GPS and Russian Glonass systems. Galileo promises realtime positioning accuracy down to one metre and is guaranteed to work under all but the most extreme circumstances. It is suitable for safety-critical systems and has the potential to run trains, guide cars and land planes. Supporters of Galileo say that it will be safer and more precise than the current GPS system and the Russian Glonass network, which were developed for military use. Approval for Galileo also has important consequences for the European space business, as a separate European network of satellites would be beyond the jurisdiction of North America. The latter aspect may have made it attractive for China to become a partner of the European Project for GPS.[41]

The China-Europe Global Navigation Satellite System Technical Training and Cooperation Centre (CENC) was inaugurated in Beijing on 19 September 2003 to train staff and organize bilateral exchanges for the Galileo project. This marks the beginning of China–Europe cooperation in the project. Today most countries have to use the GPS developed and operated by the US. This is a space technology system that was put into use by the US military in 1973.[42]

China has realized that since the 1991 Gulf War, all US military operations would have been impossible without GPS – its receivers were supplied to every combat unit and even individuals. GPS can also be widely used in people's daily life, for instance, using it on ships, planes, automobiles and mobile telephones, thus creating huge economic benefits and social effects. However, for a long time, the United States has provided its military units with precise positioning signals, whereas the ones it has provided to other users are low-precision signals (with deliberately added interference). That is to say, only the American military know the exact position of any object on the earth; other countries only have a 'general idea' concerning positions.

In order to counter this US dominance the Russian Defence Ministry launched 24 high and medium-level circular-orbit satellites in 1995 at a cost of over US$3 billion and set up its own global positioning system called Glonass for short, which has a precision better than the GPS with added interference. After five years of repeated discussion and analysis, the European Union formulated the Galileo project. In March 2002, the 15 communication ministers of EU members decided to launch the programme.[43] The US reacted strongly to the Galileo proposal and argued that there was no need for Europe to implement the Galileo programme, and in May 2000 waived restrictions on interference, raising its positioning precision for civilian purposes from the 100-metre level to 10 metres.

There are a number of reasons why China would like to see not only a successful Galileo project but also its own active participation in the programme. Firstly, China's support and participation can facilitate the EU's negotiation with the United States and Russia on cooperation. Secondly, China's huge market provides promising prospects for the programme. By now GPS technology has been fully applied to new automobiles and even mobile phones, and China's hundred million mobile users and its rapidly growing number of autos can expand Galileo's market domain tremendously. Additional reasons are that China is sure to become a space power in the 21st century, so is able and needs to have its own global positioning system. The EU-China participation would yield new knowledge in the fields of satellite launching and manufacturing, the environment of radio transmission, ground systems and radio frequencies, as well as the standards for receivers. Simultaneously China looks forward to developing its own industry for manufacturing receiving equipment and high-precision locators. The Chinese planners are concerned about the long-term future in developing the country's space technology and realize that although the Galileo project is the biggest cooperation in a science and technology programme between China and the EU, China's participation is rather limited. Actually it is equivalent to only one EU member and accounts for a scanty share of the investment. Furthermore, a major share of the funds will come from Chinese enterprises taking part in the cooperation.

China is gradually placing its own system in place and in May 2003 the nation sent into space its third 'Beidou' (the Big Dipper-1) navigation positioning satellite, forming a complete positioning system together with the two earlier Beidou satellites launched in 2000. This system offers precision reaching dozens of metres and also has communication functions. The launching of the third Beidou satellite marks the formation of a complete and sound all-weather satellite navigation and positioning system, said the experts.[44] China's Beidou navigation system is a regional positioning system mainly covering the country and its neighbouring areas, thus making vertical positioning impossible while limiting the number of users. A global positioning system is a space infrastructure that it is essential for China to possess if it is to be a space power in the 21st century, and the Sino-EU cooperation will significantly help China to attain its goal.

COOPERATION WITH BRAZIL

Another example of international space collaboration is that with Brazil. The two countries have confirmed that they will launch at least three more Earth observation satellites in the 'near future' to form a remote-sensing system that will be both competitive and compatible with the world's needs.[45] The new satellites will enable the two developing nations to reduce their dependence on the use of remote-sensing images provided by developed countries. The first China-Brazil Earth Resources Satellite (CBERS) was launched in 1999. China and Brazil pooled their technical skills and financial resources in the late 1980s to initiate the CBERS programme, with the expectation that it could become the most used remote-sensing satellite by 2010 – not only by Brazil and China, but also by many other countries. The launch of the CBERS series of satellites will lead China into an era when the bulk of its remote-sensing data needs no longer be bought from abroad.

GLOBAL POLITICS

Since the early 1990s, China's military doctrine has come to emphasize a highly capable air defence, able to carry out long-range precision strikes and not least able to control completely its information sources and structures. These factors are now shaping the restructuring of China's industrial and technological structure for the development of a variety of weapons systems. However, defence R&D and production must compete with other economic priorities. At the same time China will be able to carry out advanced research that is much less costly than prototype R&D, testing and production. China's advanced military research would provide an important instrument in evaluating and selecting systems and technologies that would support the country's strategic ambitions in the most efficient way. In the military technology domain, China has greatly benefited from the collapse of the USSR and the absence of any viable international export control regime. Although international sources of military technology have widened, China is faced with the daunting task of transforming the PLA from a people's army to a defence system in which computers and communications are coming to play a dominant role.

A new technological landscape is taking shape in Northeast Asia and China, which will in due course replace Japan as the technological driver in the region. Industrially such a change is already taking place, although China still has to scale the technological heights where the US, Japan and the EU still maintain dominant positions. China is benefiting from a tremendous inflow of foreign direct investment which has also brought significant technological inputs, although most of it is embedded in foreign industrial and management structures. The country is simultaneously expanding and supporting its domestic resources for research and technological development. It is benefiting from the increasingly global nature of commercial high-tech R&D. By combining substantial inflows of technology with its own R&D development, China is in a position to profit greatly from the ongoing globalization process, and gaining an advantage over other countries at a similar stage of development.

China has become the main beneficiary due to its vast market potential, increasing role in the global economy and its various policies and programmes to exploit the potential of the recent dynamic international situation. The global economic forces are creating a new structural relationship with the EU, as well as with other global actors of which the US is the most prominent. Thus states and regions are competing for shares of the world high-tech markets and changes take place much more quickly than in the past.

There is little doubt that China is becoming an increasingly attractive partner not only in high-tech trade but also in joint R&D activities. The US may initially have benefited from establishing close S&T relations with China since 1979, partly because of a major inflow of Chinese students looking for advanced education and training in highly recognized universities. However, the EU in recent years has been able to foster increasingly close relationships in science and technology. The Framework Programmes for Research and Technological Development have recently provided an attractive platform for promoting collaboration in science and technology. Simultaneously, almost all member states of the EU have on their own initiatives established bilateral relations in science and technology with China.

China launched its first satellite – Dongfanghong – on 24 April 1970 and had launched altogether 40 domestic satellites by the end of 1997. This number includes 17 retrievable reconnaissance satellites, 3 meteorological satellites, 8 communication and broadcasting satellites and 12 experimental satellites, the latter ones possibly being military.[46] China is using the Long March series of rockets to launch satellites while East Wind rockets are used in the intercontinental ballistic missiles programme. However, there is a consensus that the first Long March rockets used to launch satellites were derived from ballistic missiles developed earlier and there was parallel research and development for the modernization of the SLVs and ICBMs.[47]

The military and security agencies in the US have been concerned that foreign involvement in the commercial activities of satellite launch vehicles (SLV) in China would more-or-less directly support the modernization of China's ballistic missiles systems. The US concern has two targets. First, the US and its close allies want to prevent China from engaging in the proliferation of advanced rockets and satellites and related technologies. Second, the US wants to prevent any leakage of advanced technology into China's military development programme when US companies enter into commercial relations for launching satellites in China. In the latter context a number of US companies, including Loral, Motorola, Hughes and Lockheed Martin, have been scrutinized for possible transgressions.

As indicated earlier in this chapter, China's access to foreign military technology, especially from Russia, has enabled its armed forces to create a strong basis for an ongoing military-technical modernization. A testimony to the US-China Economic and Security Review Commission in early 2004 provided the following forewarning to the US:[48]

> .... weapon systems the PLA is acquiring will allow it to greatly impede a future U.S. attempt to rescue democratic Taiwan in the event of a PRC attack. Foreign military systems are also propelling what Taiwanese officials predict will be a 'crossover' in which the military balance on the Taiwan Strait will start to favour the PLA after 2005. Foreign military technology may also allow the PLA to build new power projection capabilities by early next decade.
>
> In assessing the degree to which foreign military technology is aiding PLA modernization, and the possible resultant dangers to U.S. national security, it is also possible to highlight the need for greater U.S. policy focus on the need to stem PLA access to more modern and dangerous technologies. While the United States has made clear its desire for peaceful relations with the Chinese people, the government of the PRC is actively preparing for a possible war with democratic Taiwan, as it continues to proliferate dangerous nuclear weapon and missile technologies to rogue regimes. It remains necessary for the U.S. to sustain its embargo of military technologies put in place in response to the 1989 massacre in Tiananmen Square. The U.S. should work with allies in Europe to explain the possible dangers if Europe ends its Tiananmen embargo in 2004. And as the U.S. was able to persuade Israel to end its sale of dangerous military technology to the PLA, it is necessary to make curtailment of Russia's substantial arms trade a higher bi-lateral issue with Moscow.

However, in the early 1970s and 1980s the US encouraged Israel to develop military technical relations with China with the objective of indirectly supporting China's military modernization against the former Soviet Union.[49] During the 1980s, Israel offered China its technology in the areas of tank weapons, anti-tank missiles, surface-to-air missiles, cruise missiles, military electronics and aircraft design. This burgeoning relation became a concern to the US not only because of the sophistication of technologies transferred but also China's access to weapons systems that were subsidized by US taxpayers. At the same time Israel was driven by its domestic motive to support costly arms industries that must be competitive and independent to assure national security.

China is reported to have benefited greatly from the co-development of a fourth-generation multi-role fighter – Chengdu Jian10 (J-10). This project was heavily dependent on the LAVI advanced fighter from Israeli Aircraft Industries that was terminated after the US withdrew its financial and political support. J-10 actually includes avionics, advanced composite materials and flight control specifications that originated in the US while the Chengdu plant was able to combine its domestic resources with those from Russia, Israel, and the US.

The end of the Israel–China relationship in the transfer of military technology occurred after Israel wanted to sell its very advanced Falcon phased-array airborne radar system. This triggered a bipartisan reaction in the US and in the late 1990s the US required Israel to curtail its exports of advanced military technology to China.[50] Israel has, since abandoning the Falcon sale, been pressured by the US to curb all sales of 'perilous' weapons to China, although such transactions have not been completely abolished.[51]

The Fisher testimony suggests that the US should sustain its own embargo, which is important in itself but also to 'demonstrate to Europe that its rapidly evolving policies that may soon lead to the removal of its arms embargoes will create yet another serious conflict with Washington'.[52] However, in addition the US 'should make stemming the supply of critical defence technologies to the PRC a higher strategic priority', and the earlier dialogue with Israel is seen as a success story.

INTEGRATING INTO GLOBAL AND CIVILIAN MARKETS

In the space and military sectors China is facing more serious challenges than in most other technological sectors and the reasons are twofold. First, much needed technology is not available in open global markets and must be secured through government-to-government agreements. Second, the US government views China as a strategic competitor and restricts the sales of many categories of technologies which have actual or potential military applications, and exerts strong pressures on other countries to follow suit. The Galileo project, which is supported by the European Space Agency (ESA) and the European Commission, has offered China the possibility to become equal partner in a development project that will provide advanced GPS facilities when it becomes fully operational by 2010.

Furthermore, the rapidly evolving technological capability within China's booming civilian electronics industry has important effects. First, it contains dual-use technologies which can easily be utilized in space and military sectors. Second, strong competition in the civilian sector has substantially lowered the costs for many components needed in the space and military sectors where dedicated research and development costs for the same items would have been substantially higher.

Finally, China is returning to the space sector as commercial operator and is expected to put a self-made communication satellite into orbit for Nigeria in 2007, making this African country the first foreign buyer of both a Chinese satellite and its launching service.[53]

NOTES

1. Bloom (1990).
2. Since the central government declared in 1985 that the Long March launch vehicles would provide an international satellite launching service, China has successfully launched about 30 foreign-made satellites for users in Pakistan, Australia, Sweden, the US and the Philippines.
3. Broad (2003).
4. The Chinese government also directed the CNSA to establish an indigenously built geosynchronous satellite broadcasting and telecommunications system using partnerships with Western companies to increase the level of Chinese technology. Sinosat-1, launched in 1998, was the first such cooperative project between the Chinese and European aerospace industries. The technology would be used to develop new FH-1 military and DFH-4 civilian communications satellites to form a command-and-control network designed to link Chinese combat forces. Deployment of the new constellation began with Zhongxing 22, launched in January 2000 (Futron report).
5. After joining the COSPAS-SARSAT search and rescue system in 1992, China established the Chinese Mission Control Centre, now used to operate China's Beidou navigation satellites, the first of which was launched in 2000 (Futron report).
6. 'China makes strides in space technology', *China Daily*, 6 October 2004.
7. Zhao (2004).
8. Futron.
9. 'China to launch new solid-fuel rocket', *China Daily*, 30 January 2004.
10. Li Zhenyuan (1997).
11. SCTIND, Commission of Science Technology and Industry for National Defence, is mainly responsible for production and routine technology management.
12. Futron Corporation (2003).
13. Ibid.
14. SAST originally developed the Long March 1, 2, and 3 series of vehicles, including the guidance systems.

15. Kan (2001).
16. Medeiros (2004).
17. Ibid.
18. Ibid.
19. 'Chinese Defence Industry: Chinese Puzzle', *Jane's Defence Review*, 21 January 2004.
20. Jiangnan Shipyard was established in 1865 and is regarded as the cradle of China's national industry as a first generation of industry workers in China was nurtured. It has become a significant part of the country's shipbuilding industry as China has become the third shipbuilding country only second to Japan and South Korea. The company merged with Qiuxin Shipyard in 2000 as a major step in the consolidation of the China State Shipbuilding Corporation (CSSC). Total employment exceeds 11 000 (http://www.globalsecurity.org/military/world/china/jiangnan-sy.htm).
21. *Luhu*-class Multirole Destroyer, http://www.globalsecurity.org/military/world/china/luhu.htm.
22. Medeiros (2004).
23. Ibid.
24. Xiong Yuxiang (2004).
25. 'Disarmament to cut 200,000', *China Daily*, 2 September 2003.
26. Dolven and Neuman (2003).
27. Kogan (2004).
28. There is a clear understanding within the management of the Antonov design bureau that cooperation with China is vital to its future. In the final analysis, however, it seems that the significant achievements made by the Chinese aerospace industry have allowed Beijing the room to further develop its capacities without such heavy reliance on Russian imports. (Ibid.)
29. For instance, China had previously used Russian defence systems; Russian engines, radars and missiles had also been installed in Chinese aircraft (such as the J-8/F-8, FC-1/Super-7, J-10/F-10 and the next generation lead-in fighter trainers designated as the L-15); and Russia had provided basic training programmes for Chinese workforce and aircrew.
30. The Ramenskoye-based (Moscow Region) Technocomplex Scientific Production Centre has been upgrading both single-seat and two-seat Su-27s for the PLAAF. The scope of this upgrade work is an example of the integrated thinking that surrounds China's future fighter programme and the J-10/F-10 and Su-27/J-11 in particular. Having mastered basic airframe assembly, China currently controls an upgrade package that will allow it to integrate the weapons and systems under development. Although the Moscow-based machine-building enterprise Salyut has not yet delivered to China the AL-31 engine, which features a fully variable, rotating, thrust-vector control (TVC) nozzle designed by the St. Petersburg-based Klimov Corporation, when the delivery occurs, the AL-31 can be retrofitted during an engine upgrade, turning the J-10 into a highly manoeuvrable fighter. Furthermore, it appears that the Chinese-built WS10A engine that powers the J-10 fighter can match the performance of the AL-31 engine, despite differences in design and manufacture.
31. Khazbiyev (2004a).
32. Khazbiyev (2004b).
33. At the end of 2002 – that is, 20 years after the start of the programme – the Chinese aircraft builders were able to start the small-scale series production of a clone, the J-10. True, the Chinese aircraft is relegated by its characteristics not to the fourth generation, like the F-16, but to the third. See Khazbiyev (2004a).
34. An instructive history of cloning military hardware in China occurred with Soviet aircraft carriers. Immediately after the collapse of the USSR, China acquired three aircraft-carrying cruisers from Russia and Ukraine – the *Minsk*, *Kiev*, and *Varyag* – which had been decommissioned from the USSR Navy. The Chinese programme to create their own aircraft carrier started in 1993, but China has not been able to build a single aircraft carrier. The programme has been frozen for an indefinite period. The *Minsk* is now being turned into the Minsk World floating attraction, and they are making a five-star hotel and casino out of the *Kiev* and *Varyag*.
35. Khazbiyev (2004b).

36. Ibid.
37. Ma Songde (2000).
38. Ibid.
39. European Commission (2002).
40. Ibid.
41. 'China joins EU space programme to break US GPS monopoly', *People's Daily*, 26 September 2003.
42. GPS is composed of 24 satellites orbiting around the earth, sending out continuous radio signals of a certain frequency from a height of about 17 000 kilometres above ground. These satellites can serve as objects of reference for any target on the earth, and a signal receiver could decide its position by only selecting signals from four satellites for analysis.
43. The Galileo project, involving an investment estimated at 3.6 billion Euro, is made up of 30 satellites distributed in three orbits at a height of 24 000 kilometres. Unlike the United States and Russia, the EU declared that Galileo is tailored for civilian use only, providing a precise global positioning service with an error of no more than 1 metre.
44. Both the satellite and carrier rocket were developed by the Chinese Research Institute of Space Technology and China Academy of Launch Vehicle Technology, which are under the China Space Science and Technology Group (http://english.people.com.cn/200305/25/eng20030525_117171.shtml).
45. 'China to launch trio of satellites with Brazil', *China Daily*, 23 July 2003.
46. Kan (2001).
47. Ibid.
48. Fisher (2004).
49. The formal go-ahead is reported to have come in 1979, when then-Defence Minister Ezer Weizman asked the late Israeli billionaire Shaul Isenberg to establish the Israeli-PRC arms trade. See Fisher (2004) who provides the following reference: Judy Dempsey, 'Israel considers arms dealings with China an acceptable risk', *Financial Times*, 23 April 1999, p. 8.
50. Concern had been building since the deal was formalized at the Paris Airshow in 1997 that Israel would combine Falcon with a Russian-supplied Beriev A-50 AWACS aircraft. The deal would have involved up to four aircraft for $1 billion. See Fischer (2004) who provides the following reference: 'Final RFP for Chinese AEW follow-on programme expected', *Journal of Electronic Defence Electronics*, 1 April 2000.
51. Fisher (2004).
52. Ibid.
53. Qin Jize (2005).

8. Regional innovation systems in China

REGIONAL DIVERSITY IN CHINA

In any country, and particularly in a country like China with its extraordinary size and diversity, technological innovation will take place in a number of its regions that are likely to become spatial (or regional) innovation systems. Huge amounts of innovation of a gradual and incremental nature are taking place in manufacturing firms all over China, although primarily in the dynamically evolving coastal areas. These firms have often agglomerated into geographical clusters and are found in many industrial sectors.

A number of such clusters are evolving into centres of strong innovative capability. They are still weakly linked and inadequately supported by actors within the state-level innovation systems. However, a natural formation of three major regions in China has prompted provinces and cities within them to act as midwives to create an environment that can deliver not only incremental innovations but also breakthrough innovations in future-oriented industries. A number of regional development programmes and projects play an important role in this process and have the potential to enhance necessary and strong links between clusters, foreign technology sources and national programmes.

Three regions in China, where policies and structures reflect an ongoing regionalization, emphasize science and technology as essential components. They are the Yangtze River Delta (YRD), the Pearl River Delta (PRD) and a third region – the BoHai Rim (BHR). This is also along China's coastal line towards the north, and although less clearly defined it is nevertheless very important. A very large number of units within the regions are closely linked with overseas associates through foreign direct investment and trade. Many industrial clusters have sprung up inside the regions and in many places have become budding technological springboards for indigenous technological development, supported by regional and national initiatives. The regions represent not only provincial ambitions but also attempts to forge various elements of a national innovation system that include national programmes, universities and foreign direct investment (FDI). There are great expectations in China that the industrial clusters that have mushroomed through massive FDI will evolve from being purely operational clusters into technological clusters that will fuel an innovative environment.

The success of regional innovation systems in China has its roots in the following three factors. First, the central government has strongly supported the regions by providing a framework and resources for the various types of zones, industrial parks, science parks and incubators where national science and technology programmes have often been involved. Second, foreign direct investment and increasingly closer industrial and technological links with the neighbouring countries have given strong impetus to regional development through technology transfer, management skills and extensive links to global markets. Third, the combined directed but often spontaneous development of technological and industrial clusters has provided the basis for further development.

China in the early 1980s departed from a policy of sustaining a national innovation system (NIS) and accepted a global innovation system by allowing a massive inflow of FDI. Subsequently foreign companies, which are now increasingly wholly-owned foreign enterprises, have come to play a significant role in China's high-tech industrial development. Furthermore, they now dominate China's high-tech exports, for which functional industrial clusters are essential, having sprung into existence all along China's coastal line, and currently often further inland as well.

In the long-term, China expects that the massive inflow of FDI will significantly contribute to extensive export earnings and to substantial new employment in local areas. The policy makers also expect that a large number of technological clusters will come into being and become self-generating in their support for future innovations in China. Such clusters are already appearing in a number of cities, which in their turn are integral parts of provinces. These are on their own attempting to consolidate groups of provinces into regional innovation systems. It is worth considering the future role of regional innovation systems and the evolving clusters within them. This will provide new insights into China's acceptance of entry into a global innovation system by accepting the immense inflow of FDI.

Regional development in China, with its tendency to cluster, is being influenced by three crucial forces that include commercial/economic as well as political elements. First, China has accepted and encouraged a massive inflow of foreign direct investment, by which the government and its agencies endeavour to assist overall economic as well as regional and technological advances. Second, China has accepted and encouraged close economic and industrial relations with outside areas, first with Hong Kong, to be followed extensively by Taiwan and Korea and to a lesser extent with Japan. Hong Kong, now part of China, Taiwan and Korea, have initially developed closest contacts with nearby provinces which correspond to the Pearl River Delta region, the Yangtze River Delta region and the BoHai Rim region respectively. Third, these forces have been reinforced by national and regional policies to establish different types of industrial and technological zone/parks, often complemented by provincial and municipal initiatives.

The increasing level of FDI in China and its sophistication, particularly in the IT industry, has naturally prompted the development of clusters for various types of specialization. Their character has mostly been of an operational nature in order to serve booming export markets. It is the natural expectation of technology policy makers in China that the present clusters, with high-tech dominated by foreign multinational corporations (MNCs), would transform themselves into technological clusters fuelling an innovative environment in the country.

Politicians in many countries have been fascinated by the idea of creating a duplicate of the Silicon Valley, and China is no exception. They have all been attracted by the apparent ability of clusters not only to support technological development through their innovative activity but also to generate employment and contribute to GDP growth. Furthermore results from historical studies of economic development show that national economic growth is directly related to successes in economically strong regions.[1]

Studies of clusters have in the past often revealed a distinct geographical concentration as one important characteristic, realistic assumption being that specialized skilled labour and specialized subcontractors may be attracted towards a region once a critical mass of firms within the industry is attracted there. However, clustering in high-tech sectors may take on a completely different pattern as it will, aside from a geographical concentration, also have to include sectoral or functional characteristics. Such clusters would often have to be understood as sub-clusters in networks that span not only sectoral but also geographical boundaries. Under such circumstances functional proximity takes precedence over geographical closeness, which has been seen in the past as the key cluster characteristic. Newly emerging clusters in China primarily fall into the category of functional proximity ones.

The regions and their clusters are now playing a significant role and their embedded innovation systems might greatly contribute towards China becoming a technological superpower. In its early attempts to lay the foundation for an industrialized structure, the People's Republic of China enthusiastically adopted a planned economy approach which was not only inspired by the Soviet Union but in all essentials was a carbon copy of policies and structures from that country. Changes took place after the break with the USSR in the late 1950s, with attempts at decentralization and the loosening of centralized control during the Cultural Revolution. However, China's leadership remained committed to a planned economy style until major reforms were unleashed in the late 1970s. Foreign direct investment had until then played a very minor role in China's economic development, and the provinces and cities under central direction like Shanghai had very little control of their destiny. This situation has changed dramatically since the early 1980s. FDI interests and regional interests, now bonded together, have produced a heavy demand and subsequent dependence on foreign technology.

Clusters thrive from a skilled and well-educated workforce, research and development with its resulting intellectual property rights, business infrastructure and physical infrastructure. This will require a number of policy measures. First, labour and capital must increase in quality and quantity. Second, the environment for innovation and entrepreneurship must be given special attention to renovate industrial structures constantly. Thus, the top-down policies which used to be dominant in China have had to be replaced by cooperative relations between local and central institutions, with central authorities subsequently have a diminishing role to play. The result has been that regions and also cities have been given a much larger control over various resources for R&D, including education.

Three Significant Regions in China

The early expectations that close industrial links, based on FDI, would come into existence between Taiwan and the cities across the Taiwan Straits – primarily in Xiamen and Fuzhou – did not materialize. Instead, and rather naturally, Shanghai has taken on the role, to become the natural basis for Taiwanese entrepreneurs and industrialists; a similar role to that of Shenzhen for the Chinese in Hong Kong. In early 2004 more than 400 000 Taiwan residents were living in Shanghai out of a total number of more than one million in the whole of China, not counting those who come only for shorter stays. This means that almost 5 per cent of the population in Taiwan is permanently residing on the Mainland. Although not being able to solve their political relationship, Taiwan and China have, with Shanghai as a hub, entered into a close economic, industrial and technological partnership that is expanding day by day.

Still further to the North lies the BoHai region with important industrial cities in the provinces of Liaoning, Shandong and Hebei. Tianjin, with the same status as a province, has emerged as an important technological and industrial centre with its good transportation and education facilities. As in Shenzhen and Shanghai, foreign direct investment has played an important role in the region's industrial development. Naturally, the closeness to the Korean Peninsula has attracted large numbers of investors from South Korea. After normalizing diplomatic relations in 1992, the two countries entered into a close economic partnership and China has become the number one trading partner for Korea in both exports and imports. Trade between the two countries constituted 14 per cent of Korea's total trade in 2003.

The world can already see the emergence of three great urban spheres which have become China's most highly developed areas – the lower Yangtze River Delta, the Pearl River Delta and the BoHai region, with Beijing and Tianjin. These regions have 5 per cent of China's total land area and some 20 per cent of total national population. By 2025 they could account for two thirds of China's total GDP, a development that in all likelihood would attract large-scale migration. However, in the meantime central and western China will probably create their own urban belts and concentrations of cities. A continued rapid urbanization might relieve some of the existing burden on nature, so that China's land and exhausted ecology can be allowed to rest and recover, and the environmental debts incurred over thousands of years can gradually begin to be repaid. Thus there is a need among the policy makers, to develop approaches that enable cluster industries to exploit economies of scope and scale. The need for developing new business models is more acute than in most other industrialized countries.[2]

The resulting industrialization of coastal China has generated highly efficient structures in a large number of manufacturing fields not easily to be found in other parts of the world. Manufacturers, whether they are domestic or foreign companies rely on clusters of component suppliers and China has increasingly become integrated into the global economy and an extensive supply chain draws on resources in a global setting.

The acceptance of foreign direct investment and its abundant availability combined with an acceptance of market forces has greatly contributed to the shaping of the three economic regions in China. Naturally, local governments at provincial and lower levels have contributed to the development of the three regions to a great extent. This is being done through various means of attracting FDI, developing infrastructure and allocating land and other scarce resources. China joining the WTO has contributed to a shift in the MNC investment focus, with more and more foreign investors opting for mergers and acquisitions instead of launching new plants. Thus, more and more foreign investment has become wholly-owned ventures instead of joint ventures. Some of the early investors have chosen to buy out their Chinese partner and take full control of the former joint venture. It is not yet clear to what extent this influences technology transfer and potential for absorption in the localities.

The high-tech trade statistics also reveal the relative importance of different regions in China. Shenzhen remains by far the dominant player with its exports of US$2.0 billion against US$2.3 billion in imports of high-tech goods, being followed closely by Shanghai with its export of US$1.5 billion against imports of US$1.9 billion. The Yangtze River Delta region includes Jiangsu province, with the cities of Wuxi and Suzhou having become important high-tech industrial centres. Jiangsu reported high-tech exports of 1.7

billion against imports of $1.5 billion for July 2003. Beijing reported high-tech exports of $0.4 billion against imports of $1.2 billion for the same month. Tianjin reported both high-tech exports and imports at the level of $0.4 billion.

In each of the three regions there exist a number of industrial clusters, often highly specialized and with close links to overseas investors, mainly from Hong Kong, Taiwan, Korea and to a lesser extent from Japan, the US and Europe. Information from such clusters provides useful insights into the dynamics of industrial development of the regions and its linkages with overseas industrial partners.

Provincial and local governments at city and county levels have a vital role in economic and technological development. They control about 70 per cent of the state budget and often have their own development strategies, although frequently directed from the central government. This diversity, and also autonomy, is evident from the document[3] *Formation of Shanghai Knowledge Economy Strategies*, which includes the following approaches:

1. As Beijing has a strong information technology industry and Shenzhen a strong manufacturing base, Shanghai focuses on further exploiting an advanced biotechnology base and building the Pudong Bio Science and Technology Park.
2. Preferential tax regime and financial assistance to help high technology start-ups invest in R&D.
3. Public resources to establish new basic research institutions focused on Shanghai's strengths and interests, to encourage cooperation between universities and corporations, and to reorganize old research institutes.

The Role of Development Zones

Distinct examples of different regional technological clusters may provide illustrations of specific characteristics of Chinese regional development. Throughout the past couple of decades the central government has been very influential in establishing various industrial and technological zones, which all have a strong regional dimension. The government has selected a number of 'intelligence-intensive' regions and adopted policies to transform them gradually into high-tech development zones with different characteristics. There are now 53 such zones that are expected to become bases for China's high-tech industrialization.

In 1991 China launched 24 New and High Technology Industry Development Zones (NHTIDZ) which at the time were perceived to follow the Zhongguancun model in Beijing (see below). Two years later another 27 NHTIDZs were established, with an additional one that was set up in 1997.

Of 53 zones altogether, only a single one has an agricultural focus. Using the Chinese division of national regions, the East, primarily the coastal areas, has 29 zones, the Middle, which includes the Jilin and Heilongjiang provinces in the Northeast, has 14 zones while the remaining ones are located in the West. Referring to regions, the BoHai Rim region has nine NHTIDZ, the Yangtze River Delta region has six and the Pearl River Delta also has six zones. Comparing provinces, Guangdong is the leader with six zones followed by Shandong with five and Jiangsu with four.

Aside from the high-technology industrial development zones, China has a number of other regional development programmes. These include special economic zones, open coastal cities, state-level economic and technological development zones, coastal economic open zones, export processing zones and bonded areas. Following the success of Shenzhen, Zhuhai, Shantou, Xiamen and Hainan as comprehensive economic zones, a number of coastal cities were designated to become open coastal zones – including Dalian, Qinhuangdao, Tianjin, Yantai and Qingdao. Somewhat similar to special economic zones and open coastal zones, China also established a number of coastal economic open zones in the Yangtze River Delta, Pearl River Delta and South Fujian Province Delta. They originally included 40 cities under provincial governments and 215 county-level locations. In all regional development programmes, special legal and financial provisions were made to attract FDI and advanced technology into the localities while at the same time increasing export earnings. Chapter 9 provides an overview of the various programmes which exist within the Shanghai Municipality.

Within national programmes for regional development, the New High-Tech Industrial Development Zones play the most important role in fostering China's continued drive for industrialization and technological advances. Statistics[4] from the Ministry of Science and Technology show that from 1991 until 2002, major economic indicators of the 53 high-tech development zones in the country grew almost 50 per cent on a year-on-year basis, with an increase of total turnover volume from RMB8.7 billion (US$1.06 billion) to RMB1533 billion (US$186.9 billion) in 2002. In the meantime, the number of workers employed in high-tech parks increased from 140 000 in 1991 to 3.49 million – an increase of nearly 25-fold. In its high-technology development MOST has emphasized, not only for NHTIDZ, that electronics and information technology should be given highest priority followed by bioengineering and new pharmaceutical industries.

Zhongguancun (ZGC) in Beijing has been seen as a role model for the high-technology development zones in other parts of the country. ZGC benefited greatly from being in the centre of a research environment where a number of high-level research institutes were located, mainly belonging to the Chinese Academy of Sciences. In the same area were also a number of colleges and universities, of which the two most famous are Beijing University and Tsinghua University. The loosening of the planned economy in the early 1980s provided completely new opportunities for researchers to become entrepreneurs, also compelling them to move because of reduced budget allocations. Closeness to the central government and funding agencies in Beijing has continued to favour entrepreneurs, researchers and enterprises in Zhongguancun. The result is that ZGC has become the largest high-tech R&D centre in China and the largest distribution centre for IT products in north China.[5]

The significance of this location is proven by the fact that a substantial share of grants under the ambitious national 863 programme has been awarded to ZGC enterprises and neighbouring research institutes and universities. The knowledge resources are impressive, with about half of the ZGC working force having at least a bachelor degree, although with a high rate of mobility.[6] The close relations with universities that initially favoured ZGC are undergoing evolution in the following three aspects. First, universities are increasingly focused on research and teaching and are less involved in enterprise creation. Second, research results that have a potential commercial value are moving to university science parks for incubation, and when ready for start-up subsequently shifting to industrial development parks. Third, as a consequence university teachers and researchers as individuals, rather than universities themselves, will become participants in start-ups.

Shenzhen – the cradle of coastal industrialization

Shenzhen (in Guangdon province) is just across the old frontier between China and Hong Kong, and the two cities have recently become closely linked together in all economic aspects. The recent and very rapid modernization and industrialization in modern China got off to a dynamic start in the late 1970s when a first batch of special economic zones was established along the coastline. Early on, Shenzhen took on a prominent role; several factors contributed to this and gave an impetus for a headstart. First, Shenzhen was virgin land for industrialists and entrepreneurs, with no tradition in planned economy solutions. Second, Hong Kong, which had hitherto been a significant manufacturing centre for toys, garments, electronics products and many other light industrial goods, was experiencing rapidly rising wages for workers in its mainly labour-intensive industries and thus became less competitive. Third, entrepreneurs in the Hong Kong environment of dominantly private business quickly sensed immediate prospects of doing good business on the Mainland when the central authorities started to grant special status and privileges to the zones.

The government in Beijing offered two important provisions. First, Shenzhen as well as other special economic zones were given almost complete liberty in establishing new ventures, a boon when compared to earlier very troublesome procedures. Second, China offered labour that was relatively well skilled, disciplined, and most important at the time, at wages that were only a fraction of those in Hong Kong. With hardly any migration across the border and with wages only slowly rising, Shenzhen provided a golden opportunity for the industrialists in Hong Kong. At the same time, China was given a dynamic injection of industrial entrepreneurship in Guangdong, good management, employment and industrial upgrading of worker skills. Thus, an exceptional synergetic relationship between China and Hong Kong occurred. Within little more than a decade the industrialists in Hong Kong had created more than three million jobs across the border, all of which were more-or-less directly serving their immediate interests. The manufacturing base in Hong Kong, previously quite substantial, almost completely disappeared. In the transformation process the economy of Hong Kong became geared to providing services, among which finance, logistics and transportation are most prominent. The early success of Shenzhen as a springboard for injecting a new spirit of modernization set the precedent for dynamic change.

This transformation has by now encompassed the complete coastline of China, although Guangdong province, where Shenzhen is located, set the stage for what would happen in many parts of China.[7] Xiamen, further to the north, also benefited from one of the next special economic zones. The city, which is located just across from Taiwan, initially attracted a number of investors from overseas. Investors from Taiwan were mostly channelling their capital though other countries. However, Xiamen and Fuzhou, the capital of Fujian province still further to the north, did not provide the same dynamic conditions that had favoured the economic relationship between Hong Kong and Shenzhen. Also, and possibly more important, was the complete lack of political understanding between Taiwan and China in how to foster their economic relations.

From early on in the south the political leadership and planners in Beijing quickly expanded the scope of industrialization. This was based on an almost unlimited acceptance of foreign direct investment followed by access to the country's pool of low-cost labour. After completing the first stage, namely providing special economic zones, the central government went on to create special zones for high-tech development. The purpose was to wean away the localities from depending on labour-intensive activities when they accepted foreign investors. A later stage has seen a rapid proliferation of incubators for exploitation of research and high-technology innovations. They are usually connected to government research institutes or to universities, which underwent a major reform and a very rapid expansion of enrolment since the late 1990s. Both have simultaneously come to play a very important role in establishing spin-off companies, and some of the most well-known high-technology companies in China owe their origin to this phenomenon.

The industrial strength in Shenzhen lies in the following five areas. First, the majority of the industrial high-tech companies are private ones, with strong participation from abroad. Second, R&D is carried out within companies. Third, incubators play an important role, with more than 1000 enterprises currently in incubators. Fourth, all required support services for the electronics sector are available. Fifth, Shenzhen gives strong support to intellectual property rights (IPR), with many companies having their own IPR sections. Shenzhen is number three in China in terms of IPR. Furthermore Shenzhen has organized annually, since 1999, a High-Tech Fair, held in mid-October.

Shenzhen Special Economic Zone was established in 1980 after the City came into being the preceding year.[8] Foreign companies have located in Shenzhen because of labour which is not only attractive for cost reasons but also for its increasing competence levels. Another reason for the increasing attractiveness of Shenzhen lies in the dominance of private companies. As a consequence, more than 90 per cent of all R&D in Shenzhen is carried out inside companies. This is very different from other major cities in China. Shenzhen has a multitude of technology development structures. Established in 1996, Shenzhen High-Tech Industrial Park (SHIP) is one of the five national science parks. SHIP serves as the base for high-tech industrialization, R&D, incubator and high-tech talents training. Twenty-two per cent of total industry output comes from this park.

By taking advantage of the environment and resources of Shenzhen Special Economic Zone, SHIP has formed a number of professional and interactive incubator groups which combine activities of government, venture capitalists, scientific research institutes and overseas Chinese students and enterprises. They include the Virtual University Incubator, the Overseas Chinese High-Tech Venture Park, the Software Park and the Bio-Engineering Gene Incubator, all located inside the Shenzhen Special Economic Zone. The SHIP incubators also include an IC Design Park. The park has 80 000 staff employed in some 1500 firms and more than half have university or college degrees. More than 10 000 have masters degrees and more than 1000 have doctoral degrees.

Shenzhen Software Park (SZSP), located inside Shenzhen High-Tech Industrial Park, is the national base for the Torch Plan Software Industry, as designated by the Ministry of Science and Technology (MOST). The Software Park has a total of 225 registered firms and 11 are among the top 100 software firms in China. SZSP is dependent on SHIP for resources and interaction with companies. The software companies employ more than 6000 people and generated software with a value of RMB2.5 billion in 2001. Shenzhen City expects SZSP to become an internationally well-known centre for software development, technology cooperation and enterprises in incubation. A set of policy measures was formulated in 2001, based on 'The Policies for Encouraging the Development of Shenzhen Software Industry based on the National Programme'. The policies include measures on funding, taxation, export promotion and personnel development.

SHIP is one integral part of the Shenzhen High-Tech Industrial Belt (SHIB) that comprises 11 parks including 9 high-tech industrial parks, a university town and one ecological agriculture park. SHIB also includes the Virtual University with some 8000 postgraduates and 6000 undergraduates (2003).

Shenzhen will soon have seven million people and is today the most multicultural city in China, developing at high speed with great flexibility. Forty per cent of the present inhabitants come from Guangdong. The rest come from Hong Kong, Taiwan and other parts of China. Route 128 and Silicon Valley in the US are often referred to as models for Shenzhen, although they were not planned but developed and expanded because of a favourable environment.

New materials and biotechnology are also expected to become prominent from today's very low levels of activity. These national ambitions can clearly be seen in Shenzhen but are more evident in cities like Shanghai and Beijing. Here the central government has more leverage through government research institutes and a large number of state-owned enterprises. Shenzhen is looking into the future and has to consider its development of human resources and R&D. It is not yet very well endowed with institutions of higher learning and government research institutes.

The Shenzhen University has 12 000 students and Shenzhen also has a Normal College. Furthermore, the national government has established a Medical Instrument Research Institute and local government has established some 20 laboratories altogether. In addition Beijing University has established its presence in Shenzhen with the Beijing University Medical Hospital, and Tsinghua University has set up the Shenzhen Institute of Tsinghua University. Other related activities are the Shenzhen Institute of International Technology Innovation and the Shenzhen-Hong Kong Institute of Industry-Education Research – PKU-HKUST.

To improve higher education, Shenzhen relies partly on the Shenzhen Virtual University (SVU), established in 1999, which is a novel approach. A university town was inaugurated in December 2003 with 10 000 students, of which 70 per cent will be full-time while the rest are local part-time students who will study in the evening and at weekends. The university town will in particular draw on resources from Tsinghua University in Beijing, Beijing University, Harbin Engineering University and Nankai University in Tianjin. SVU is located inside the Science Park. Almost all the well-recognized universities in China plus some from Hong Kong are members of the SVU. Beijing University and Tsinghua University initially played an important role together with Harbin University of Engineering, the presence of the last one being explained by the fact that one of the earlier mayors of Shenzhen City came from Harbin.

Operational and technological clusters

It is apparent that the emergence of a knowledge society is creating a very unequal distribution within the geographical space of a country or region. Obviously relatively small territories – cities and their close surroundings – are becoming the key production areas in an emerging knowledge economy. This also means that there is a shift away from the national perspective to the regional in understanding the welfare of a nation. It is possible to identify three types of regional agglomeration. First, in a number of locations there is already a strong industrial base which naturally provides good conditions. This corresponds to what could be termed 'traditional clusters'. A second category is locations where high-tech activities have started to agglomerate and where policies support frontier research. A third category is locations which have a weak industrial and technological base. In some cases the attempt to establish R&D agglomerations runs counter to major trends in the process of ongoing restructuring of R&D and industrial activities.

In China the second category prevails. There is an inherent conflict between regional or localized development on one hand and the rapid concentration of industrial and technological activities on the other. In the second category industrial and technological stimulation in a region will require a multitude of resources and cannot rely only on local companies. The science and technology industrial parks are not evenly distributed in China, and do not represent the actual distribution of population or economic activity. They will no doubt advance technology-led growth that by necessity requires precedence in obtaining scarce resources such as capital, engineers and scientists in research institutions and land which is attractive for foreign MNCs. This development also favours businessmen and professionals over the agricultural population, which still provides the labour for the many process and assembly enterprises that presently play a dominant role in China's modernization.

The manufacturing process of many industrial products, for which electronics provides highly visible examples, has been divided into a number of discrete stages. This has given manufacturers considerable opportunities to choose locations for each stage to meet demand for access to low-cost labour, closeness to markets and supporting industries and easy access to research and development. Such opportunities explain why a number of production stages have migrated to developing countries, including China – in particular for final stage labour-intensive assembly. The manufacture of hard disk drives (HDDs), which are used in all computers, provides a good example. American companies, which played a dominant role in the sector, assembled almost all HDDs in the US in the early 1980s. This share was reduced to 5 per cent by 1995 with South-East Asian countries taking on more than 70 per cent of the assembly operations.

Product development and assembly used to be concentrated in one location in the 1980s. However, functional differentiation has resulted in assembly taking place almost exclusively outside the country of origin and it has become clustered predominantly in Singapore, with additional assembly operations in Malaysia and Thailand and more recently in China. Product development that used to be almost completely maintained at locations where HDDs are manufactured has been modified through mergers and acquisitions, exemplified by Hitachi acquiring the HDD division from IBM.

The disaggregating of functional activities has led to global production networks (GPN) that can freely choose locations for various functions in the production chain. This has generated industrial clusters of different sizes and functional orientation in a substantial number of locations. The GPN for hard disk drives, as for many other products, has bred two different types of spatial concentration: operational clusters and technological clusters.[9]

The OECD provides the following definition of clusters.[10]

> Clusters can be characterized as being networks of production of strongly interdependent firms (including specialized suppliers), knowledge producing agents (universities, research institutes, engineering companies), bridging institutions (brokers, consultants) and customers, linked to each other in a value adding production chain. The cluster approach focuses on the linkages and interdependence between actors in the network of production when producing products and services and creating innovations.

Earlier studies of HDD manufacturing clearly indicated that clusters in South-East Asia came into existence based on

> economies of proximity in input-output relations: speed of throughput, product changeovers, increasingly specialized engineering and assembly labour. Operational clusters may on occasion be sources of new product ideas, but their principal goal is to achieve operational efficiencies, and any new technologies they create are meant to improve production processes of supply chain management.[11]

Technological clusters signify the other end of a technology chain and represent places where innovations take place. This would include the co-location of activities that lead to the recognition of new market opportunities, the development of new technologies and the design of new products. Such clusters change over time as new firms enter into the technological field and new designs offer or demand major changes in global production networks.

Technological clusters, as observed in the manufacture of hard disk drives, are of two different kinds. One involves design coordination between component makers and enterprises that are in charge of final products, while the other involves coordination between designers of critical product components. In the former type there is usually intensive interaction during the design stage for new products and the design and production of new product prototypes. In the latter type complementary technical changes may be needed in the manufacture of several components when new technology is introduced in one particular component.

Naturally, it happens that technological clusters lose their innovativeness, which may reflect major changes in the market or management shortcomings or a combination of both. Similarly, operational clusters continue to undergo changes and may occasionally become redundant. Thailand, still the world's second largest producer (assembler) of hard disk drives, is facing serious competition from operational and possibly technological clusters in China. *The Business Times* of Singapore in June 2004 reported that Thailand will have to strengthen its education and research if it wants to compete with regional rivals, which today include not only Singapore and Malaysia but also China.[12] An advantage of China, it is reported, is that makers of components for hard disk drives are establishing themselves in China, bringing them closer to their customers. All the major makers of HDDs, which include Seagate, Maxtor and Hitachi, have increased their production capacities in China.

THE FUTURE OF REGIONAL INNOVATION SYSTEMS IN CHINA

Industrialization leads to concentration. However, intelligent and learning regions tend to develop outside early industrialization concentrations, which happened both in Europe and the US. The primary industrial concentration in Europe has undergone major changes in recent decades and new technology and knowledge centres have emerged in new locations. Regional concentrations with respect to population, higher education, research-intensive industry, small and medium enterprises and diversified services are important factors in generating regional development. This alters the conditions for traditional industries and also requires new roles to be played by universities and other learning institutions.

Planners and policy makers in all countries are today eager to promote regional development. In both Europe and North America we can observe a development pattern that shows increased concentration of knowledge, capital and people. In recent decades similar changes have taken place in China, although still on a much more modest scale. Figure 8.1 provides a crude presentation of various important actors and factors in China's regional innovation systems.

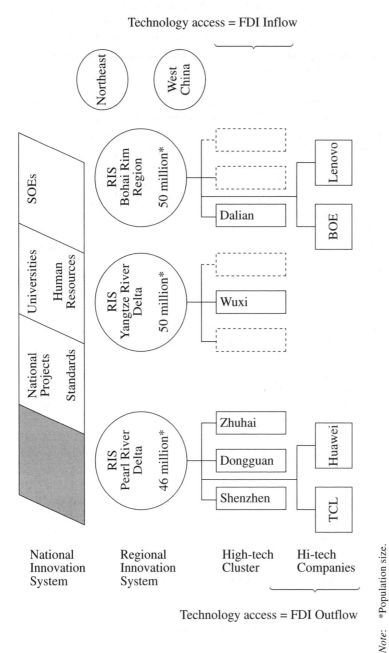

Figure 8.1 Innovation system structure in China, 2004

Note: *Population size.

Looking back into earlier stages of industrial development it is important to understand that industrial plants as such are no longer the focus. It is rather the enabling structures and information flows that determine the efficiency of a system. Thus, knowledge flows and linkages have become key elements in knowledge-based production. The following characteristics define production of advanced knowledge today.

1. Rise in the number of locations and environments where new knowledge can be generated. The traditional universities and colleges are increasingly complemented by centres or research and knowledge institutes, as a growing share of new knowledge is being generated in corporate laboratories. Basically all organizations, both public organizations and private companies, are progressively using highly trained experts and consultants.
2. Locations and environments are connected through well-functioning communication networks, which are electronic and organizational as well as social.
3. Knowledge becomes increasingly differentiated to suit various needs and conditions. Traditional specialities are redefined and combined into new knowledge areas, which indicates a move away from traditional disciplines.

In the light of these changes it is occasionally argued that the modern state today might be vulnerable because it is too small or too big. Certain states are too small to incorporate all the networks and linkages that are required for successful development of domestic industry and research. This profound change reflects the disappearing territorial symmetry between industrial, scientific and political interests. Fully developed democracy requires trust and commonality, which can only exist in tight social networks.

Viewing China's extraordinarily rapid economic development many observers have emphasized that China remains a fragmented federal system, although unified by a single political party.[13] A corollary is that domestic companies are not able to exploit fully the huge Chinese market as they are too reliant on local structures and officials. This drastically reduces their capability to develop unique and proprietary technologies, which has the following two consequences. First, by allowing a major inflow of foreign direct investment, China's high-tech exports are dominated by foreign companies. Second, their Chinese counterparts are highly dependent on designs, crucial components and manufacturing equipment, which are imported in major quantities. Thus it has recently been argued that China 'has joined the global economy on terms that reinforce its dependence on foreign technology and investment and restrict its ability to become an industrial and technological threat to advanced industrial democracies.[14]

China needs to continue its reform of economic and political structures and considerably improve its corporate governance in order to reap fully its fruits of modernization. Today the car industry provides an example of such changes where until recently a foreign company – Volkswagen – was the dominant car manufacturer with some 50 per cent of the market. Until recently the local manufacturers, often in partnerships with foreign makers, were scattered all over the country. Their scale of production was small and they were dependent on foreign technology, a situation that has gradually changed as more and more foreign component makers have been lured to set up shop in China.

Social and Cultural Networks

Let us look at the relation between territory and networks by considering the characteristics of two important types of social and cultural network: physical and institutional. A territory is perceived as an interconnected part of the globe while networks relate to individual geographical positions, and are characterized by their ability to link those points (nodes) into a structure that has economic, social and other functions. Social and cultural networks connect individuals, who are the smallest units in a nodal network. These networks are essential for transmitting new ideas and approaches and exist, in ways that are similar to nervous systems, in all social life.

Physical networks

These include structures such as roads, railways, electricity and telecommunication lines which provide the means for transportation of materials, people and messages. The character of physical networks becomes obvious when comparing networks which have sparse nodes, such as air communications, with road networks where nodes are more amply distributed. Cities are often major nodal points for physical networks.

Institutional networks

These connect units and structures that are involved in all types of societal activities – whether they produce goods, are responsible for administration or provide other services. There is an important distinction between internal and external institutional networks. The former exist only with a preferred organization and facilitate control of internal flows of resources, and can be either hierarchical or flat. The external institutional networks link various environments that exist in different organizations. Given the dominance of networks it is increasingly argued that a networked society is taking shape in our new world.

China's current regional development poses one very important question that can only be tentatively answered. Does the fragmented and federated nature of China's economy delay the restructuring of the corporate sector and make it more dependent on foreign technology?

In 1949 the population in China's cities and towns had only reached 57 million or 10.6 per cent of the country's total population. This is actually three percentage points lower than the world average level of urbanization in 1900. Since 1949 the urbanization process has passed through four stages.

1. The first stage was characterized by the rehabilitation of the economy and the construction of a number of large industrial projects. Some cities expanded during this period, ending in 1957, with an urbanization level of 15.4 per cent.
2. During the second stage, from 1958 to 1966 the economy experienced great fluctuations, which directly affected the level of urbanization. Initially, during the Great Leap Forward there was a move into the cities and the urbanization level reached 19.7 per cent before a reversal took place – to be slightly modified before the Cultural Revolution started.
3. The third stage covered the period from 1966–1978, which almost brought economic development to a standstill. In order to reduce the pressure on cities a large number of young students, cadres and intellectuals were transferred to the countryside and urbanization remained at a low level.
4. The fourth stage, since 1978, has seen a stable and rapid economic development with a gradual transformation of the planned economy system. A number of new cities were established and the urbanization level reached about 40 per cent (see Table 8.1).

Table 8.1   Urbanization level in China, 1949–98

	1949	1957	1965	1978	1984	1990	1994	1998
Urbanization (%)	10.6	15.4	18.0	17.9	23.0	26.4	28.6	30.4

Source: Liu He et al. (2000).

Considering that there is a direct correlation between the level of economic development and urbanization, it has been calculated that the level of urbanization should have reached 40 per cent in 1996. The existing gap is primarily due to the fact that the natural connection between industrialization and urbanization has been severed on many occasions, as indicated above. The highly centralized planned economy relied on administrative directives and basically all economic activities were organized according to strict plans.

This actually enabled industrial development without the services and infrastructure that exists in a market economy and the expansion of service industries was severely contained. It should also be observed that the delayed urbanization made it possible to allocate more investment to industrialization as such.

Furthermore, the residence control system that started in 1958 was able to secure a separation of the cities from the rural areas. Thus, the expansion of urban areas only relied on the natural growth of the urban population. The priority in economic development for a long period of time remained in fostering heavy industry while at the same time ignoring service industries. As a consequence, the efficiency of industrial investment was far below the optimal level due to the lack of services, even though this had also created greater flexibility in industrial production. In sum, urban centres were not able to provide their full contribution to industrial and economic growth.

The existing industrial structure poses two serious problems – both within town and village enterprises and within the large state-owned enterprises. The former have allowed the surplus rural labour force to become employed in industrial development, which also played an important role in increasing peasant incomes. However, this dispersed industrialization, today involving more than 100 million rural people, suffers from a number of shortcomings. One serious problem is the small-scale operations in many types of industries that actually require economies of scale. The other is the paucity of linkages to necessary services that could improve the efficiency of rural enterprises.

An equally serious problem lies in the still existing 'kombinat' character of many state-owned enterprises. In many cases they still maintain service functions and provide infrastructure facilities that could more efficiently be provided by separate entities. Thus corporate governance is a major policy issue. This problem has been partly solved through combining state-owned enterprises into a substantial number of large group companies, while at the same time shedding many of their peripheral activities like printing, transportation and canteen facilities.

Education and Innovation in the Regions

From various Chinese statistics it becomes apparent that a relatively small number of locations control disproportionally large shares of resources which are critical for further economic development. The three cities Beijing, Tianjin and Shanghai, together with the Jiangsu province, provided approximately 15 per cent of the national enrolment into institutions of higher learning in 2002. Their share of the enrolment in specialized secondary schools was equally large, although their population constituted only little more than 5 per cent of the country's population (Table 8.2).

Table 8.2 Enrolment in institutions of higher education and specialized secondary schools by region (2002)

	Institutions of Higher Education			Specialized Secondary Schools		
	Number of Graduates	New Student Enrolment	Student Enrolment	Number of Graduates	New Students	Student Enrolment
National Total	1 337 309	3 204 976	9 033 631	1 441 539	1 553 062	4 563 511
Beijing	67 958	127 597	395 713	31 645	32 893	113 870
Tianjin	27 295	69 453	196 892	26 397	20 743	74 690
Shanghai	55 198	109 184	331 649	29 418	39 331	126 883
Jiangsu	104 079	222 880	700 210	117 444	160 058	451 377

Source: *China Statistical Yearbook 2003.*

Table 8.3 Three types of patent applications examined and granted by region (2002)

	Number of Patents Applications Examined	Inventions	Utility Models	Designs	Number of Patent Applications Granted	Inventions	Utility Models	Designs
National Total	205 544	39 806	92 166	73 572	112 103	5 868	57 092	49 143
Beijing	13 842	5 785	5 920	2 137	6 345	1 061	3 721	1 563
Shanghai	19 970	3 968	4 952	11 050	6 695	341	2 805	3 549
Jiangsu	13 075	1 940	7 002	4 133	7 595	334	4 304	2 957
Zhejiang	17 265	1 843	6 390	9 032	10 479	188	3 860	6 431
Guangdong	34 352	3 819	9 972	20 561	22 761	352	6 395	16 014

Source: China Statistical Yearbook 2003.

A similar pattern appears when examining the geographical location of patent activity. The two cities of Beijing and Shanghai, together with the three provinces of Jiangsu, Zhejiang and Guangdong, are responsible for almost 50 per cent of patent applications as well as granted patents, although populations share is less than 15 per cent (Table 8.3).

URBAN–RURAL DIVIDE AND URBANIZATION IN CHINA

Rural Employment

The development of China's rural areas has a direct bearing on the country's continued economic development and its future as a technology-based nation. However, China's rural employment problem cannot be easily solved and will require completely new policy approaches. This is expressed in the following way by a representative from a leading think tank in Beijing, where Mr Wang Jian says:[15]

> It would be just wishful thinking if anyone were to believe that China can provide a modern off-farm job to every working-age citizen. The solution lies far beyond that. It lies also beyond China's existing small town model, which means just a factory town plus a city government to take care of the industrial infrastructure, and little room for services.

The total number of manufacturing jobs in developed countries is around 90 million. China has been the favourite destination when investors have relocated manufacturing operations to low-cost countries. However, a continued total relocation of manufacturing to China, unrealistic as it is, would be far from providing enough jobs for China's some 300 million underemployed farmers. Wang Jian mentions that the proportion of manufacturing to services is 1 to 1.5 for the overall economy in China, while it is 1.5 to 1 in the small towns. This indicates, he says, that there are serious limitations to such a crude urban development model. Thus, he proclaims that the most effective way to create jobs is to engineer a massive effort to turn rural towns into massive cities, as presently 'a mainland small town has nothing to distinguish itself from the rustic way of life in the old countryside'.

China has allocated large funds in recent years to the building of public infrastructure such as modern highways, S&T parks, industrial parks and a number of other initiatives, although there has been little discussion regarding the purpose of this large expenditure. Critical voices now raise concern and argue that city building should have a much broader significance than providing infrastructure. This leads to the following conclusion which would have far-reaching consequences for China's future development.[16]

In addition to infrastructure development, city building requires knocking down the rigid household registration system, which results in impossible hassle for rural citizens to move to the cities; and also requires providing a welfare shelter to them – otherwise they wouldn't be willing to surrender their old plots in the countryside.

Official statistics indicate that there are now 130 million migrant workers in Chinese cities. This is almost equivalent to the total combined population of Germany and France. It also means that China has more migrant than urban workers, and that they today constitute the main Chinese industrial workforce.[17] China's population, presently reported to be around 1300 million, includes 900 million rural dwellers, with 500 million being the rural labour force. About 100 million engage in agricultural production, and tens of millions work in township enterprises. However, this leaves a surplus in the region of 300–400 million. Thus the current figure of 130 million migrant workers in urban areas constitutes only the tip of an iceberg that is China's employment challenge within the coming decades.

Migrant workers earn an average RMB660 per month, according to a survey conducted by the labour and social security department, based on data from 2600 enterprises in 26 Chinese cities, including Beijing, Tianjin and Shenzhen. This is lower by about RMB300 than that earned by urban industrial workers.[18] In another survey, conducted by a China Urban Labour Employment and Labour Flow research team, 50 per cent of the floating rural population wanted to stay in the city, and less than 10 per cent expressed a desire to return home.[19]

However, migrant workers contribute greatly to the rural economy. For example the seven million migrant workers from Anhui province contributed an amount equal to the annual provincial GDP. They remitted RMB30 billion annually, an amount higher than the entire provincial revenue. Such funds are used to improve living standards of family members, build houses and roads, establish schools and generally advance rural economic development. Returning migrant workers often start their own businesses, bringing with them new concepts that advance local economic development.[20]

Migrant labour forces have contributed about 1.5 percentage points to the 9 per cent annual average growth rate of the country's economy in the past two decades, and will continue to be an important factor in pushing forward economic growth in the years to come.[21]

Ongoing Urbanization

China's urbanization rate is expected to rise from the current 39 per cent to 75 per cent within the next 50 years. During this period, the population in cities and towns could reach between 1.1 billion and 1.2 billion – making

China into an urbanized nation.[22] It is estimated that around 55 per cent of the country will be urbanized by 2020. The large-scale migration of farmers into non-agricultural areas in the process will gradually change the old pattern of the divided urban–rural dual economic structure.

A team producing an urban development report for the Chinese Academy of Social Sciences (CASS), has drawn up a framework of three central metropolitan areas, seven economic belts and a number of central cities with common features to cope with China's rapid urbanization. The metropolitan areas are the groups of cities around BoHai Bay, the Yangtze River Delta and the Pearl River Delta. They will serve as economic centres boosting the development of their surrounding areas. The areas will contain more than half of China's population while accounting for 80 per cent of the national economy and 90 per cent of China's industry output value. About 95 per cent of the nation's trade volume will be produced there.

Table 8.4   Statistics for China's five city regions (2003)

	Gross production RMB billion	Local GDP RMB billion	Gross Production Increase (2003) %
Beijing	361.19	About 350	10.5
Tianjin	238.69	About 230	14.5
Shanghai	625.08	>600	11.8
Chongqing	225.01	About 230	11.4
Guangzhou	346.66	About 350	15.0

Note: Findings published by Niu Wenyuan, head of 100-plus experts who prepared an in-depth urban development report for the Chinese Academy of Social Sciences (CASS).

Source: 'Shanghai ranks 1st among metropolitan economic giants', *China Daily*, 4 March 2004, http://www.chinadaily.com.cn/english/doc/2004-03/04/content_311723.htm.

The Chinese mainland has 668 cities and more than 20 000 towns,[23] where about 40 per cent of the nation's 1.3 billion people live. After analysing the comprehensive competitiveness of 200 Chinese sample cities at or above the prefecture level, the CASS team found that China's top 10 most competitive cities were Shanghai, Beijing, Shenzhen, Guangzhou, Suzhou, Hangzhou, Tianjin, Ningbo, Nanjing and Wenzhou. Statistics for the five city regions can be found in Table 8.4. Guangzhou, Tianjin, Nanjing, Hangzhou, Wuhan, Qingdao, and Chengdu are among China's biggest cities.[24] In terms of learning capability, Beijing beats Shanghai because of its group of renowned universities, research institutions, think-tanks and international research and development centres.[25]

Statistics indicate that the Yangtze River Delta (YRD) region – accounting for less than 2 per cent of China's total land area and about 6 per cent of its population – contributed nearly 20 per cent of the country's gross domestic product (GDP) in 2003. Meanwhile, the area's actual overseas investment increased to nearly 50 per cent in 2003, reaching some US$25.5 billion, and its export volume now accounts for over 30 per cent of China's total. In terms of private economy, Suzhou of Jiangsu Province recorded a leading 18 per cent GDP growth in 2003, and even the slowest-growing area – Shanghai – saw an 11.8 per cent growth in its GDP in 2003. It is estimated that the annual per capita GDP in the region could reach US$15 000 by 2020, compared with the present US$2500.[26]

Urban residents have witnessed dramatic improvements in their living standard over the past two decades. By 2002, on average, every 100 urban families owned 126 colour TV sets, 93 washing machines and 87 refrigerators.[27] A majority of China's population – rural residents – are seeing widening gaps between themselves and urban residents in terms of income and consumption growth. China's 500 million urban residents account for only about 40 per cent of the nation's population of 1.3 billion, while their consumption demand accounts for 65 per cent.[28]

It has been argued:[29]

> It will be ideal if this rate maintains growth of 1.5 to 2 percentage points annually for the next few years and reaches 10 per cent by 2010. With the urbanization rate going up 1 percentage point, more than 15 million rural residents could every year move into cities and towns. The consumption created in this process would amount to tens of millions Yuan and boost the GDP growth by 1 to 2 percentage points.

THE COMING OF RURAL URBANIZATION

There are several likely effects of China's extremely rapid development of coastal areas, which are the most advanced regions in the country. First, China will have a number of mega-cities located in its coastal areas where vigorous industrialization started and they will not only remain centres of continued development but also constitute captivating magnets for people and activities from inland. Second, partly as a consequence, people who are now referred to as floaters will become residents in coastal areas and will be joined by many more from inland. Third, also as a consequence, a substantial shift of China's population will take place away from inland areas. Thus, the emerging industrial and economic structure in the inland provinces is more difficult to predict. However, it is not unlikely that the economic and industrial structures of the three major coastal regions – Pearl River Delta, Yangtze River Delta, and BoHai Rim – will eventually evolve along different lines, linked strongly to the global economy. There is apparent competition among them. The question is whether this will fuel efficiency and faster development.

The Minister of MOST has argued strongly that domestic firms should step up their innovations to earn the advantages of intellectual property, while integrating themselves with advanced technologies and overseas capital. He has also called on domestic high-tech firms to intensify the integration of financing and science and technology, and to improve the environment of investment and financing. 'For this purpose we should set up multiple capital markets, encourage the development of venture investment, and make it a means to drive the industrialization of high technology,' Xu has said.[30] China has set up high-tech parks in Britain and the United States with the purpose of playing an active role in giving Chinese enterprises experience of developed countries to promote their high-tech products in world markets.[31]

The initial success of regional innovation systems along the coastal areas of China is based on having created favourable conditions for a large number of clusters to come into being. Many of them, possibly most, are clusters that require functional proximity, for which FDI has offered great possibilities by linking local clusters into Global Production Networks. This special characteristic poses immense challenges in bringing a similar approach to the western parts of China and making it successful there. This further challenges politicians and policy makers in the already successful coastal areas to provide the necessary conditions for transforming operational clusters into technological clusters.

Given the diversity and size of China it is not surprising that a regional development policy framework has evolved rapidly. The planners have seen the creation of development zones and parks as one of their major objectives. The result has been that there existed in 2004 6741 development zones of various kinds in the country.[32] The number of diverse programmes, each one usually having a distinct focus, may occasionally have squandered scarce resources in the huge number of industrial development zones, high-tech development zones, science parks and incubator structures to serve industry and universities.[33] They are supported and financed both by the central government and by the localities, with varying concepts and objectives for developing regional economies. This has provided an infrastructure to innovate, discover and manage knowledge and contributes to its transfer where needed. Knowledge creation and its efficient use is a central issue with a focus on the acquisition of knowledge, innovation and management of brainpower.

The ability to accept and employ new ideas is a determining factor in success and competition for resources will under many circumstances be contributing to improving efficiency. The role of universities and other institutions of higher or specialized training have taken on an increasingly important role. When embedded in local partnership they will serve industrial development and further technological upgrading.

The explosive growth of development zones and parks has also created a number of serious problems. One is the indiscriminate use of land, which has recently led to strong countermeasures. Thus, the central government in April 2004 announced serious control measures on land to be used for development zones and parks. The Ministry of Land Resources reported that the government had suspended the establishment of new economic development zones in 2003 and invalidated 3763 of the total 6015 economic development zones which had already been approved – a remarkable progress made in the country's land market regulation.[34]

The character of China's three major regions will naturally influence the speed and direction of rural urbanization. There are great differences between the three major regions and also a number of important and striking similarities. Shanghai is dominated by state-owned companies and government agencies and has in recent years seen a great influx of companies from Taiwan.

The private sector completely dominates in Shenzhen, a fact that owes very much to its very close relations with neighbouring Hong Kong. IT dominates in Shenzhen's high technology with 91 per cent, followed by new materials with 4.2 per cent and biotechnology with a paltry 1.4 per cent. Attention is given to optical, space, ocean and new energy technologies. Although industry plays a dominant role in Shenzhen it constitutes only 39 per cent of the economy with only 1 per cent for agriculture and 50 per cent in services, where financial services play a very important role. While undergoing structural changes Shenzhen will continuously update its IT industry, support for development of new materials and biotechnology. Special attention will be given to a future semiconductor park, thus challenging the dominance of Shanghai in this field.

The BoHai Rim region, which covers parts of Hebei Shandong and Liaoning provinces, has a good physical infrastructure which includes good transportation and telecommunications and easy access to ports and airports. Tianjin and other cities in the region provide an almost unlimited supply of low-cost labour and are well endowed with universities, colleges, and research institutes.

Successful and innovative firms are seldom alone, and networking and clustering has become more important in recent years as the character of competition in market-based innovation systems has undergone fundamental changes. The approach to clusters has changed from direct intervention to indirect stimulus, for which supporting infrastructure remains important, and many clusters are now increasingly market-led. An early narrow sectoral approach has been replaced by support for knowledge flows within wide production networks.

The problem of lagging urbanization has become very serious in China, where industrialization didn't start until the era of the planned economy, when the central government was using most of the nation's resources to build a substantial heavy industry sector. This led to an artificial, almost regimented divide between the cities and the country that lasted for most of a 30-year period.

Wang Jian[35] is among the strong advocates that argue that the time has come to embark on a large-scale city-building programme, as China has already become a factory for the world. He has pointed out:[36] 'when China's manufacturing power has been able to match a middle-income nation (with per capita GDP at some US$3000), its consumer structure is still like a poor nation (with more than 60 per cent of the population living in the countryside and per capita GDP hovering at the US$500 level)'. He suggests that a nation with per capita GDP of US$3000 would be one where 60 to 70 per cent of the population is living in the cities, and where the service industry would account for around half of the total economy. There is a special reason why cities make up such a small proportion of the population in China. Primarily it is a heritage from a dual economy – with a distinct split between a small industrial sector in a few modern cities and a vast underdeveloped and even antiquated agricultural system.

Tremendous changes have taken place since the start of the economic reforms, when thousands and thousands of farmers launched their privately-owned workshops. Clusters of small industrial towns emerged in what used to be paddy fields, especially in the Yangtze and Pearl River deltas. However, Wang Jian argues that these changes have not yet been profound enough. He says that unless they manage to continue to grow, small towns will not be able to compete with the major cities in terms of economic sophistication and growth potential. Only by becoming major urban centres will the towns provide advanced services and create an environment that will promote the development of technological clusters that has already taken place in many locations in the until now more favoured coastal areas.

The rural towns need to grow beyond their present rustic character, and he notes that since the mid-1990s there has been a marked decline in the growth of small towns while the number of new off-farm jobs has been limited. Wang Jian says that both small industries and farming are becoming more capital-intensive – with their growth depending more on capital input than labour. From 1988 to 2000, the labour input required for the production of each unit of farm products – from rice to fruit – was reduced by about one-third. 'Even small town industry has become more capital-intensive, as reflected by the annualized growth in their investment in fixed assets (such as machinery) per worker. In the last few years of the 1990s, small town industry was even more aggressive than large city industry in spending on fixed assets.'[37]

City mayors in China's Central and Western provinces argue that their provinces should attract more foreign investment in manufacturing as they have improved their infrastructure, especially given rising production costs in coastal regions.[38] However, this argument neglects a number of important factors that may have changed the conditions for further expansion of manufacturing in China.

First, the relocation of manufacturing to China and other low-cost countries since the 1980s occurred when foreign manufacturers gave up their production base – a process which is now coming to an end. Thus, any shift in manufacturing investment from coastal regions to the hinterland would primarily reflect an expansion in production capacity. Unless production costs are rising considerably in the Pearl River Delta and other coastal regions, the producers would have little reason to move their manufacturing operations inland. Second, China's Central and Western regions have limited advantages in labour and production costs, as skilled workers have moved from Western and Central regions over the past 20 years, mainly to Guangdong and Jiangsu provinces. The result is that the migration of rural labour from interior regions has reduced salary levels in coastal regions and contributed to reducing labour costs for the manufacturers in these locations.

The present imbalance might actually widen the gap between affluent coastal regions and poor provinces in Central and Western China. Professor Wei Jie of Tsinghua University in Beijing argues:[39] 'A more realistic option is to abolish all the obstacles that stand in the way of talented staff and labour migration while banning all discrimination against migrant labourers so people can benefit from the advances in coastal regions'.

REGIONALIZING CHINA'S INNOVATION SYSTEM

In many locations policy makers have created economic space for new markets which is shaping the behaviour of entrepreneurs and new companies. Innovations are heavily influenced by institutional path dependency, to varying degrees relying on the early industrial base which until recent years was dominated by SOEs. These enterprises, along with other important actors, are socially and politically embedded. Thus, policymakers, businesses, universities and research institutes have different roles in different parts of China. Subsequently this requires an understanding of how their roles facilitate or hinder innovations.

Huge amounts of innovations – of a gradual and incremental nature – are taking place in manufacturing firms all over China, although primarily in the dynamically evolving coastal areas. These firms have often agglomerated into geographical clusters and are found in a number of industrial sectors. In many locations clusters are evolving into centres of strong innovative capability. They are still weakly linked and inadequately supported by actors within the central government.

China can no longer be considered a centrally planned economy being ruled from Beijing and there has emerged a strong need to delineate the responsibilities of individual ministries and agencies at both national and sub-national levels. Many cities in China's coastal areas, like Hangzhou, Suzhou and Wuxi, are all pursuing their own export-oriented development strategies, taking advantage of the external economies of the region as a whole. At the same time, each city also devises its own development plans and pursues its own individual strategies, competing for more MNCs and more DFI. Thus, forces of competition and complementarity are working hand in hand.

Until the late 1970s the innovation system in China was a basically self-contained structure which had self-sufficiency as its leading principle. All structures existed with a political technological space and all major decisions on R&D strategy and its implementation were taken within a domain in which state agencies exercised a political monopoly. This system has undergone and continues to undergo fundamental changes in two parallel ways. First, the innovation system controlled at the national level is now joined by three semi-independent systems which include regional innovation systems, the universities and corporate innovation systems with their laboratories. These structures are growing in relative importance, although the national innovation system is also expanding. Second, the power and control of resources have shifted towards an innovation system where the private technological space is becoming more important. These changes continue and will eventually result in a Chinese innovation system that is dominated by the regions and by the corporate sector which is likely to have a significant share of foreign ownership (see Figure 8.2).

Government agencies at the local level have been able to play an important role in attracting FDI with a focus on high technology and/or R&D by establishing science parks and offering specific incentives and facilitating availability and mobility of highly qualified personnel. It appears that Chinese policy makers have acquired a deepening understanding of how to promote interaction between FDI and domestic companies, and between the state sector and the private sector. Important instruments are the development of science parks, public R&D laboratories, incubators and related categories of infrastructure.

In many locations it has become possible to discern close interaction between universities, R&D units and high technology FDI – often within

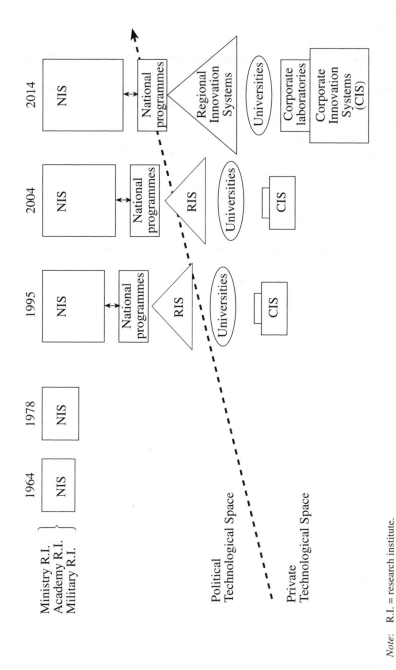

Note: R.I. = research institute.

Figure 8.2 NIS evolution in China

confined geographical domains where investment in human resources plays an important role. The industrial dynamics of structural change has been well understood by many municipal and provincial planners who in a timely fashion have expanded college education with considerable leeway to private initiatives in higher education, although supervised by the national Ministry of Higher Education.

Development zones within regions have been a very important planning instrument since the mid-1980s. They exist in different forms and are initiated and controlled at different government levels. A major objective has been to attract national, local and FDI enterprises to cities or specific areas to stimulate industrial and technological development. Land is generally made available at preferential rates and buildings and facilities are usually erected at high speed. A number of preferential conditions are generally offered such as low taxes and easy access to loans at low rates. The development zones in China have a number of similarities with the favoured concept of industrial estates that was prevalent in the 1970s and early 1980s. However, in China they have evolved in a dynamic planning instrument in which a number of economic factors are brought together. The names and orientation of development zones reflect different levels of ambition such as 1) new high technology development zones of which there exist only 53; 2) industrial development zones at national, provincial and city level; and 3) general development zones which exist at lower levels and for specific purposes.

Major cities have often combined a new high technology development zone with an industrial development zone and occasionally promoted co-location with a major university which would often have its own university science park with an attached incubator system.

NOTES

1. Pollard (1981).
2. Tamada (2000), pp. 2–17.
3. Shanghai Science and Technology Committee (1999).
4. Xiao Cao (2003).
5. Rowen (2004).
6. Ibid.
7. Vogel (1990).
8. Reflection: one day when passing from Hong Kong into China at Lowu there were only four passengers in the international group and only some tens among the rest – against an hourly passage of 4000 at Huang Kang which is now a major border crossing between Hong Kong and the Mainland.
9. McKendrick (1998).
10. Roelandt and den Hertog (1998).
11. Ibid.
12. 'Thai disk drive goal needs support', *The Business Times*, 16 June 2004.
13. See for example Gilboy (2004).

14. Ibid.
15. Zhang, Ed (2003).
16. Ibid.
17. 'Migrant workers: urban underclass', *China Today*, 14 March 2004, http://www.chinadaily.com.cn/english/doc/2004-04/14/content_323279.htm.
18. Ibid.
19. Legal experts have pointed out that the present system makes it difficult for migrant workers to enjoy their civil rights in cities, as urban social organizational systems are, at present, not open to them and that they are in many places constrained by discriminatory local laws and regulations.
20. 'Migrant workers: urban under class', *China Today*, 14 March 2004.
21. Comments by Wang Mengkui, Director of the State Council Development Research Centre, at the Fourth Meeting of the China-UK Forum, as reported by Wu Yixue (2003).
22. Findings published by Niu Wenyuan, head of 100-plus experts who prepared an in-depth urban development report for the Chinese Academy of Social Sciences (CASS), Source: 'Shanghai ranks 1st among metropolitan economic giants', *China Daily*, 4 March 2004, http://www.chinadaily.com.cn/english/doc/2004-03/04/content_311723.htm.
23. China has five large municipalities directly under the Central Government. Provinces and autonomous regions are divided into prefectures, counties and cities. Counties are divided into townships and towns.
24. Fu Jin (2004).
25. Ibid.
26. Liang Yu (2004b).
27. Qi Jingmei (2004).
28. Ibid.
29. Ibid.
30. Xiao Cao (2003).
31. Cui Ning (2003).
32. Qin Chuan (2004).
33. Soufun.com Academy, the research department of China's largest property portal, will soon release its recent research on the competitiveness of China's development zones. The research was conducted in early 2004 by Soufun.com Academy, the Real Estate Institute of Tsinghua University and the Institute of Enterprises under the Development Research Centre of the State Council (DRCSC). The study covered 53 state-level high-tech development zones and 54 economic and technological development zones. The report was to be released during the China Development Zone Competitiveness Forum, planned for June 10–12 in Beijing's Diaoyutai Guest House Hotel. Source: 'Development zone research to be released', *China Business Weekly*, 24 May 2004.
34. '3,763 economic development zones cancelled', *Xinhua*, 10 April 2004.
35. Wang Jian is director the China Macro-Economics Web, an online service for research institutions that distributes scholarly reports for circulation among central government agencies.
36. Zhang, Ed (2003).
37. Ibid.
38. Jia Hepeng (2003).
39. Ibid.

9. Shanghai: from development to knowledge city[1]

CREATING A KNOWLEDGE METROPOLIS

The visitor to Shanghai immediately becomes aware of an expansive city skyline of skyscrapers and the expressways, and will soon become aware of the fact that Shanghai has built the first maglev train to exist anywhere in the world. However, Shanghai is transforming itself in many ways that are less noticeable.

Shanghai's economic development is based on the twin pillars of knowledge creation and knowledge application. The latter has been strongly supported by the city's attraction to foreign investment in a wide range of industrial activities. The former is coming to the forefront by a rapid expansion of higher education and scientific research, where foreign investors have also become active by setting up research laboratories.

The city enrols more than 50 per cent of its senior high school students into colleges and universities. Three major universities among a total of some 60 colleges and universities are set to become recognized research universities by rapidly expanding their graduate training programmes and attracting research funds.

Zhangjiang, the large high-technology park in Pudong across the river from old Shanghai, is known to almost everyone. In addition Shanghai has a handful of expanding technology parks, aside from science parks attached to major universities. Each of the city's 11 districts has its own industrial park. Shanghai has 28 incubators, with at least one in each district, housing a total of some 20 000 companies.

These changes in Shanghai's technological and scientific landscape are not just top-down, driven from the local government supported by the central government in Beijing. Forces of change also originate from major foreign companies present in Shanghai with regional headquarters, production plants and more recently also with their R&D units. Shanghai has already registered more than 140 R&D units established by foreign companies. Changes are also strongly driven by entrepreneurs, scientists and researchers in companies and universities who want to see their Shanghai emerge as a world metropolis for knowledge creation and knowledge application within the next couple of decades.

The role played by R&D is becoming important in Shanghai. Total investment in R&D reached RMB16.3 billion in 2003, which corresponds to 2.1 per cent of Shanghai's GDP. In absolute terms this is twice as much as in 2000. The estimate is that R&D as share of GDP would reach 2.3 per cent in 2004 and the objective is that it should reach 2.5 per cent in 2005. Officials of the Development and Reform Commission indicate that investment in R&D would stabilize at this level as the economy in the region might in itself continue to increase annually up to 10–15 per cent.

Shanghai has 59 colleges and universities altogether, of which ten are included in the national programme of 100 universities that will be supported to become high-ranking institutions of higher learning. Three among them – Fudan University, Dongji University and Shanghai Jiaotong University – have been selected to become world-recognized research universities. These have also established large science parks combined with incubators. Incubators have also been established throughout all Shanghai and each of the city's 11 districts has at least one, usually with a focus on a distinct industrial or technological field. The first industrial park, Zhangjiang Hi-Tech Park, was established in 1992 when Shanghai was given special economic zone status, and has been followed by a wide array of district-based industrial parks and a number of specialized zones.

These far-reaching developments in creating and applying knowledge have endowed Shanghai with resources and capabilities that in many ways are comparable to those in highly developed regions in Europe or in the US. Furthermore, a similar development has taken place in many cities in the neighbouring provinces of Jiangsu and Zhejiang, with the expected result that Shanghai will emerge as a leading metropolis in an economically booming Yangtze River Delta region, and will challenge the dominance of Beijing in many areas.

SHANGHAI UNIVERSITIES

Twenty-nine of Shanghai's institutions of higher learning offer four-year training programmes, while many of the others want to upgrade their curricula from three to four years. Twenty-three of the institutions are available for foreign students. Shanghai also has 27 independently organized junior colleges for adult education.

The total number of students in institutions of higher education is close to 600 000 which corresponds to 3.5 per cent of the population in Shanghai. There are 16 privately-run colleges and universities. The number of university and college students in these institutions has reached 40 000 students, which accounts for almost 7 per cent of total students in higher education institutions. In 2003 the gross enrolment ratio was 53 per cent, which means that more than one half of all students leaving senior high schools continued their studies in colleges or universities. A few years ago the enrolment rate was slightly above 40 per cent; Shanghai calculates that enrolment of high school students to the college and universities will increase to 60 per cent by 2010 and that the city will have a total enrolment of 900 000 students in higher education. Graduate programmes are rapidly expanded and major universities expect an annual increase of some 15 per cent, which includes an ongoing tendency to train more PhD students.

Another source says that in 2003 one out of every 10 university students in Shanghai was pursuing studies at private universities.[2] A municipal education department official said that Shanghai now has over 1800 private and non-government funded educational institutions, including 40 private institutes and several hundred colleges providing higher education without diplomas. Most private universities want to recruit higher management staff and faculty from the prestigious local public universities to make their universities more attractive to students. One example is the Shanghai Fuqiu Vocational Studies College, which was approved in mid-2003 and is located in the Pudong Lujiazui financial trade zone. Its President Hu Aiben was formerly the assistant to the president of Fudan University. Many teachers at Fuqiu come from Shanghai Jiaotong University, Shanghai University of Finance and Economics and Shanghai University. It is planned that students will get professional training diplomas after three years' study, on the assumption that they pass examinations organized by Fudan and Tongji universities. Fuqiu College has applied for certification to offer four-year degrees.

Entrepreneurs investing in education today are doing it as a business enterprise, rather than as a donation. Says Xu Xincai of Fuqiu College:[3] 'The education industry has a bright future, although it takes a longer period of time to receive a return on the investment'. The Law on Private Education Promotion, passed at the end of 2002, stipulated that 'private school investors can get a reasonable repayment after deducting schooling costs and reserving development funds and other expenses'.

Talent recruitment is very important for the university system and Shanghai is scouting all over the country as well as overseas. This is facilitated by additional funds coming from society to fund additional costs. Rebuilding campuses is also an important task for which funds are flowing not only from the government but also from society. A recent educational policy has included the acceptance of private funds, often from overseas Chinese, to attract highly qualified staff to key positions – with the Zhangjiang Honour Professorship as an important scheme.

Zhangjiang professors are seen as important to link key personalities with key subject areas in a university. These include lectureships as well as programme leaders, with a dominance of the latter category. The contract period is often five years, and many have remained within the programme since it started in the mid-1990s – earlier than the 985 programme (see below). There are some 40–50 Zhangjiang professors in Shanghai, recruited both domestically and overseas. There are a number of challenges for Zhangjiang professors as they have to adjust to a new environment where cultural clashes may occur. They may be successful, though perhaps not leaders in their scientific fields. An example of the Zhangjiang Professor Programme is given below for the School of Microelectronics at Fudan University.

The central government has also carried out the 985 project since 1998; this covers 30 universities and aims to make them fundamentally very strong. Three universities in Shanghai are included in this programme – Fudan University, Tongji University and Shanghai Jiaotong University.

Fudan University

Fudan University is composed of 17 schools offering a complete range of majors and disciplines, with undergraduate, masters and doctoral programmes. The university has five national key laboratories, 57 institutes and 80 research centres. Eight of the undergraduate programmes have been designated as national centres for basic scientific research and teacher training.[4] Total enrolment at Fudan University is close to 40 000 students. The 2100 faculty members of Fudan University include more than 1300 professors and associate professors, of which 24 have been admitted as members of the Chinese Academy of Sciences or the Chinese Academy of Engineering. The aim is to build Fudan into a high-level research-oriented university with a comprehensive range of academic disciplines, to make it one of the influential universities in the world.

Fudan University is today a well-recognized university that started in 1905 on a very modest scale as Fudan College. In 1917 Fudan College started its undergraduate programme and was renamed Fudan University, and the number of students increased gradually. Subsequently it founded art, science, business and preparatory departments as well as a high school. The number of students reached almost 500 in 1921, which was also the year when the Institute of Psychology was founded. It was at the time a private university. In 1929, Fudan University carried out a reorganization and established departments of journalism, law and education. Under the educational reforms of 1952 certain areas of training were abolished while art and science departments that originally belonged to another ten colleges were merged into Fudan. After the reform of 1978 Fudan developed into an international comprehensive university covering humanities, social sciences, natural sciences, engineering sciences and management sciences. A further consolidation took

Table 9.1 Fudan University enrolment

	Undergraduate students	Masters and PhD students
2000	12 559	5 115
2001	13 241	8 193
2002	14 194	9 400
2003	15 170	10 148
2004	14 998	11 029

Source: Foreign Affairs Office, Fudan University.

place in April 2000 when it merged with the Shanghai Medical University. Fudan University is now under the direct jurisdiction of the Ministry of Education. Recent student enrolment figures can be found in Table 9.1.

The Fudan University is jointly funded by the Ministry of Education and Shanghai Municipal Government. The total income in 2003 was RMB1581 million, an increase of 30 per cent over the preceding year. Income sources were:

- Ministry of Education 27 per cent
- Shanghai Municipal Government 6 per cent
- Educational undertakings 27 per cent
- Research funds 19 per cent
- Donations 6 per cent
- Return on industrial spin-off 1 per cent
- Interest on investment 2 per cent
- Other sources 12 per cent.

Faculty and staff salaries used 26 per cent of funding with another 15 per cent for their welfare, administrative costs required 27 per cent and 12 per cent was used for equipment.

Aside from its own research institutes Fudan has four sets of key laboratories which are directly supported by the national government.[5] Major key laboratories, supported by national agencies, are listed below.

1. State Key Laboratories
    - Genetic Engineering (SKLGE)
    - Applied Surface Physics
    - Application-Specific Integrated Circuits (ASIC) and System ICs
    - Medical Neurobiology
    - Advanced Photonic Materials and Devices

2. Ministry of Education Key Laboratories
   - Applied Ion Beam Physics Laboratory
   - Biodiversity Science and Ecological Engineering
   - Non-Linear Mathematic Model and Methods
   - Molecular Engineering of Polymers
   - Medical Molecular Virology
   - Molecular Medicine
   - Studies of Carcinogenesis and Invasiveness
   - Wave Scattering and Remote Sensing Information
3. Ministry of Health Key Laboratories
   - Glycoconjugate Research
   - Antibiotics and Clinical Pharmacology
   - Functional Reconstruction of Hands
   - Medical Audiology
   - Viral Myocardial Diseases
   - Study of Myopic Eyes
4. Ministry of Education – Key Research Bases of the Humanities and Social Sciences
   - Centre for Ancient Chinese Studies
   - Centre for Chinese Historical Geography Studies
   - Centre for Contemporary Western Marxist Studies
   - Centre for Chinese Socialist Market Studies
   - Centre for American Studies
   - Research Centre for Information and Mass Communication
   - Institute of World Economy.

Department of Microelectronics

The School of Microelectronics at Fudan University is a big department with a focus on advanced IC technology. The department has around 600 graduate students and good contacts with semiconductor foundries in the Shanghai region, such as Grace and SMIC in the Zhangjiang Semiconductor Industry Park. The ability to do excellent research is exemplified by the success of an integrated circuit,[6] known as Zhongshi No.1, for which the School was the independent designer. This is an illustration of China's ambitions to capture leading-edge technologies with significant commercial potential.[7] Analysts say that mass production of the chip will be an enhancement to China's digital TV industry; it is reported to be adaptable to international as well as domestic standards and cost-effective compared with similar international products. The chip embodies the core technology for the new generation high definition television (HDTV) that has been the focus of research and development, not only in China but worldwide since the early 1990s. The domestic importance lies in the fact that, being the world's most populous nation,

China has more than 370 million TV sets and an average 40 million sets are being sold each year. Furthermore, China plans to broadcast the 2008 Beijing Olympics with digital TV and to popularize digital TV nationwide by 2015.[8]

The attraction of Fudan University and this Department in particular was confirmed by Professor Dian Zhou of the University of Texas at Dallas (UTD) who in late 2003 was appointed Dean of the School of Microelectronics at Fudan University. He was offered a Zhangjiang Scholar Award to Dallas, where he received his physics BSc and MSc degrees in 1982 and 1985 respectively. Including his doctoral studies he spent 19 years in the US and is on leave of absence from UTD to develop and expand microelectronics at Fudan University. Zhou was appointed by Fudan University in agreement with the Ministry of Education and the Zhangjian Honour Programme, which has made such appointments in several other Chinese universities.

The award rules require the holder to have a doctoral degree and an outstanding record in research, and be internationally recognized and able to develop a first-class research programme. The award is sponsored by the Ministry of Education with additional funds from an industrialist in Hong Kong, Mr Jiacheng Li, and other generous sponsors. It is an ambitious programme to develop China's research universities rapidly. Fudan is one of them and has been given the task of developing advanced microelectronics research capabilities.

In October 2004 the School of Microelectronics received a special grant of RMB100 million to establish a 'Micro-Nano Electronic Platform'. This is part of ambitious government objectives and is related to the National Key Laboratory on Application-Specific Integrated Circuits (ASIC) already in operation at Fudan University. This laboratory has been in operation for more than ten years and has been responsible for the development of 32 bit CPUs. The equipment and resources of the Platform will be open to researchers from all over China who will be accepted on the basis of highly recognized proposals and will bring their own financial resources. The platform activities will have five focal areas which include telecommunications and HDTV, System-on-Chip (SOC) high-level clean rooms, and electron-beam technology. Furthermore, substantial financial support has also come from commercial companies, including Applied Materials and Novelius Systems (Shanghai).

Professor Zhou perceives the development of the semiconductor industry in China to be in a very exciting stage. Although IC design capability is still low, Chinese engineers and developers are rapidly catching up, and there are several favourable conditions enabling China to move forward. First, the timing is ripe as advanced foundries are moving to China. Second, the earlier IT bubble in the US and elsewhere has led to a recession within the electronics industry. As a result, many Chinese engineers and researchers who have been studying and working in the US, maybe for some 20 years, are not only willing but also eager to return to China. This cohort of willing returnees includes highly qualified engineers who can seize exciting and challenging tasks in China. Professor Zhou sees the development of the HDTV chip at Fudan as a good illustration.

However, Professor Zhou emphasizes that the returnees are also facing serious challenges and hardships. First, the scientific and technological infrastructure still remains weak. Although China offers ample talented manpower (researchers), the infrastructure for advanced research is still weak and is poised at a level somewhere between a developing country and an advanced industrialized nation. Second, the management culture in China not only creates difficulties, but is often the root of serious cultural clashes and conflicts. However, there are many opportunities for carrying out exciting projects in China by introducing returnees in teams that get generous funding from national and local sources. Zhou has already recruited six faculty members from overseas – four from the US, including one former Motorola employee, and two from Europe. If everything goes well Zhou expects to have 20 returnees out of a total faculty of some 60 staff members.

Professor Zhou stresses that the project that he is engaged in, as well as many others, is part of an ongoing experiment that is, so far, only taking place at a small number of the leading universities in China. He anticipates that culture and management are going to change rapidly under the influence of the returnees.

Fudan Graduate School[9]
The Graduate School of Fudan University is the successor of the graduate schools of the former Fudan University and the former Shanghai Medical University, both of which were among the first 22 universities authorized to establish graduate schools after the graduate education system in China was formally organized according to a nation-wide plan in 1984. Graduate education was resumed in 1978 and has expanded rapidly. Between 1978 and 2001, the University admitted 24 852 graduate students, among whom 19 854 were Masters students and 5898, doctoral students. The university had awarded 2991 PhD and MD degrees and 12 313 masters degrees by the middle of 2001.

A multidisciplinary structure for cultivating graduate students has been formed, and it includes the disciplines of humanities, social sciences, sciences, engineering and medicine. Doctors and Masters degrees can be granted in 103 and 148 secondary disciplines respectively. Five types of professional degrees – Masters of Business Administration, Juris Masters, Masters of Public Administration, Masters of Engineering and Masters and Doctors of Medicine can be granted. There are 22 postdoctoral stations, 29 state key subjects and five key state laboratories which support graduate training.

Fudan University Science Park
The university science park[10] concept was initiated in 2001 when 22 university science parks were approved jointly by the Ministry of Education and the Ministry of Science and Technology. Another 14 university parks were approved in 2003, and there are now 36 such parks altogether in China. There are also science parks in another four universities – Shanghai University, Donghua University, Shanghai Jiaotong University and Tongji University. The main purpose of the university science parks is to support technology transfer and commercialize technologies, in a close relationship with the universities. The Fudan University Science Park (FUSP) resembles Oxford Science Park in the UK, where private investment is also encouraged.

The Park is built in the surroundings of Fudan University and is located in the midst of one third of Shanghai's universities and colleges, and surrounded by more than 100 research institutes. Its six main investors include Fudan University, a Shanghai investment company and the Technical Innovation Centre. None of the investors have a controlling power. The Park at an early stage attracted companies such as Fudan Microelectronics, Fudan Zhangjiang, Fudan Kingstar, Fudan Tianchen, Fudan Biomedicine, and Fudan Water Engineering.

The university science park concept differs in several important ways from the high-technology parks that constitute an integral element of the Torch Programme. First, they attract smaller companies which are utilizing or developing novel technologies, often in close collaboration with university departments. Second, university parks only have a limited number of foreign-invested companies and are less oriented towards export markets. However, there is frequent interaction between the research universities and their science parks on the one hand and with the Zhangjiang High-Tech Park in Shanghai on the other. Fudan university has established two related research institutions in Shanghai Zhangjiang Hi-Tech Park located in the Pudong area across the river:

1. Microelectronics Research Institute.
2. National Microanalysis Research Centre.

Yangpu district, where the Fudan Science Park is located, is a heavily developed part of Shanghai City. Consequently, the Park is not geographically concentrated but has its activities in different places around Fudan University. The present structure includes

- an on-campus industrial research institute
- Handan Road Incubator
- Siping Road Headquarters

- Fenglin Road Campus, which also contains Shanghai Medical University
- Shanghai Qingpu Industrial Base
- Jiangsu Kunshan Industrial Base, actually located in a neighbouring province, to provide industrialization of products that have been developed within the Park.

Most university science parks in China are managed by the universities themselves by providing capital and manpower. Fudan Science Park is different as the University does not have the same controlling power as in other places. Fudan Science Park has from the very beginning been market-oriented with the objective of making profits and is paying annual dividends on investment. As an organization, it resembles a property development company that is looking for tenants that meet certain criteria. FUSP[11] pays an annual dividend of 10 per cent on invested capital of RMB100 million. Shanghai Fudan Science Park Co., Ltd is a sponsorship-type equity limited company that was founded in September 2000 by joint investment from:

1. Fudan University.
2. Shanghai Wujiachang Hi-tech Combined Development Company Ltd.
3. Shanghai Shangke Technical Investment Ltd.
4. Shanghai Lujiazui Finance and Trade Zone Development Ltd.
5. Shanghai Municipal Technical Venture Centre.
6. Shanghai Yangpu Construction (Group) Ltd.

The development of Fudan Science Park has passed through several stages. In Phase I property development was carried out by a local construction company that transferred the ownership to Fudan Science Park in 2002 for RMB85 million. In Phase II the Park bought land for expansion from the government at a cost of RMB300 million. During this phase a 24 storey building was begun, to be completed before September 2005 when Fudan University officially celebrates its 100 Year Anniversary. Expansion is primarily financed by bank loans, for which the Science Park pays interest in the region of 5–6 per cent. When Phase III and Phase IV are completed the Fudan Science Park will cover a total area of 100 000 square metres.

In early 2005 FUSP housed approximately 300 companies inside the Park, although only half of them have a direct relationship to Fudan University. A new high-rise building in Phase II can potentially house another 200 companies, although newcomers may be limited to about 20 enterprises in order to provide expansion possibilities for existing companies. Total employment of companies located within the Fudan Science Park is in the region of 3000–5000.

The Fudan Science Park, operating as a profit-making company, has greatly benefited from the booming property market. Since its inception property prices have risen from about RMB5000 per square metre to more than 10 000, and it is expected that land prices will continue to rise. The company is responsible for management, providing services to tenants and making investments. These are focused on microelectronics, biomedicine, new materials, information engineering, education business and environmental engineering. The Park has set up venture capital funds in collaboration with foreign companies such as the SK Group in Korea.

A high-tech company has several options for its location in Shanghai. Zhangjiang Hi-Tech Park is an obvious choice but the threshold in terms of size and required investment is quite high. Furthermore all 11 of Shanghai's districts have also established their own industrial parks. Like Fudan, another four major universities in Shanghai have also established science parks. It is relatively easy to be accepted in Fudan Science Park, without any requirement of being associated with the university as such, although it is emphasized that new companies should bring original technology. Furthermore, Fudan Science Park has also established its own Software Park in the vicinity – Fudan Fuhua (Fudan Forward) – which has already been listed on the stock market.

INDUSTRY AND TECHNOLOGY IN SHANGHAI[12]

Shanghai has experienced a double-digit economic growth since 1992 and its GDP in 2003 reached RMB625 billion, with a per capita income that exceeded US$5000. The economic structure is almost equally divided between industry and services, with a miniscule 1.5 per cent in agriculture. Shanghai covers an area of 6340 square kilometres and by the end of 2003 had a resident population of 13.4 million and a total population of 17.1 million.

The national government in China approved the first four special economic zones (SEZ) in 1979. They included Shenzhen and Zhuhai, located just across the borders of Hong Kong and Macau respectively, which were to attract foreign investors with advantageous conditions such as low-cost land, good infrastructure and favourable tax rates. The chosen places were at the time only small rural sites and have since then become booming industrial centres. This centrally-directed initiative was followed in 1984 by opening 14 larger and older cities along the coast to foreign investment, with basically the same conditions that applied for the first batch of four SEZs. In 1990 Shanghai, already having the administrative status of a province, became the next site with SEZ privileges, although with more flexible regulations than the earlier ones. Two years later the central government decided to provide similar status for another 23 cities further inland. Shanghai approved more than 4000 FDI projects in 2004, with a total investment value of US$11.6 billion, an increase of 12.65 per cent over the preceding years, with companies in Hong Kong, UK and Japan being responsible for more than one half. By the end of 2004 Shanghai had accumulated some 36 000 FDI units with approved contract value of US$86 billion.

The history of Shanghai helps to clarify the conditions that enabled the city to take a leading role in China's economic and industrial development in recent years. Its previous development as China's foremost transportation centre benefited from its dual function both as a sea port and a river port. Shanghai was in the past the first port that was opened to trade with the West and for a long time it dominated China's commerce. After being defeated by Great Britain in the Opium War in 1842, China was forced to open Shanghai to European trade and concessions, which gave a foreign imprint on its continued expansion. After 1949, when the communist government came to power in China, Shanghai became a major industrial centre to support national development plants. The city also became one of China's leading centres for scientific research and higher education with numerous colleges and universities. Technical personnel now constitute one in twenty of the population. There are today 800 institutions for scientific research and technological development, with 152 persons who are members of the Chinese Academy of Sciences or the Chinese Academy of Engineering Sciences, and three members belonging to both academies.

The technology structure and industrial production is constantly changing towards information technologies, biotechnology and new materials. Statistics for 2003 indicate a total output value for high technology and new technology industry of RMB298 billion (a growth of 43.5 per cent compared with 2002). Their share of total industrial input was 26.5 per cent. The IT industry has increased annually during the past six years by more than 20 per cent and has become the number one pillar industry of Shanghai. The sales value of IT production had reached RMB334 billion by the end of November 2004, which is an increase over the same period of the previous year of 53.7 per cent.

The information service industry, which includes the software industry, registered sales of RMB 59 billion, which is an increase of 31.6 per cent over the same period of the previous year. Estimates indicate that for the whole of 2004 the IT industry constituted 11 per cent of the GDP in Shanghai. Foreign direct investment has played a major role in this development and makes up approximately one third of total investment in the sector. A substantial number of the world's leading IT companies are currently established in Shanghai with some 40 having set up R&D centres and ten having their regional headquarters in Shanghai.

Shanghai already has seven 8 inch wafer production lines, with manufacturing technology that corresponds to international levels. Integrated circuit design and innovation ability is rapidly improving and Chinese language digital signal processors are already produced, as are chip sets for the new TD-SCDMA mobile system. This development is supported by Shanghai IC R&D Centre and by Shanghai Silicon Intellectual Property Rights Transaction Centre, as well as other organizations, to provide necessary support for technological development. In concrete terms Shanghai has given priority to the development of the chip design industry followed by technological advances in the chip manufacturing industry. The immediate goal is that IC industry output in 2005 should reach a value in the region of RMB35–40 billion, which would correspond to 3–4 per cent of the world market. Based on strong IPR and rapid technological development Shanghai should have become one of the major IC industry bases in the world by 2010.

The biomedicine industry is one of the priority areas for developing high-technology and new industries in Shanghai. There are some 70 academicians within the domain of life sciences, biotechnology and medicine, who are the key leaders for 863 projects. Shanghai is already using national-level R&D centres as the core for developing modern life sciences and medicine technology. The sector reached an estimated production value of RMB23 billion in 2004, which amounts to 20 per cent of the national total. The expected value for 2007 is RMB35 billion, which would correspond to 30 per cent of the national total. By 2010 Shanghai expects that a comprehensive and innovative cluster of biomedical industry will have an influence on the rest of the world.

Software Industry in Shanghai

Software development has become exceptionally strongly supported by Shanghai authorities, with the Ministry of Information Industry (MII) playing an important role in meeting demand from companies in the US, Japan and Europe. An Annual Software Summit is held in Shanghai, as the city has several advantages in software development, including human resources, technology and support from the national government.

The Shanghai software industry has experienced rapid growth as its industrial scale expands and technology improves, and Shanghai has become one of the country's leading exporters of software. During the past three years Shanghai has become a centre for R&D, production, services and export of software. By 2005 Shanghai expects its software industry to have sales of RMB35 billion, having established a number of backbone enterprises, and with IPR controlled by Chinese companies – employing altogether between 80 000 and 100 000 people.

Shanghai Software Development Park,[13] established in 1999, is a core element of Shanghai's software industry. It consists at present of three subsidiary parks, Pudong Software Park in Zhangjiang Hi-tech Park, which started operations in May 1998, Fudan Software Park and Shanghai Jiaotong University (SJTU) Caohejing Software Park.

The Shanghai Pudong Software Park (SPSP) company has a complex ownership structure with joint investment from the China Electronic Corporation and Zhangjiang Hi-tech Park Development Corporation, representing the Ministry of Information Technology – the predecessor of MII – and Shanghai Municipality respectively. SPSP is being developed in three phases; the first development phase was completed in 2000 and all premises are fully occupied by companies. A second development phase was completed in August 2002. The mission of SPSP is twofold. First, it is expected to establish a favourable environment to attract good software companies, ranging from research, development, production, sales and services, to foster a group of world-class IT software enterprises with the ability to capture IPRs. Second, it aims to establish itself as a hub for global and local software companies, becoming a centre for software product and technology development on the Western Pacific Coast.

The development of software has very different requirements compared with the expansion of the semiconductor sector. The latter needs much capital, for which FDI is required in order to secure the technological base, while software development depends on human resources, of which China has an ample supply. The major factors enhancing software development are creative human resources, new products and markets. Production of software in Shanghai reached RMB10 billion in 2002, of which RMB3 billion was exported – mainly as outsourcing for US and Japan companies. New software game development is creating close relations with counterparts in Korea, while the domestic market is growing in importance.[14]

The future development of the software industry will require closer relations between enterprises and research institutes and further development of human resources. Important future areas for software development include industries such as biotechnology, automobiles, semiconductors and aviation. China is suffering from a number of bottlenecks as competencies among software developers are still comparatively low. Furthermore, there is a lack of team leaders, project managers and company managers. Shanghai planners suggest that the local software industry must emphasize flexibility and productivity, which have a bearing on company optimal size. China's software industry is lacking in scale and finance as large loans are not available to small software companies, and hardly any companies so far have the desire to go global.[15]

An important dimension of Shanghai Science and Technology Commission policy is attracting talents from other countries. The system has certain similarities with the Zhangjiang Honour Professor Programme for which the Ministry of Education in Beijing has the responsibility; this is discussed in the next section. In the early 1990s Shanghai City formulated three programmes to attract talents to come to Shanghai through providing such incentives as accommodation, education for children and higher salaries. There are three such programmes:

1. *Star* aims to bring new talents to universities and research institutes in Shanghai. Thirty-six such positions are awarded annually.
2. *Leader* also aims to bring people to companies with approximately 50 positions annually.
3. *Returnees* also supports both companies and the academic sector with more than 30 positions every year.

HIGH TECHNOLOGY PARKS

Zhangjiang Hi-Tech Park is considered to be the most successful among the 53 national high-technology parks arising from the Torch Programme initiative. Preliminary statistics show that total revenues for 2004 reached RMB200 billion, with an exported value of US$10.3 billion. Shanghai has another four high-technology parks and altogether 28 incubators which are located throughout the city. A large number of the city's most important enterprises are located inside the Zhangjiang Hi-Tech Park, with many of them engaged in advanced knowledge creation.

The Science and Technology Commission prepares annual plans which are directly related to the Five Year Plan for Science and Technology, allowing adjustments for changing conditions and environment. Four regional park centres will play an important role in meeting the objectives of these plans.

A substantial number of industrial and high-technology parks, often with attached incubator systems, play a very important role in Shanghai's industrialization efforts. Although the number of parks is overwhelming, and their structure and governance occasionally confusing, the following examples attempt to provide a certain amount of clarification. Zhangjiang Park is located in the Pudong area, which is also the site for two large free trade and export processing zones. The other three technology zones mentioned below – Caohejing, Songjiang and Zizhu – are located in the old parts of Shanghai.[16]

- Caohejing National Industrial Park (Caohejing Xinxin Jishu Kaifa Qu) was established in the early 1980s, with a focus on information and industry, and is the oldest high-technology park in Shanghai. It is a technology development zone of importance for the nation as a whole, and the State Council has authorized the formation of bases for microelectronics, photoelectronics, computers, software and new materials to support new major industries. By combining domestic advances with international exchanges the Park will develop into a High-Science and Technology Park along the Pujiang River. The total number of companies is more than 1700, which includes 450 foreign firms. Annual revenue has reached RMB40 billion.

- Zizhu Science-Based Industrial Park (Minhang district) was established in 2001 and will combine three forces in its development – private companies, institutions of higher learning and government agencies. It will rely on market forces and use educational and research resources to make Zizhu into a modern science park. With six leading sectors including microelectronics technology, optoelectronics, digital technology, software technology, nano-technology and life sciences, Zizhu expects to attract high-tech enterprise headquarters and research and development centres, as well as venture capital companies.
- Songjiang Science and Technology Park (Songjiang district) was established in July 2002. With this most recent technology park, Shanghai is striving to set up a third microelectronics industrial base after the Zhangjiang integrated-circuit (IC) industrial base and the Caohejing IC industrial base. This endeavour has been strongly supported by the Shanghai Semiconductor Industry Association, which estimates that IC businesses in Shanghai will earn more than half of the total revenue for the sector in China. At present a number of foreign companies which include TCMC (Grace) from Taiwan as well as domestic companies have decided to make a total investment of US$1.15 billion, for which construction has already started.

The Zhangjiang Hi-Tech Park

Zhangjiang Hi-Tech Park (SZHTP) was established in 1992, and has become well-known to foreign visitors. SZHTP aimed from the very beginning to learn from the Hsinchu Industrial Park in Taiwan. SZHTP in Shanghai and Zhongguancun High Technology Park in Beijing are generally seen as the two most successful examples of this kind of institutional structure in China. In August 1999 Shanghai City government made a strategic decision to make Zhangjiang Hi-Tech Park into a focal centre for biological medicine, integrated circuits and software – as three leading industries. Within the ongoing tenth Five Year Plan the area must undertake the task of becoming the home for high technology and new technology industry and is expected by 2010 to have become a world-class high-technology park.

SZHTP is different from many of the other 53 hi-tech parks in China as the content and value of science is much higher here and the Park has attracted more foreign investment in R&D facilities than other locations. Furthermore, it has attracted Fudan University in Shanghai and many other advanced teaching and research institutions to set up branch campuses, and there are now altogether more than 20 educational units.[17]

Park officials argue that there is considerable interaction within the software and microelectronics fields and that SZHTP will by 2010 be of world class and compare favourably with Hsinchu and other similar sites around the world. The ambition has been to establish 'innovation chains' in the three fields of the semiconductor industry, the software industry and the biotechnology and pharmaceutical industry, and also provide support for small and medium-sized enterprises through its incubator systems. In addition SZHTP also provides support for the bankcard and optical electronics industries.

A number of companies, both domestic and foreign, have been attracted to set up R&D centres within the Park. The largest single R&D unit within the SZHTP is operated by the Zhongxing Technologies (ZTE) Corporation with a total number of staff exceeding 3000. Other important national centres include:

1. Shanghai Super Computer Centre.
2. The National Light Source Project in Shanghai. This is a third-generation synchronous radiation device at the energy level of several GeV with the objective of becoming a multidisciplinary research centre to support the industrial development of advanced materials and provide a basis for fundamental research in a wide array of knowledge fields.
3. Shanghai High Polymer Material R&D Centre. This is an open and new research institute involved in R&D and its commercialization in the area of high polymer materials.

Employment within the Park had reached more than 40 000 by 2004, of which 10 per cent have Masters degrees and another 3 per cent have doctoral degrees. Returnees have come to play an important role in further developments within the Park. According to incomplete statistics, there are about 3500 returned overseas Chinese students having started businesses or working in the Park – with companies established by returned overseas Chinese students reaching close to 500.[18] Thus, Zhangjiang Hi-Tech Park has become an attractive choice for returnees.

The Park has its technological focal points in three industries – semiconductors, software, and biotechnology and pharmaceuticals. This is reflected in the physical layout of the Park and its further expansion and naturally also reflected in the choice of companies invited or desiring to settle. The early choice of focusing on three sectors has also directly influenced the location of university branch campuses and the decision to locate key technology centres or Key Research Facilities in the Park. We provide a short description below of the characteristics of the three focal areas and for each an example of a major high-technology company that is active in the Park.

The semiconductor industry sector and the SMIC foundry
Semiconductor Manufacturing International Corporation (SMIC) employs some 7700 people, of whom 86 per cent work in China. One hundred and forty have PhD degrees and another 12 per cent have Masters degrees with 31 per cent with bachelor degrees. Some 2700 are contract workers. SMIC does not have an expatriate system but supports education and provides accommodation. This has facilitated moving some 1000 staff to China.

SMIC has a training programme in collaboration with three universities in Shanghai – Fudan University, Tongji University and Shanghai Jiaotong University. SMIC (in Shanghai only) has 600 R&D staff members in projects that were approved by the Ministry of Commerce in June 2004. SMIC has a number of IC foundry clusters in Shanghai, Beijing and Guangdong, and its foundries can produce chips with line width down to 0.13 micron. In 2005 SMIC starts a mask making collaboration with Toppan in Japan, while chip packing is done in Chengdu.

SMIC is a unique implant in China, being completely owned by foreign investors, with Richard Chang playing a very important pioneer role. Huawei, ZTE and many other companies in China received substantial planning support. SMIC people spent a lot of time in discussions with officials, before finally agreeing to establish its major semiconductor foundry operation in China. The company wanted to move a substantial number of overseas employees without giving them expatriate status. One challenge was that SMIC wanted to have its own school system and officials queried this. However, eventually an agreement was reached on this and a number of other issues. After much persuasion the Shanghai government provided substantial support and the SMIC Government Relations Team has turned out to be very important.

SMIC provides a one-stop solution for its customers from design to final testing and its business model is similar to that of TSMC in Taiwan. SMIC plants located in Shanghai, Beijing and Tianjin are pure-play foundries that offer 0.35–0.13 micron technologies for logic, mixed signal/RF, high-voltage circuits, memory, system-on-chip and LCOS (liquid crystal on silicon). Marketing and services are located in Milan, Tokyo, Fremont and Dallas. The main facilities in Shanghai also include mask making and testing. The newly created SMIC structure will have several foundries all based on 12 inch wafers in the future. During 2005 SMIC will enter into a joint venture with

Toppan to manufacture on-chip colour filters and micro lenses, and another joint venture will be established in Chengdu for packaging and testing.

SMIC delivers various IC components for digital still cameras, HDTV, DVD players, smart phones and PDAs. It has a strong R&D team with more than 700 specialists, and has established close partnerships with Toshiba, Fujitsu, Chartered, IMEC, Infineon, Elpida and Toppan. SMIC is delivering more and more high-end ICs. The share of ICs with a line-width of 0.18 micron increased from 22 per cent of total sales in 2003 to 46 per cent during the third quarter of 2004. Logic wafer production constitutes the majority of SMIC services. Partners include Elpida for DRAM down to 0.10 micron, and with Infineon down to 0.11 micron. SMIC has taken over the Motorola IC production plant in Tianjin after discussions that were completed by 1 January 2004 after lasting two years.

SMIC identified a number of cost advantages to operating in China, which include the following. First, cost advantages include water recycling and natural cooling. Second, lower utility costs with water fees being one third of those in Taiwan. Third, SMIC has access to a large talent pool and lower labour costs by having access to engineering graduates from leading universities in China.

Shanghai is a major location for semiconductor foundries. China's IC manufacturing capacity is growing and there are presently 55 units in production. Thirty-eight units are working with wafer sizes of 3–5 inches and with line widths of less than one micron. China has nine foundries at present using 8 inch wafers, with line widths in the region 0.35–0.11 micron, of which four are operated by SMIC, which has also built a 12 inch wafer unit in Beijing.

The ambition of Shanghai is that SZHTP Park should have 20 production lines for wafer production and 20 enterprises in photo mask production, packaging and testing that would give an expected output value of US$10 billion by 2010.[19] By March 2003 the Park's industrial structure included 96 IC enterprise – 3 wafer fabs, 44 fabless units, 16 photo mask, packaging and testing enterprises, 10 R&D educational institutions and 23 vendors.[20]

Software industry and Zhongxing Technologies R&D Centre
The Shanghai R&D Centre of Zhongxing Technologies (ZTE) is located in the Innovation Zone of the Shanghai Zhangjiang Hi-tech Park. It is the largest R&D centre operated by a hi-tech enterprise in Shanghai, with more than 3000 research staff. Its work is focused on R&D for 2G, 2.5G and 3G mobile systems, and mobile handset and network products. ZTE operates its own 'university', located in Dameisha, Shenzhen, to provide training for international and domestic customers on technology, products and management.

ZTE, with headquarters in Shenzhen, was established in 1985 and listed on the Shenzhen Stock Exchange in 1997. At the beginning ZTE was only a simple trading company and entered the telecom sector in the early 1990s when domestic demand emerged and the company found that it had matching engineering competence in the field. ZTE has a total of 13 R&D centres of which eight are located in China. Shenzhen R&D centre is mainly involved in the development of CDMA systems, while also doing IC design, optical communications, teleconferencing systems and power supply systems for telecom equipment. ZTE has three R&D centres in the US, located in San Diego for CDMA2000, in Dallas for optical transmission and in New Jersey for soft switches. The company also has a centre in Korea where development of handsets is carried out, and another small centre in Sweden. ZTE invests more than 10 per cent of its annual revenue in R&D.

ZTE has worldwide employment of 19 500. This includes more than 500 PhDs and some 6000 Masters degree holders. The share of bachelor degree holders is around 70 per cent. Employees engaged in R&D consist of 42 per cent of the total ZTE workforce, with 19 per cent in production and 32 per cent in marketing.

The company has entered into agreements with a number of foreign companies. One important relationship is with Motorola for joint development of ICs for speech digital signal processors. ZTE also maintains close relations with Intel and TI for other semiconductor categories. Another collaboration is with Agere in the development of 3G platforms. ZTE has also established working relations with some 50 academic institutions, almost all of them in China. One important aspect of this network is to be able to identify and recruit talented engineers and scientists. Another very important element is joint research on frontline technologies such as 3G technology, optical communications, soft switches and data communication.

ZTE is a global network solution provider and China's largest listed telecom equipment manufacturer. The company has grown very rapidly in recent years. In 2003, sales revenue reached US$3 billion, an increase of some 50 per cent over 2002. During the first half year of 2004, ZTE doubled its revenue over the same period in 2003. By the end of 2004 ZTE had deployed 35 million lines of GSM systems in 20 countries. It is China's largest CDMA equipment supplier in international markets, with 18 million lines of CDMA system adopted by over 30 countries. Contracted sales for 2003 in overseas markets reached US$610 million, which is a year-on-year increase of 100 per cent. ZTE recorded a very large expansion of its overseas sales during the first half of 2004. In 2004 it raised capital on the Hong Kong Stock Exchange for its future expansion.

Pudong National Software Park is located within the Shanghai Zhangjiang Hi-Tech Park and has attracted a substantial number of companies, domestically and from abroad. By the end of March 2003, more than 1000 companies[21] had been registered, of which 228 had entered the Park by mid-2004.[22] Their total employment reached 7000 and covers a number of fields such as system integration, e-commerce, chip design and information security. This group of enterprises include Microsoft.net, Citibank, Synopsis, Sony, BearingPoint, Kyocera, UnionPay, Kingdee, Tata, Infosys and Satyam.

The software industry in the Park also includes the China Eastern 863 HiTech Information Security Base, with the following components:

- The National Information Security Engineering Centre
- The National Public Security Technology PKI Research Centre
- The National Computer Virus Technology Research Centre
- The Information Security Engineering Institute of Jiaotong University.

A closely related activity is the Shanghai Bankcard Industry Park, with the stated objective of becoming the design and R&D centre for financial information products, and a data processing centre for financial organizations. The designated area comprises three square kilometres where the following companies will be present: China UnionPay, China Bank of Communication, Ping An Insurance Company and Bank of China.

Biotechnology and Pharmaceutical Industry, and ChemExplorer working for Lilly

Shanghai ChemExplorer Co. Ltd is one of the pharmaceutical companies within the Park structure that is promoted by the Zhangjiang Biotech and Pharmaceutical Base Development Co. Shanghai ChemExplorer started on a very small scale with only five people in July 2002. The company, doing contract research on organic compounds for Lilly in the US, was founded by Mr Hui who used to work for the Institute of Organic Chemistry in Shanghai, which is under the leadership of the Chinese Academy of Sciences. After leaving the Institute Mr Hui served as Shanghai Vice-Minister for Science and Technology before he founded ChemExplorer. The company can only carry out R&D activities which are directly related to the interests of Lilly. Mr Michael is managing director.

ChemExplorer has concentrated its research on three areas, of which one is identification of active substances in the screening of natural samples. Every two weeks meetings are held with Lilly counterparts. Lilly has a number of research sites in China – in Hangzhou, Qingdao, Shenzhen, Dalian and Shanghai. After its early start ChemExplorer has expanded rapidly and presently consists of some 200 people of whom 15 per cent have Masters degrees and another 15 per cent have Doctorates. The company expects to recruit 20–30 more staff members during 2005. Talented people are sought from other places and the company has been on recruitment missions to Hangzhou, Nanjing, Dalian and Tianjin, an activity which is also boosting the image of the company. The company will increasingly look for qualified people.

Location in the Park has a number of advantages. First, an environment for pharmaceutical research has been created within the Park, to which the presence of the Shanghai College of Traditional Chinese Medicine (TCM) has contributed. Second local government support has also been important.

The aim of Zhangjiang Hi-Tech Park (SZHTP) in the biotechnology and pharmaceutical industry is to create a cluster of activities that will provide the platform for industrial development in biotechnology and traditional Chinese medicine as well as new drugs and medical devices. To support its objective a specialized development company was created in 1996 – Zhangjiang Biotech and Pharmaceutical Base Development Co., Ltd – established jointly by the Ministry of Science and Technology, the Ministry of Health, the Chinese Academy of Sciences, the State Food and Drug Administration and the Shanghai Municipal Government.

SZHTP aims to attract recognized pharmaceutical companies from overseas and from China, and to expand the R&D structure and improve its contents by attracting talents. More than 120 small and medium-sized companies have already been established in the Park, which also hosts the Shuguang Hospital and the Shanghai University of Traditional Chinese Medicine. Research activities include the following:

- The Shanghai Institute of Materia Medica (CAS)
- The National Human Genome Centre at Shanghai National Centre of Drug Screening
- The National Centre for New Drug Safety Evaluation and Research
- The National Centre for TCM Innovation
- The National BioChip Engineering Research Centre.

Foreign pharmaceutical companies with a presence in the Park include Roche, Amersham, GlaxoSmithKline, Boehringer Ingelheim, Kirin, Sankyo, Tsumura and Medtronic. The domestic ones include Pioneer, Greenvalley, Celsar, Sanjiu, Tasly and Taiji. A number of them, including Roche and Dupont, have a substantial R&D presence in the Park. The US firm DuPont started to build its China Corporate Research and Development Centre in the Zhangjiang High-Tech Park in late 2003 with a planned investment of RMB124 million. The centre will eventually accommodate up to 200 scientists and focus on technical marketing, with the aim of localizing the production of existing products in the Asia-Pacific market. The R&D centre will open in early 2005 and will become DuPont's third comprehensive R&D facility outside the United States, the other two being in Europe and Japan.[23]

INDUSTRIAL PARKS IN SHANGHAI – EXAMPLES

The high-technology zones, as exemplified by Zhangjiang Hi-Tech Park, the incubator system and the university science parks, are primarily involved in knowledge creation. In addition Shanghai has established a number of industrial parks which can be illustrated by the examples below. Thus, Shanghai has taken on the responsibility of using certain planning mechanisms that were prevalent in the formerly nationally planned economy. The most important remaining planning instrument for long and medium-tem planning is the government's control of land, which has been the basis for creating industrial, technological and scientific parks in various parts of the City to meet specific expectations. This instrument can also be used to select industries, for which water resources and other environmental considerations also have to be taken into account. A more general influence, although very significant, is the expansion of the educational systems and its orientation – to support specific industries and technologies. An important dimension for Shanghai and many other old cities in their development of industries is to integrate new technology into traditional industries.

Shanghai Chemical Industrial Park (SCIP)

This ranks as the largest industrial project in Mainland China.[24] It is located in Jinshan District in the southern part of Shanghai Municipality and major multinational chemical companies such as Bayer, BASF and British Petroleum have decided to locate there. This industrial park is the first industrial zone specializing in the development of petrochemical and fine chemistry businesses, and will be one of the four key industrial bases of Shanghai according to the master development plan for the city. It will provide investors in the Park with a good investment environment through public utilities, logistics, environmental protection and administration services. SCIP aims to be one of the largest and the most integrated and advanced world petrochemical bases in Asia.

Shanghai Comprehensive Industrial Development Zone (SCIDZ)

This is one of Shanghai's nine municipal-level industrial zones. The major industries encouraged in SCIDZ include information technology, new materials, high-tech industries, green products, real estate and other tertiary industries.

Fengpu Industrial Park[25]

This was approved as a Municipal Industrial Park by the Shanghai People's Government in 1995. In 2000, the Shanghai People's Government Industrial Investment Group took control of it, and decided that it should be Shanghai's only comprehensive industrial zone, and one of the city's three main industrial parks. In 2003, the State Council approved the establishment of the Shanghai Fengpu Export Processing Zone located within the Fengpu Industrial Park on an area that measures three square kilometres.

Songjiang Industrial Zone (SJIZ)

This was the first municipal-level industrial zone approved by the local government, in July 1992. It is located southwest of Shanghai. By June 2003, more than 400 foreign-funded projects had been approved, with a total investment of over US$5 billion.

Songjiang Export Processing Zone (SJEPZ)[26]

This was established by the State Council in April 2000. Key industries encouraged within the zone include companies that are active in information technology, bio-pharmaceuticals, new building materials, fine chemicals and light machinery. Investors in the zone can enjoy special tax incentives. Like other zones of this kind investors can finish all export procedures within the zone, as services and institutions include customs, goods inspection, commercial transactions, taxation, banking and foreign trade.

In addition Shanghai has a number of specialized zones in the Pudong area. This includes Waigaoqiao Free Trade Zone (WFTZ), a location for Intel and where IBM in 2002 established a large plant for sealing and packing integrated circuits, with easy access to the global market from the free trade zone. In between WFTZ and Zhangjiang High-Technology Park lies Jinqiao Export Processing Zone (JEPZ), which started operation in 2002 and is one of the largest export processing zones in China. Shanghai expects it to become an integral part of a 'chip belt' in the Pudong area.

Closer to the city centre, also in Pudong, lies Lujiazui Finance and Trade Zone (LFTZ), now seen as an important part of Shanghai's central business district (CBD), with the expectation of becoming a global business district. According to plans LFTZ will eventually offer 4 million square metres of office area, including both the 337 000 square metre World Finance Centre to be completed in 2007, and the 220 000 square metre Sun Hung Kai Project to be completed by 2008.[27]

Incubators in Shanghai[28]

Shanghai established city-wide incubators in the early 2000s and each of the city's 11 districts has at least one incubator, usually with a distinct industrial or technological focus. Altogether Shanghai had created 28 incubators by early 2005, which cover a total area of 650 square kilometres, with some 1500 companies; up from under 400 companies in 1999. Total employment is around 20 000 and revenue reached RMB6 billion in 2004. Two hundred companies have (successfully) left the incubators. An enterprise within the Shanghai incubator system would typically have around 15 employees and an annual revenue of around RMB4 million. Thus the incubators in Shanghai serve as intermediaries between universities or other sources of new technologies and industrially more advanced enterprises. The latter would normally be found within one of the industrial parks for which Shanghai offers a wide variety as illustrated above.

The Shanghai City government promotes incubator development in various ways which, aside from policy directives, includes investment promotion through proper introduction, and fund allocation to individual companies. The Science and Technology Commission has a total budget of RMB1.4 billion, part of which can be used to support the incubator system in the city. The Commission hosts annual meetings with more than 1000 participants to collect ideas, identifying promising projects and companies worthy of specific support. Based on such meetings the commission makes a public release to invite applications.

The first incubator in Shanghai was established in April 1988 – Shanghai Technology Innovation Centre (STIC) – and later on authorized by MOST to be a State-Level High and New Technology Innovation Service Centre. STIC accommodated 52 companies with more than 1000 employees in 2003. In early 2004 there were altogether 28 incubators in Shanghai under various administrations. Five of them had been set up by state-level hi-tech parks and another seven by universities. One of the incubators is operated by Shanghai Science and Technology Commission while four have been created through various investment schemes.

The incubator system in Shanghai has developed through four stages. First, hi-tech parks were the natural location of incubators as exemplified by Caohejing Technology Innovation Centre. Second, incubators were established near recognized universities, and one such example is Yangpu Technology Innovation Centre near Fudan University. After that Shanghai created district-based incubators with a specific industrial focus, such as the Hongkou Technology Innovation Centre in Hongkou District. Finally, following industrial advances in Shanghai, specialized incubators were created in different technology fields, as exemplified by the Integrated Circuit Design Incubator.

A Shanghai Hi-tech Business Incubator Network includes an intermediate agency that provides various services, and also performs a catalytic role to get access to seed funds and venture capital, for example, from Special Incubation Funds, the Shanghai S&T Innovation Fund for SMEs, the State Innovation Fund for SMEs and loan guarantees for incubator companies, as well as loans from commercial banks.

To support internationalization, in 1997 Shanghai, with approval from MOST, established the Shanghai International Business Incubator (IBI), with headquarters at STIC. IBI supports internationalization by drawing on the resources and experience of six incubator bases:

1. Caohejing Technology Innovation Centre
2. Shanghai Technology Innovation Centre (STIC)
3. Zhangjiang Technology Innovation Centre
4. Withub Hi-tech Business Incubator
5. Yangpu Technology Innovation Centre
6. Shanghai University Science and Technology Park.

Another initiative is to make the various incubators more specialized to promote clustering with a concentration of qualified team members, and thereby reduce the costs of innovation. Such specialized incubators include the following: the IC Design Technology Innovation Centre for integrated circuit design, the China Torch Internet Innovation Centre for Internet-related innovations, the Kehui Technology Innovation Centre, for nano-technology materials, the Fengxian Technology Innovation Centre for modern agriculture, and Xuhui Software Park for programme development. Shanghai has set the objective that an additional number of incubators should be specialized, as follows:

- Environment protection in Yangpu District
- Multimedia technology in Changning District
- Urban industry design in Huangpu District
- Biotechnology in Xuhui District
- Fine chemical engineering in Jinshan District
- Manufacturing equipment in Nanhui District
- Software design in Minhang District
- Engineering software at Tongji University
- Modern agriculture in Fengxian
- Urban industrial design in Huangbu.

Some of these incubators should be seen as seeds for rapid urbanization and continued industrialization in Shanghai, and the plans for Fengxian provide an interesting illustration. These aim to change the district from a rural district into part of the urban area of Shanghai by 2020.[29] Some 9000 villages scattered around the district will gradually be merged, and 22 towns will be reduced to eight with Nanqiao New City becoming the new central place for a district that currently has a population of one million. In a press release it is mentioned that the district government is working hard to encourage some

100 000 farmers into the district's new towns, which will help cut unemployment in villages and increase urbanization, which could reach 84 per cent of the population by 2020. Industry would take a prominent role in the district's economic development in the future, with four industrial zones for electricity distribution equipment, fine chemicals, logistics and manufacturing. These four zones would house most of the enterprises in the district. Agricultural industrialization is ranked high in the district and Fengxian Modern Agriculture Park, with an area of 20 square kilometres, is designed to become a model to develop modern agriculture, where produce is grown in large quantities in unpolluted areas based on unified standards.

Minhang District, which is also the location of Zizhu Science-Based Industrial Park, has attracted Intel with its focus on software design. The company has indicated that this would become its third site in Shanghai in 2005. This plan is connected to an agreement with Shanda Networking to develop digital entertainment technology. In an announcement it is mentioned that China's online gaming and entertainment industry is experiencing rapid growth, with more than 54 million gamers and 187 million Internet users expected in the nation by 2008.[30] The two companies will focus on developing total online interactive entertainment solutions for a variety of devices including PCs, TV set-top boxes and mobile phones based on Intel chip architecture, and will also establish a digital home and game testing and innovation centre. Intel employees engaged in research and development will move to Zizhu in 2005, which will offer space for future expansion as Intel participates in global research projects.

Xuhui District is already the location of several colleges and academic institutes, including Shanghai Jiaotong University, the Medical School of Fudan University and the Shanghai branch of the Chinese Academy of Sciences. There are also nine high-level hospitals and a dozen other medical treatment units within the district. Xuhui officials announced in September 2003 the goal of becoming an international business area. Medical treatment, information technology and the materials industry have been identified for its future development.[31] An indication of the ambition of the district is its development since 1992 of an optical fibre network for which it has invested RMB6 million.[32]

KNOWLEDGE CREATION AND INTELLECTUAL PROPERTY RIGHTS (IPR)

Shanghai gives a high priority to intellectual property rights and attempts to strengthen their protection as an essential core of the city's knowledge system. Patents from Shanghai in 2003 amounted to 22 374 applications sent to the national patent office (SIPO), and 16 571 were granted in the same year. These numbers include all categories of patents – real innovations as well as utility and design patents. However, the position of Shanghai in real innovations is still rather weak. In 2003 5936 patents were applied for while in the same year only 880 real innovation patents were granted.

China has, since joining the WTO, revised its intellectual property rights (IPR) laws, to make them consistent with the rules of the Trade-related Aspects of Intellectual Property Rights (TRIPS). Shanghai is anxious to have international biotechnology firms entering China. However, concerns remain about China's implementation of IPR-related laws, as IPR may mean everything for biotechnology companies, which are very cautious about making major investments in China. Shanghai has mapped its local strategy for establishing an IPR protection system by 2010 that would be consistent with laws regulating the national market economy and international practices. Under this strategy a Shanghai official suggested that 'an average million people will be required to invent 150 authorized patents annually by 2010, topping other cities'.[33]

The 2004 Shanghai International IPR Forum discussed intensively the relations between IPR and the city's competitiveness, with the participation of scholars, government officials and economists from China as well as Europe, the US, Japan and Korea. Shanghai's efforts are supported by the State Intellectual Property Office, which will sum up the experiences gained from overseas and domestic practices and seek ways to protect IPR in order to reinforce the competitiveness of Shanghai. The City will need to become an example for other parts of the country by following the world's IPR development trend, if it wants to get a foothold in the global economic competition.[34]

Shanghai Intellectual Property Administration (SIPA) started to cooperate with international organizations in 2004 in order to develop further an IPR protection system. Furthermore, SIPA has launched a training programme with counterparts in the US, which will provide Shanghai with a number of IPR teachers, judges and government officials. Shanghai has about 500 IPR professionals working in government departments, courts, universities and intermediary service agencies, with 10 per cent in the local courts and another 10 per cent as IPR teachers in the universities. Realizing the need for more staff competence, Shanghai will train 50 senior IPR experts in the United States by 2010. They will be chosen from local government departments, research institutions and universities.[35]

SHANGHAI LEAPFROG INTO HI-TECH DEVELOPMENT

A natural formation of three major regions in China have prompted provinces and cities within them to act as midwives to the birth of an environment that can deliver not only incremental innovations but also breakthrough innovations in future-oriented industries. A number of regional development programmes and projects play an important role in this process and have the potential of enhancing strong and necessary links between clusters, foreign technology sources and national programmes. Shanghai has naturally come to play a dominant role in the Yangtze River Delta (YRD) region, surrounded by dynamic cities such as Ningbo to the South and Hangzhou, Wuxi and Suzhou further inland. Simultaneously, increasingly close industrial and technological relations with Taiwan have boosted Shanghai's knowledge potential.

Shanghai is on the verge of becoming an innovative knowledge region with no parallel in China except Beijing. Although this development is guided from above, and with substantial local or national budget resources, it is strongly supported by local initiatives and entrepreneurship. This dynamic character is likely to be duplicated in many other locations in China. The regions with their dynamic cities and technological clusters are now playing a significant role, and their embedded innovation systems might greatly contribute to China becoming a technological superpower.

NOTES

1. The preparation of this working paper has been made possible through a one-week study visit to Shanghai in early 2005. All my visits were facilitated through the office of Vice-Mayor Zhou Yupeng, who kindly instructed the Foreign Affairs and Trade Commission of Shanghai Municipality to arrange the visits and interviews on which most of this chapter is based. Any mistakes or misunderstanding of information given is solely the responsibility of the author.
2. 'Growing role for private education', *China Daily*, 9 July 2003.
3. Ibid.
4. The establishment of the 'Student Scientific Research Fund' and the 'Student Summer Fieldtrip Fund', together with a hundred high-level lectures, a hundred recommended books, a hundred social fieldtrips and experiments in laboratories, constitute the 'three hundred plan' which dominates campus activities.
5. Fudan University brochure, undated.
6. 'China makes its first self-designed digital TV chip', *China Daily*, 12 December 2004, http://www.chinadaily.com.cn/english/doc/2004-12/27/content_403686.htm.
7. There are three competing groups that are vying for the successful design of chips to be used in future digital HDTV sets. One is the microelectronics researchers at Fudan University who are collaborating with colleagues at Tsinghua University in Beijing. The other two are Shanghai Jiaotong University and the Shanghai Video and Audio Electronics Group, commonly referred to as SVA, which is a state-owned company.
8. 'China makes its first self-designed digital TV chip', *China Daily*, 12 December 2004.

9. Fudan University website, http://www.fudan.edu.cn (28 January 2004).
10. Information on Fudan Science Park is primarily based on an interview with Mr Jia Wei, Deputy General Manager (11 January 2005) and Shanghai Fudan Science Park brochure, undated.
11. The management team includes specialists from finance and political circles, law, management and education. The President of the Board is Dr Yang Yuliang, who is Fudan university Vice-chancellor, also Special Professor of Ministry of the Education 'Zhangjiang' Project, and also Chief Scientist of the Ministry of Science and Technology 973 Key Fundamental Study Project.
12. The information in this section is extracted from a PowerPoint-based briefing in Chinese by Deputy Chief of the High Technology Department of Shanghai Development and Reform Commission, Mr Ji Zhaoliang (14 January 2004).
13. http://www.spsp.com.cn.
14. Only six of the largest software companies in China are certified at the level of four or five of the Capability Maturity Model (CMM), which contrasts with India where all of the top 30 software companies have achieved these rankings. This is a reflection on process control and product management that the McKinsey analysts say remain weak, with limited attention to personnel upgrading programmes.
15. A comparison between the software industries in China and India reveals a number of fundamental differences. A recent survey shows that the top ten IT service companies in China have only about a 20 per cent market share compared with 45 per cent for the top ten in India. Another fact is that China has more than 8000 providers of software services and close to three quarters of them have less than 50 employees, with only five of them having more than 2000 employees. India has fewer than 3000 companies and at least 15 have more than 2000 workers. Some of the Indian companies – including Infosys Technologies, Tata Consultancy Services and Wipro Technologies – have established a strong position in the global market for software services.

    Although revenues from IT services are increasing rapidly they are presently only one half of India's. The growth in China is primarily driven by domestic demand from SME companies that want customized solutions. The share of outsourcing business amounts to only about 10 per cent of total revenues, against 70 per cent in India. Most outsourcing is done for Japanese companies that account for two thirds of mostly low-value application development contracts. The analysts found that Chinese software companies indicate little interest in growing bigger through mergers and acquisitions and are generally not yet eager to enter into the global market.
16. Furthermore Shanghai also has another four high-technology parks:
    1. Jingqiao Modern Science and Technology Park (Jingqiao Xiandai Keji Yunqu), with a focus on electronics, was recently established. It has some 40 companies which are mostly foreign, including Bayer from Switzerland.
    2. University Science and Technology Park(s) System includes parks at eight universities in Shanghai: Fudan, Jiaotong, Tongji, Shanghai, Donghua, Huagong Ligong, East China Normal and Maritime.
    3. Jianding Minying Science and Technology District (Jianding Minying Keji Qu) is a 'returnee' park that was established in 2001 and already has 500 companies of which 300 have been established by returning Chinese. Most other companies have a focus on energy and new materials.
    4. China Textile International Science and Technology City (Zhongguo Fangzhi Guoji Keji Cheng) was established in the early 2000s and has a total of 145 companies.
17. These include:
    - Microelectronics Research Institute of Beijing University
    - Microelectronics Research Centre of Tsinghua University
    - National Software Institute of Fudan University
    - Microelectronics Research Institute of Fudan University
    - Information Security Engineering Institute of Jiaotong University
    - Shanghai Research Institute of Xi'an Jiaotong University
    - Shanghai R&D Centre of the University of Science and Technology of China.

18. Shanghai Zhangjiang Hi-Tech Park, Shanghai Zhangjiang Group Co., Ltd brochure, Shanghai 2004.
19. *Shanghai Zhangjiang Hi-Tech Park Investment Guidebook.*
20. IC companies within SZHTP include the following:
    - Fabless: Infineon, ISSI, SST, Pixelworks, LSI Logic, Cadence, Synopsis, Sunplus, Via, Elan, Xirlink, Spreadtrum, SmartAsic, Huahong, Marvell, nVdia, Conexant, Onsemi
    - Foundry: SMIC, GSMC, HHNEC
    - Assembly and Testing and Photomask: ASE, GAPT, STATS, Photronics, Ellepsis
    - Equipment vendor: Applied Materials, TEL, Novelius, ASML, Screen, LamResearch, KLA-Tencor, TOWA, Praxair, Saes, Bolch
    - System application: Lenovo, Amoi, ZTE, Huahong, UTStarcom.
21. *Shanghai Zhangjiang Hi-Tech Park Investment Guidebook.*
22. Shanghai Zhangjiang Hi-Tech Park, Shanghai Zhangjiang Group Co., Ltd brochure, Shanghai 2004.
23. Zhang Yong (2003b). More recent information suggests that Dupont is making more significant investment in R&D facilities within the park, amounting to US$100 million, and will employ 250 people.
24. SCIP website, http://www.scip.com.cn.
25. Fengpu website, http://www.fengpu.com.
26. http://www.sjepz.com.
27. Shanghai's present use of space by international financial corporations is around 160 000 square metres while that in Hong Kong's financial district is 1 000,000 square metres, and over 20 million square metres on Wall Street in the United States. Source: 'Lujiazui seeks comprehensive development', *China Daily*, 27 December 2004.
28. Information extracted from PowerPoint presentation 'Shanghai Hi-tech Business Incubator Network – Networking, Internationalization, Specialization', prepared by Mr Wang Rong, Director of the Shanghai Technology Innovation Centre (STIC).
29. 'Fengxian's blueprint for beautiful urbanization', *Shanghai Daily*, 17 September 2004.
30. Chen Qide (2004).
31. Xin and Tian (2003).
32. Ibid.
33. Guo and Chen (2004).
34. Ibid.
35. 'Round table looks at IPR protection', *China Daily*, 29 October 2004.

10. China regaining its position as a source of learning

SCIENCE AND CIVILIZATION IN CHINA

Towards the end of the 1700s China had similar opportunities to start an industrial revolution comparable to those in England. One important difference was that the essential new energy source – coal – was widespread close to the manufacturing sites in Europe, while distances were far greater in China. In addition, Great Britain and Europe were to benefit from the flow of raw materials, like cotton, from the United States. This is the main argument advanced by Kennneth Pommeranz to explain why industrialization did not start in East Asia.[1] However, Benjamin Elman argues that China was in a high-level equilibrium trap in which non-industrial methods were efficient enough to prevent industrial methods.[2] Underlying his argument is an assertion that the older and the modern traditions embody fundamental differences and that the former do not permit changes. In a related discourse Ellen Chen has suggested that the problem of tradition and modernization is much more serious and disturbing for original civilizations.[3] Thus Chinese, Indian and Islamic civilizations face greater difficulties than civilizations that have been constantly attuned to borrowing from other traditions. As a consequence China's path towards modernization has been particularly tortuous as, like Japan, it had limited borrowing experience and nor did it have any of the colonial guidance from which, for example, India benefited. Chen sees China's acceptance of the Marxist ideology as a complete system that would in its totality replace the totalistic Confucian system that had failed to direct China on the path towards modernization. By adopting Marxism China gradually drifted into a dream state that reached its climax during the Cultural Revolution, and at its extreme attempted to obliterate all cultural memories.

There has been considerable interest in the fact that China did not develop modern Western scientific methods, and an important explanation lies in understanding the basic difference that has existed between the traditional Chinese and Western way of regarding the natural world.[4] Elman argues strongly that:[5]

Unlike the colonial environment in India, where British imperial power after 1700 could dictate the terms of social, cultural, and political interaction between natives and Westerners, natural studies in late imperial China were until 1900 part of a nativist imperial project to master and control Western views on what constituted legitimate natural knowledge. Each side made a virtue out of the accommodation project, and each converted the other's forms of natural studies into acceptable local conventions of knowledge. Europeans sought the technological secrets for silk, porcelain, and tea production from the Chinese. Chinese literati borrowed algebra, geometry, trigonometry, and logarithms from the West. Indeed, the epistemological premises of modern Western science were not triumphant in China until the early twentieth century. Until 1900, then, the Chinese interpreted the transition in early modern Europe from new forms of scientific knowledge to new modes of industrial power on their own terms.

However, China was at the same time one of the great non-western domains of human wisdom in all aspects of science, technology and wisdom. Joseph Needham and Nathan Sivin have restored a measure of respect to traditional Chinese sciences, and Elman says[6] that 'the history of modern science can never again be what it once was, the history of a select group of early modern Europeans who supposedly discovered the natural laws of the universe and then passed them on with good will and cheer in the name of progress to the rest of the world'. Still, it has been the conventional wisdom that China could never have produced science because it had no industrial revolution and had never produced capitalism.

In China's ongoing hectic efforts to catch up with the industrialized countries in all aspects of science and technology, it is of considerable interest to understand why this process was so long delayed. During the Tang and Sung dynasties, works on mathematics and astronomy were banned for security reasons. However, Ming dynasty examination records reveal that candidates were tested for knowledge of astronomy and other aspects of the natural world, although knowledge of classical text remained pre-eminent. The examiners expected candidates to place technical learning within the context of classical narration, as statecraft was assumed to be based on the direct linkage between classical learning and political competence.[7]

Elman presents a view on the development of a ban on Chinese literati studying astronomy by the early 1700s, saying that although effective in the examination system the ban did not carry over to literati learning. In an essay prepared in the early 1860s there are suggestions that the civil examination system should be reformed in order to create a new balance in the selection process with regard to the needs of the future. The same author also called for a widening of the selection process for officials, to include recommendation and promotion of clerks who had demonstrated their administrative abilities. He also suggested that civil examination system should have two paths – with one group required to master machinery and physics. However, not until

1887 were candidates specializing in mathematics allowed to pass provincial examinations under a special quota, although they also had to fulfil the same classical requirements.

Although the Chinese literati took an interest in the theoretical aspects of science and technology for a couple of centuries until the late 1880s, they showed very limited interest in European science from a practical perspective that would require laboratories to reproduce experiments and to confirm or reject past findings. However, Elman notes that:

> Recent research indicates, however, that the various arsenals, shipyards, and factories in the treaty ports were important technological venues for experimental practice where, in addition to the production of weapons, ammunition, and navies, a union of scientific knowledge and experimental practice among literati and artisans was first forged. Indeed, it is likely the case that the 'techno-science' of late-Ch'ing China was an important building block for the rise of both dynastic and private industry in the late nineteenth century treaty ports where most of the arsenals were established. The Chiang-nan Arsenal and Fu-chou Shipyards, for instance, were acknowledged by Europeans and Japanese to be more advanced than their competitor in Meiji Japan, the Yokosuka Dockyard, until the 1880s.

Thus, the events of the Sino-Japanese War (1894–95) came as a surprise when the Chinese navy was defeated by the Japanese navy with its basis in Yokosuka technology. It became evident that Chinese naval technology in Fuzhou and Shanghai did not match that of Japan. Subsequently, uncritical views of Western schools and educational reform were widely accepted and the 1898 Reform Movement required fundamental changes in the education system in China. This would only be possible if the civil examination system was reformed, but it actually survived until 1905.

A lasting result of the Reform Movement was the creation of the Imperial Peking University. It was established to be at the apex of an empire-wide network of schools, and designed like the Translation College to train civil degree holders in Western subjects, suitable for government service. The experiment, although promising, came to an end during the 'Boxer Rebellion', when everything of the new university was destroyed in the summer of 1900, soon after which Western and Japanese troops arrived in Beijing. However, one of the results was a rapidly growing interest in institutions of higher learning that would stress modern science. The process was further accelerated by the perception of many radicals, after the 1911 Revolution, that a scientific revolution was also required. Culminating after 1915, the advocates of the New Culture Movement strongly supported modern science and medicine. The ten years between 1905 and 1915 saw a complete demolition of an educational regime based on classical learning that had been reproduced by the civil examination system since 1370. The mistaken conclusion, based on a surprising intellectual consensus among Western and

Chinese scholars, was that China had failed to develop science before the Western influence.

China is gradually and systematically taking steps to become a knowledge-based economy. It is equally evident that the country has yet to become a technological superpower with a noticeable presence in all important research fields. The evidence for a future technologically strong nation can be gleaned from its space programme, new structures that are being established in biotechnology and nanotechnology and the substantial resources allocated to these and other sectors. Reaching the status of a technological superpower will require a number of vital conditions to be fulfilled.

First, China must considerably expand universities and other institutions of higher learning to meet the demand from a knowledge-based economy. A number of important steps have already been taken and the universities and colleges will soon produce more than one million graduates annually in engineering and scientific disciplines. Furthermore, the education system has to be upgraded and increasingly become the hub of research, which will require giving up its strong links to the business sector. In comparison with Japan, where the universities have taken a pride in remaining above the commercial sector, many universities in China have become closely enmeshed in business activities, a situation that was partly forced upon them when state funds were withdrawn or reduced.

Second, China's industrial success in recent years has depended heavily upon the infusion of foreign technology and will continue to do so for a long time until China completes its catch-up phase, as happened in Japan by the early 1980s. National security, which includes both important economic as well as military considerations, has led Japan occasionally to adopt a posture of techno-nationalism. This happened in the aftermath of the two oil shocks of the 1970s. China's posture has for more than two decades been one of techno-globalism. Present policies and understanding indicates that this situation will remain unchanged for the foreseeable future. Still, China is faced with the challenge of incorporating all its imported technology so that it seeps, and eventually flows, out of the enclaves where most of it is embedded. When successful in this endeavour and coming to the end of the catch-up phase, China is likely to find that foreign companies are less willing to transfer technology than they used to be. Furthermore, China will have to develop its own military technologies as the United States will as a global power obstruct any attempt by China to have a military capability that would in any way threaten the American dominance. Aside from such considerations, China must have an absorptive capacity in all scientific and technological fields to follow what is going on in the rest of the world.

Third, China has in the past employed a dominant majority of its high-talent manpower in various sorts of government research institutes. The Chinese

Academy played and still plays an important role in organization and institutes are primarily involved in basic research. However, official statistics suggest that more than 60 per cent of all R&D personnel have been transferred to the corporate sector as a result of R&D reform. Previously most of them worked in research institutes belonging to various industrial ministries. They may still remain in their metal towers in the same way as many scientific researchers still sit in their ivory towers – without being able to make any real contribution to science and technology in China.

The Cultural Revolution left the educational system in complete disarray and naturally it took longer for the graduate programmes to recover. The country has been faced with the problem that most talented young Chinese who left for undergraduate or graduate studies abroad have not returned. This remains a problem, although the size and coverage of national graduate programmes have greatly expanded in recent years. However, there could be no doubt that it would be desirable to attract the most talented PhDs, now residing overseas, to basic research and other challenging tasks in China. Ongoing expansion of the Chinese university system could drastically change the situation in a relatively short period and a *Science* article in May 2004 suggests that Chinese expansion, based on present trends, would greatly contribute to the following results: 'By 2010, 90% of all PhD physical scientists and engineers in the world will be Asian, and half of them will be living in Asia'.[8]

Fourth, the increase in funding for R&D has expanded more rapidly than the rate of economic growth during the past ten years. Thus, official statistics indicate that Gross Expenditure on R&D (GERD) as share of GDP increased from 0.6 per cent in 1996 to 1.3 per cent in 2002. However, this R&D expenditure would also include activities in the large chunks of ministry research that have been transferred to the corporate sector, mainly state-owned enterprises. Available information suggests that this part of the system only contributes in a minor way to economic growth.

Fifth, institutions and intellectual culture must evolve in a major way to advance science and technology in China. The legacy of the past, including the planned economy borrowed from the Soviet Union and the Cultural Revolution that followed will take another full generation to eradicate.

China is capturing high-technology market shares through FDI increased value-added is being generated through diffusion effects and; subsequently, the gradual growth of technological clusters. Substantial amounts of innovations – of a gradual and incremental nature – are taking place in manufacturing firms all over China, although primarily in the dynamically evolving coastal areas. Firms have often agglomerated into geographical clusters which can be found in many industrial sectors. A number of clusters are evolving into centres of innovative capability. They are still weakly linked and inadequately supported by actors within the state-level innovation system.

Financial and human resources for R&D are rapidly expanding, although accompanied by a gloomy view on the need to restructure the old formal system. A much more optimistic landscape of technological advances is emerging in the manufacturing sector. The position of Chinese companies in Intellectual Property Rights (IPRs), as revealed in patents, still remains weak, although it is relatively strong in design and utility protection. Setting standards, particularly in the electronics sector, has received much attention from the Ministry of Information Industry (MII). Initiatives have been taken to form a Northeast Asia joint support forum for various standardization schemes.

Realizing the inherent difficulties of catching up in microelectronics, China has decided to allocate substantial resources to emerging scientific fields. These include areas such as biotechnology and nanotechnology, with subfields such as bioinformatics and nano-biology. China has decided to join forces with international partners. One example is the full membership in the ITER nuclear fusion project together with Russia, the EU, Japan, South Korea and the US. Another is the Galileo Project undertaken by the EU, where China will gain access to advanced space technology.

China 2020

The year 2020 is the target for China's ambitious goals in science and technology, for which massive planning and preparations started in 2003. This is also reflected in many comments and statements on China's future advances in technology and industrial development, for which examples are provided on the following pages.

China will launch its second manned spacecraft Shenzhou VI, carrying two astronauts, in 2005. The Chinese government has also announced its plan to launch a satellite to orbit the moon in 2007. It is to be followed by the landing of an unmanned vehicle on the moon by 2010 and collecting samples of lunar soil with an unmanned vehicle by 2020.[9] Looking into the future, the President of the China Academy of Space Technology has declared that China will 'establish a sound mechanism to commercialize its space technology'. However, at the same time he made the statement that 'currently China's space industry mainly serves its national economy and national defence'.[10]

China has declared the nuclear power industry as a priority in its high-technology research and development plan and requested in late 2004 that the industry should grow at an annual rate of 15 per cent in the following five years. China already has eight commercial nuclear power stations with 19 reactors altogether, of which nine are in operation. Four of them are based on Chinese design, and China has established a complete nuclear fuel system from resource exploitation to disposal. By the end of 2003, nuclear energy in China accounted for 1.6 per cent of the country's total power generation capacity. The country's nuclear power generation capability will, according to plans, increase threefold and account for 4 per cent of total power generation by 2020.[11]

China's next phase of economic expansion envisages a rapid development of the automobile industry as one of the driving forces. Production was expected to reach five million units in 2004; this makes China the third largest manufacturer after the US and Japan. This is one million more than the previous year, when the number of private cars increased by 80 per cent, reflecting the increasing affluence of the middle class. The Ministry of Communication has predicted that the number of cars in China will eventually reach 250 million as an assumed maximum, which would correspond to about 150 cars per 1000 people.[12] However, planners in China expect that by 2020 there will be 140 million cars on the road, which would be a seven-fold increase compared to 2004.[13] This would no doubt put tremendous pressure on the transportation infrastructure, although the Ministry suggests that the current construction of highways between provinces in western China, as well as those in the Yangtze River Delta and the Pearl River Delta would be completed by 2010.

The number of private cars already exceeds ten million and the auto industry is now seen as 'one of the growth engines for China's gross domestic product (GDP)', which is expected to reach US$4000 billion by 2020.[14] The domestic passenger car market is dominated by foreign companies, which control 90 per cent of the manufacture and sales of passenger cars in China, while all the big state-owned car makers are basically assembling foreign brand cars. The expectation has been that China, by opening up the domestic market, would benefit from the transfer of advanced automobile technology. However, the Chinese firms have on the whole failed to assimilate enough technology to create an indigenous technological capability.[15]

A major reason for this may lie in the character of the former state-owned companies (SOEs), which still maintain not only earlier bureaucratic tendencies in management but also preserve close ties with political power holders at state, provincial or city levels. This may be even more of a problem for the 189 important SOEs that are under the direct control of the State-owned Assets Supervision and Administration Commission (SASAC). These companies belong to China's pillar industries and should be compared with overseas multinationals. However, these major Chinese companies on average spend about 1 per cent of their sales on R&D and many of them even less.[16] Some SOEs have become export-oriented companies and the number of Chinese companies in the Fortune Global 500 list has increased from three in 1995 to 18 in 2004. Still, the government is far from satisfied and has mapped out a plan to help to set up hundreds of multinational enterprises of varying scale. The aim is to have 50 Chinese enterprises among the Global

500 by 2015. At the same time the government has suggested that another 500 medium-sized companies and 5000 small firms should become fully established multinational companies.[17]

The general opinion in China today is that enterprises with distinct advantages should be encouraged to set up processing or assembly plants and improve their sales network. This would eventually encourage a number of Chinese enterprises to engage in multinational operations with their own brand names. Such approaches already include household appliance firms that have set up plants in developing countries, while others have ventured into high-tech cooperation with enterprises in industrialized countries, or have started to explore oil and mineral resources overseas.

In a number of industrial sectors the ambitions and interests of China and its policy makers clash with those of foreign partners, for which the auto industry provides a good illustration. In early June 2004 the National Development and Reform Commission issued a long-awaited new policy document for China's rapidly growing automobile industry, replacing a State Council decree of 1994. At this time there were altogether some 120 vehicle plants in China. The large Chinese car makers would be encouraged to join forces with foreign partners to merge domestic and foreign vehicle production. The Shanghai Automotive Industry Corporation (SAIC) – one of the three major domestic makers – had already joined hands with General Motors in the US and Suzuki in Japan in late 2002 to take over Daewoo Motors of South Korea.[18] SAIC signed a memorandum of understanding in July 2004 with creditors of the ailing Ssangyong Motors to buy a majority stake in the South Korean automaker.[19]

The policy will create barriers for new domestic non-auto investors as new auto projects will require an investment of at least RMB2000 million and include a product R&D organization of not less than RMB500 million.[20] The policy will also tighten the restrictions on foreign investors in the auto industry in establishing new joint ventures although majority control is allowed if they are located in China's export processing zones. A related expectation is that some internationally competitive Chinese car makers will join the Global 500 MNCs by 2010.[21]

China's car industry has become one of the biggest growth engines for China's economic development with a total workforce of more than 12 million employed directly or in related services, and a turnover of RMB680 million in 2003. However, China has basically failed to create its own development capabilities and brands, and 90 per cent of passenger cars made and sold in China are brands of foreign automakers. So far only a few small Chinese manufacturers are struggling with domestic brands, such as Geely, Chery and China Brilliance Auto, although with limited shares of the domestic car market. They are small and have received only scant support from the government.

The government planners had expected foreign automakers, when allowed to produce vehicles in China, to transfer advanced technology to their Chinese partners. This expectation has hardly been fulfilled, one reason being that Chinese firms have failed to assimilate enough technology to enhance greatly their development capabilities. The underlying reason is that the three major automakers seem unworried about innovation as they can get many products from foreign partners and operate profitable joint ventures with two or more foreign partners. The Chairman of Geely, the sole privately-owned passenger car producer in China, says that 'China's car industry cannot be at the mercy of foreign giants any more, and we should cast away the illusion that they will really help boost our development capabilities'.[22]

As exemplified by SAIC's purchase of a majority stake in Ssangyong Motors, described above, automakers are going abroad, although in a much smaller way than foreign massive forays into China.[23] Chinese automakers have three main routes – mergers and acquisitions (M&A), building plants and direct exports. However, Chinese companies still have limited experience in international mergers and acquisitions, and are exposed to many new challenges such as handling trade union relations and a variety of legal issues.

Advances in Coastal China – Ningbo Example

Ningbo is one of China's important coastal cities, located to the south of Shanghai and like Shenzhen in the South, was one of the first cities to be targeted for rapid economic development. Its population exceeds six million, which also includes the rural population in an area that covers some 10 000 square kilometres. Ningbo has a number of port-based industries including logistics, based on an excellent deep-water harbour, and major industrial plants in steel, paper and petrochemicals. Ningbo also has highway-based industries that include the manufacture of cars, car components and advanced moulding machinery. The city also has significant IT industries, of which the most significant example is Ningbo Bird, which has evolved as a leading domestic manufacturer of mobile handsets trailing closely behind Nokia. An early and successful industrial and economic development has propelled Ningbo into a leading position in China with a per capita income of US$3500, which is three times the national average.

Ningbo offers an interesting example of China's expansion of its higher education. The recent annual enrolment is roughly 30 per cent of school leavers while the national average has reached 20 per cent. The majority of this expansion has taken place in various types of professional colleges (Table 10.1).

Ningbo still requires qualified personnel from other parts of China and organizes Talent Fairs on an annual basis. Graduates in Northeast China

Table 10.1 Institutions of higher learning in Ningbo City

Institution	Supervision	Number of Full-time Students		Annual Enrollment	Date of Establishment
		Normal education	Adult University		
1. Ningbo Vocational and Technical College	Ningbo Municipal Government	5 523	5 522	2 500	1959
2. Zhejiang Light Industrial and Textile College	Ningbo Municipal Government	3 034	692	1 700	1979
3. Ningbo Radio and TV University	Ningbo Municipal Government	1 828	1 347	1 500	1979
4. Ningbo Engineering College	Ningbo Municipal Government	6 978	2 661	2 500	1983
5. Ningbo Teachers' College	Ningbo Municipal Government		7 892	2 500	1984
6. Ningbo University	Ningbo Municipal Government, provincial and municipal level	21 807	13 268	6 500	1985
7. Public Security and Maritime Police College	Ministry of Public Security	1 029	779	500	1999
8. Zhejiang Pharmaceutical College	Zhejiang Provincial Government	4 368	73	1 800	1999
9. Zhejiang Wanli University	Ningbo Municipal Government	14 341	1 548	5 500	1999
10. Zhejiang Business Technology Institute	Zhejiang Provincial Government	5 099	1 049	2 000	2001
11. Ningbo Institute of Technology (Zhejiang University)	Zhejiang Provincial Government	6 942	607	3 000	2001
12. Ningbo Dahongying Vocational Technical College	Ningbo Municipal Government	3 810	239	3 500	2001
13. Ningbo Garment Vocational College	Ningbo Municipal Government	3 607	176	1 700	2002
14. Ningbo City Technology College	Ningbo Municipal Government	1 520		2 500	2003
15. Ningbo Tianyi Vocational and Technical College	Ningbo Municipal Government			1 500	2004

Source: Data provided by Ningbo Municipal Foreign Trade and Economic Cooperation Bureau.

would get an annual salary in the region of RMB12 000–18 000, while they would earn considerably more in Ningbo. However, a number of initiatives are significantly changing the higher education sector, which already comprises 14 institutions, with Ningbo University playing a major role as the first institution of higher learning aside, from engineering colleges, established in 1985 in the early stage of China's modernization.

China's University Leap Forward

China's future development will depend on its consistent upgrading of human resources, in which an expected leap forward in colleges and universities will play a major role. Foreign observers and Chinese economists are equally uncertain when questioned whether China's continuous boom since 1979 can continue through 2020. Still, many commentators suggest that China could see an annual economic growth of between 7 and 8 per cent in the coming years and gross domestic product (GDP) could amount to RMB38 000 billion by 2020, with a per capita GDP of RMB26 000.[24] This would still leave China with a per capita GDP of less than 10 per cent of many industrialized countries. However, an expected leap forward in higher education would propel China into a leading position as a knowledge nation. The assumption is that the entrance rate for higher education will reach 35 per cent by 2020. This would make China a country with the richest human resources in the world, which, together with low labour costs, would provide a strong driving force for the country's continued economic development.

The university system in particular, as well as education in general, has received great attention as a source of China's future as a knowledge-based economy. Universities have undergone dramatic changes in recent years and it is still too early to judge their performance fully, which must be assessed with regard to undergraduate teaching, graduates studies and research. Universities in China have not been involved in advanced research until recently, and graduate studies were only introduced after the major reforms started in the late 1970s.

Project 211, the national programme for higher education, is mainly oriented to economic development in China and emphasis will be given to supporting institutions and key disciplinary areas which are closely related to 'pillar sectors of industry' where high-level professional manpower is needed. Priority will be given to some 25 universities which have a concentration of critical disciplines. These universities are expected to reach high international standards in both teaching and research and become models for other universities in China. An underlying objective is to break away from the narrow disciplinary orientation that existed in the former university system, to broaden the coverage of various disciplines and to foster the emergence of cross-disciplinary teaching and research.

Assuming that China's higher education will continue to expand with a national annual intake higher than that of Ningbo, by 2020 China could have a total of more than 120 million citizens who have received education in universities and colleges. Simultaneously it could be expected that 25–30 per cent would have received degrees or diplomas in engineering or science disciplines (Figure 10.1). However, there can be little doubt that this expansion would require not only substantial financial resources but also daunting requirements on teaching staff to raise quality standards.

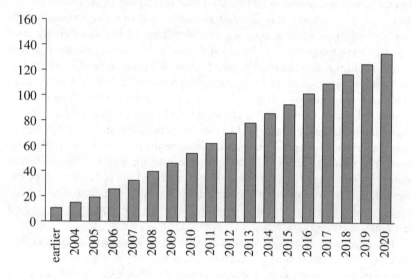

Figure 10.1   Estimate of China's accumulated graduates up to 2020 (million)

China in the Global Innovation System

China has also set its eyes on technologies that have the potential to yield results in the long-term future. Participation in a global innovation system (GIS) has recently become highly valued. It shows itself for example in the increasingly close relations with the European science community. China has reached an agreement to become a full member of the Galileo Project. This is carried out within the sixth Framework Programme and will involve close collaboration with the European Space Agency, giving China access to future advanced satellite technology and systems. Another example is China's decision in 2003 to become a full member of the ITER nuclear energy fusion project.

The migration of electronics production to China and its subsequent upgrading and increasing sophistication have yielded substantial benefits to the defence sector in its requirements for military technologies. These are now often dual-use technologies and many components and sub-systems are becoming available on the shelves of electronics companies inside China. However, such parts have to be integrated into systems and built into platforms. The US refuses to sell advanced military systems and platforms and is prompting the EU to do the same. This has left China with Russia as the only willing supplier of advanced military systems and platforms as Israel and other potential suppliers are constrained by the US.

China has accepted the trends of globalization. Its technology strategy includes three pillars. First, the reform of corporate R&D, which massive infusion of FDI and acquisitions abroad are combined with remnants of the earlier system. Second, national programmes have been initiated to support present and future capability. Third, China is committed to international collaboration wherever the need arises and without jeopardizing military and economic security.

China is able to move forward rapidly as it has embarked on its present stage of development under very favourable conditions. First, by accommodating globalization, the size of the Chinese market has attracted and will continue to attract huge amounts of FDI, which is becoming embedded in constantly upgraded industrial and technology structures. Second, backed by substantial foreign exchange reserves, entrepreneurial Chinese companies are on the way to establishing themselves in global markets through marketing and technological acquisitions. Third, the development of human resources with a strong emphasis on engineering and science will, together with national programmes, provide the basis for future advances in science and technology.

Major operations by Chinese companies have been noticeably helped over a short period of time by the evolution of regional innovation systems, for which provincial and municipal governments play an important role. Embedded in their promotional activities are clusters which originally provided only operational efficiency for foreign investors. These clusters have gradually evolved into technological clusters which are becoming hubs of not only economic but also innovative activity. The decision of Hitachi to locate a significant portion of its hard-disk drive (HDD) activities to Shenzhen after taking over the HDD division from IBM, might indicate that a HDD technological cluster is being created in that city.

The weakness of China's innovation has been realized by its policy makers and planners and their actions can be interpreted in the following way. Although being highly successful in export markets for many high-technology products, China has realized that it will never be able to compete on an even playing field with the well-recognized global players in the present range of technologies. China thus decided to allocate substantial resources to science and technology fields where the gap with advanced countries is much smaller. This includes areas where the research front has recently been established in sub-fields within nanotechnology and biotechnology. This offers much better entry possibilities for China than microelectronics, where a research front was already established in the late 1940s. This would in no way indicate that China is giving up ambitions in the electronics sector. A majority of final assembly of all global production of electrical and electronics products will soon be carried out in China. The earlier simple processing and assembly industries have rapidly evolved into higher value-added operations and require two important inputs – integrated circuits and software. The first requires complex and highly capital-intensive plants for the manufacture of memory circuits (DRAM), central processor units (CPU) and application-specific integrated circuit (ASICs).

Manufacturing plants for all three categories of semiconductors are migrating to China, which will soon surpass Japan and the US as the major market for semiconductors. CPUs and DRAM will be dominated by established global leaders such as Intel and Samsung, although it was rumoured that one industrial group in China had actually started negotiations to acquire the ailing Hynix memory-manufacturing company in Korea. China has determined through its Ministry of Information Industry (MII) that it should concentrate on software development, which is required for all types of products, and thereby indirectly controlling or influencing the ASIC sector. This semiconductor segment is increasing in relative importance to memory and CPU. A number of ASIC plants already exist in China and ambitious plans for high-end production plants are being implemented. Taiwanese companies are playing an important role in this process, partly to maintain and further strengthen their own interests on the Mainland. An important technological evolution, system-on-chip (SOC), suggests the integration of all three categories of semiconductors on a single chip.

Although the size of the market is a major attraction for chip makers to locate on the Mainland, its comparative advantage lays in future software development which has high-value content, and also offers the possibility of garnering IPRs. Until recently language proficiency and earlier advances had given India an edge over China in software development. However, the migration of electronics production to China is offering a golden opportunity when foreign companies are sourcing not only production but increasingly product development. Part of the production chain is highly software-intensive for many electronics products. This being the case it offers great costs savings, especially when software engineers are being paid salaries at levels that correspond to 30 per cent of those in the US and Japan while industrial

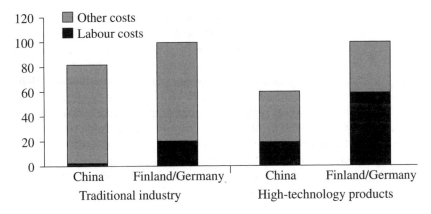

Figure 10.2 Salary comparison between China and Finland and Germany (percentage)

workers receive only 10 per cent of the wages in highly industrialized countries (Figure 10.2).

China's attention to the role of the universities indicates that improved conditions for research and development are being created. However, this does not initially offer China the opportunity to engage in research in areas which it prioritizes. To jump on to the software wagon China has to train increasing numbers of engineers. The rapid expansion of the tertiary education system indicates that is already being done, although still only at an early stage.

China's concern about its future in electronics is clearly reflected in its attention to standard setting. Standardization has become a very important element of China's technology strategy in recent years. There are basically two trends to consider in the light of the tectonic shift of electronics to East Asia. One is the increasing attempts to harmonize interests and development among three countries in East Asia – China, Japan and Korea. The other and more important trend is that China wants to establish its own technological platforms, in as many areas as possible, in order to gain independence from foreign high-tech companies and drastically reduce the level of licence fees.

Thus the Chinese government is today strongly supporting the development of various industrial technology standards in a number of areas, including digital TV and the controversial Chinese (WAPI) protocol for W-LAN, which offers higher security but has highly angered both the US and Japan. Another sensitive area is openware software, where Japan and China have common interests. China wants a Linux version for the Chinese language. Although,

China, Japan and Korea have different interests there are prospects for cooperation in East Asia that could reduce the market dominance of Microsoft.

The demographic transition that has taken place in China during the past couple of decades has the potential for a substantial positive effect on savings and investment. The impressive rise in savings and investment rates in neighbouring countries since the late 1960s was a very important factor, combined with increasing labour participation, in propelling economic growth in East Asia. China was not itself involved in the process of economic growth at the time as it had entrenched itself within a planned and autarkic economy and was preoccupied with its Cultural Revolution. As a latecomer, it is now following an earlier established pattern of high savings and high investment and is presently enjoying a favourable ratio between working population and dependents.

Policy makers are given the opportunity to use the demographic dividend to start shifting resources toward increasing access to more advanced forms of education. With lower birth rates, demand for primary education will go down, and this will later on be repeated at the secondary level. Thus, improved opportunities will be given to those leaving secondary education to enter into universities and colleges for training and degrees that will be increasingly demanded in the society.

TECHNOLOGICAL SUPERPOWER PERSPECTIVES

The Japanese Experience

In 1990 The US Department of Commerce released a report that carried the title *Japan as a Scientific and Technological Superpower*.[25] A summary says that Japan has a continuing demand for young engineers that may not be met in the future because of an ageing population and a declining interest in engineering as a profession. Japan graduates far fewer PhDs in the sciences and engineering than does the US but devotes much more attention to in-service education and training of technical personnel after they have entered employment. The funding of academic R&D in Japan follows a markedly different pattern from that of the US. The prestigious national universities conduct a disproportionate share of all academic R&D and are funded essentially by the central government. The report further states that[26] '(T)he economic success being enjoyed by Japan has resulted in large surpluses of government income over expenses. However, the surpluses are not being used to materially increase government support of R&D or hence to increase the ratio of government to private spending on R&D.' The author makes the following notable comment:

Knowledgeable people in both the US and Japan unanimously agree that Japan's technological successes have depended heavily upon the infusion of foreign science and technology. However, Japan has moved from a 'catch-up' mode to one of relative technical equality or equilibrium with the West, so that current transfers of technology to Japan are much more sophisticated than in the past. Japan has become a major exporter of medium-level technology to developing countries and is gradually increasing exports of advanced technology to the West.

Following from that the author says '(F)rom the point of view of industrial competitiveness, lessons can be taken by the US from the Japanese model to increase technical quality and literacy of the American industrial work force, to accelerate the process of converting scientific knowledge to industrial practice, and to be more alert to the results of R&D being carried out in Japan and elsewhere'. A most important part of the summary says: 'The worst mistake that U.S. policy makes and industrial leaders can make is to underestimate Japanese technical capability and determination to succeed. Attempts to isolate Japan from Western science and technology will also act to spur successful indigenous development.'

EU Views on China's Competitive Advantage

Less than 15 years later similar concerns were voiced in Europe, and the EU published a Competitiveness Report in late November 2004 that raised major concerns about China's emerging technological prowess.[27] In a press release Günther Verheugen, Commission Vice President, makes the following striking comment with reference to the report:[28]

> [We] need to enhance competitiveness and innovation in the EU to respond to the challenge of countries such as China, which is turning itself into a low-cost competitor in high-skill industries. One of the priorities of this Commission will be to face this challenge in the framework of the Lisbon agenda.

He notes that import competition from China used to focus on labour-intensive goods and low-skill industries, although China's active industrial policy is turning the country into a low-cost competitor in high-skill industries. The rapid growth of skill-intensive imports from China epitomizes a serious challenge to the EU, although the growth of trade has offered major trade and investment opportunities for European companies. It has become more and more apparent that China has been successful in selectively attracting FDI in technology-intensive industries in order to benefit from embedded technology and organizational skills. Simultaneously China has actively promoted domestic companies as pillar industries, eventually to become leading global companies, which together has strengthened China's industrial structures.

The Commission discussed at some length the challenges to the European car industry, which is seen as too fragmented and suffering from high labour costs while at the same time being challenged by major technological developments such as fuel cell technology, where China has already taken major initiatives. However, the EU does not see China's automobile industry as a major threat because of its fragmented structure, continued preference for local suppliers and the limited ability to export into the EU market. However, expertise on implementing fuel-saving technologies will become a competitive factor in the Chinese market.

Another major area of concern is future research policy within the EU, for which the following perspective is provided.

> In 2002, the EU spent roughly 2 per cent of GDP on research and development, the US 2.5 per cent and Japan (in 2001) 3 per cent. Public sector R&D in the EU is comparable to the levels in the US and Japan, while business sector R&D in the EU lags significantly behind our competitors. The EU target, set at the Barcelona summit, of raising R&D expenditure to 3 per cent of GDP by 2010 is motivated by this weak R&D effort in the EU and the need to boost innovation and growth.[29]

As for China, the start of the Open Door Policy in 1978 has allowed a highly productive, abundant labour supply that is not only available at low cost but also has a high rate of literacy and increasingly high skills. The Chinese economic system has been gradually transformed and is able to respond quickly to market signals about scarcity and profits. The result is the establishment of performance-oriented institutions that are constantly undergoing changes to meet domestic requirements and adjust to market needs and possibilities abroad.

As a result China had captured 35.7 per cent of the OECD trade in textiles and garments in 2001, and an even larger share in leather and leather products – 43 per cent.[30] More astonishing is that China's export share of office machinery and computers had reached 22.4 per cent of OECD trade in 2001, with an even higher share in radio, television and communication equipment – 28.5 per cent.

Is There a Chinese High-technology Threat?

The EU report makes a mistake by stressing that mature industrialized countries face a challenge from emerging economies which can in particular enjoy a labour cost advantage. Thus, technological progress and investment in human resources are key strategies for industrialized countries. Although correct at the aggregate national level, this attention to the factor endowment of labour overlooks the great diversity that exists in China. A recent report from the Inter-American Development Bank[31] argues that:

large urban centres such as Shanghai are comparable to the fast-growing capital- and skill-abundant Tiger economies of Asia or the more developed countries in Latin America. This diversity of endowment within China may be at least as wide as variation across countries in Latin America, suggesting that China could begin to export skill- and capital-intensive products long before the aggregate economy seems able to do so.[32]

All segments of the electronics industry in China have developed spectacularly since the early 1990s, a development for which the domestic market has become increasingly important. Until recently China's success in mobile handsets, personal computers and peripherals has been most spectacular. However, the IT industry is still dominated by industrialized countries which maintain technological leadership and capture a large share of value added. An international division of labour has favoured low-cost countries and China has benefited tremendously in recent years from the creation of global production networks. China's share of value-added in IT manufacture was a miniscule 8.1 per cent in 2001, but is expected to reach 33 per cent of the global total added value in the manufacture of IT products, which would be almost all of the combined added value on IT production in developing countries.[33] Available evidence indicates that the IT industry in China is very efficient in comparison with other manufacturing industries, with a labour productivity that is 50 per cent higher than the average Chinese manufacturing productivity.[34]

Government agencies and companies in China are not comfortable with the existing high dependency on foreign technology and marketing strategies and are making great efforts to support domestic technological development not only for components and capital goods but also to create more-or-less completely new system technologies that could become the basis for Chinese IPR. Consequently we can expect that China will be less and less involved in IT assembly and that whole value chains will increasingly be established in China – not only to meet domestic demand but also to serve export markets. China will be dependent for a number of years on foreign imports of processing equipment for the manufacture of semiconductors, although it will probably erode more quickly the European and US dominance in infrastructure for telecommunication operators.

At the very end the EU report provides the following very positive comment.[35]

> Currently, institutional conditions favouring innovations are better in the EU-15 as well as in new Member States than in China. But the vast, inexpensive work force, combined with the development of knowledge economy provides an excellent basis for offshore centres for the manufacture of a broad range of products and services. The competitiveness of Europe, its capacity to cope with the challenges as well as its ability to make use of the opportunities brought forward by a rising China economy, will, to a great extent, be determined by its innovation performance.

This positive note is based on the assumption that FDI flows into China are increasingly turning into market-seeking rather than resource-seeking flows. Consequently, fears of job losses to the low wage destination of China should be minimized across Europe as, given China's large market potential, the long-term utilization of the opportunities ought to be exploited above all by indigenous production. Related to this the following conclusion is offered:[36]

> Faced with China's competition, the new Member States and candidate countries have so far been able to offer more attractive near-shore centres. But further improvements in the performance of knowledge economies in the new Member States as well as a better synchronization of traditional EU-15 and new Member States structures are crucial to cope with China's challenge.

A Threat to Japan

There are indications that China is aspiring to become a world research centre. Naturally, such ambitions pose a direct challenge to Japan, which perceives itself to be the leading technological nation in the region. This concern is fuelled by inexpensive goods that have flooded the Japanese market in recent years, thus contributing to deflation. This is partly a reflection of the fact that many industrial goods can no longer be competitively manufactured in Japan and Japanese companies have, like companies in many other countries, relocated production to China – a shift that is partially replacing an earlier preference for Southeast Asian countries or Mexico.

An expanding horizontal division manufacturing inside Global Production Networks, primarily practised by US companies, has recently rekindled the debate on hollowing-out, this time with a focus on advanced technological competencies.[37] This concern is shown in various ways by Japanese companies. The following information from Japan provides a few examples which indicate a homeward retreat in terms of cutting-edge technology:[38]

1. A Kameyama plant in the Kansai region is the most important production facility for wide-screen LCD displays and for LCD televisions for the Sharp Corporation. Sharp is investing 150 billion yen in the plant in 2003 and 2004 to focus on developing and manufacturing LCDs in the Kansai area.
2. Matsushita Electric Industrial plans to build the world's largest plasma TV manufacturing plant in Amagasaki in Hyogo prefecture.
3. Canon will locate its production facilities for new digital cameras in Oita prefecture. The Canon President, Fujio Mitarai, expresses the position in the following words[39] 'Let's make things in Japan'.

Japan is benefiting for the time being from the growth of the digital consumer electronics market as products such as digital cameras, DVD recorders and flat panel TVs have sufficiently high added value to justify their manufacture in Japan, where high labour costs can be combined with highly advanced manufacturing skills. Unless new technology is completely integrated into 'black boxes', constantly upgraded and quickly brought to the market, there is nothing to hinder the increasingly skilled engineers and scientists in China from rapidly catching up with the industrialized countries. In an attempt to reduce the competitive threat Sony has decided that it will increase its internal production of semiconductors from the current 20 per cent to 40 per cent by 2007.[40] Sony's new strategy for semiconductors will help the company to differentiate its digital cameras, computers, LCD TV sets and other products from those of rival companies, while at the same time cutting costs.

China is No Longer a Strategic Partner to the US, But a Strategic Competitor

However, a very different perspective on China's development has developed in the US. The future challenge of China to American technological, industrial and military interests is obviously feared. China had been perceived as a strategic partner before the collapse of the Soviet Union and before it emerged as the factory of the world through attracting large-scale FDI from the early 1990s onwards. The latter aspect eventually triggered a concern among workers and policy makers, but it was not until the period between the inauguration of President Bush in January of 2001 and the attack on the World Trade Center that the conceptualization of China changed from strategic partner to strategic competitor. The coining of the term was never officially announced by the administration but has increasingly become prevalent in media reports on China. The US concern or rather fear of a technologically rapidly evolving industrial superstructure in China has its basis in three important factors that fall outside normally accepted policy interventions.

First, the US is no longer the main focus of Taiwan exports as it now exports twice as much to China as it does to the US. Simultaneously, imports from the US are falling while imports from China are rapidly increasing. What is more important than sheer numbers is the fact that Taiwan's exports to China have an increasingly larger share of high-technology products, which is accompanied by a huge flow of Taiwanese engineers and managers. They contribute in a major way to technological upgrading on the Mainland, which has been followed in recent years by investment in advanced semiconductor production facilities, and they have set up substantial research laboratories in many sectors, with ten of them located in Beijing and an even larger number in Shanghai. In sum, Taiwan and China are increasingly integrated in economic, industrial and technological terms.[41]

The US-China Economic and Security Review Commission, under the Congress, has been established to analyse important changes and early in 2004 organized a hearing in San Diego.[42] On this occasion much attention was given to the rise of China as a player in Asia and its consequences for the US in the areas of the economy, politics and the military. One of the invited speakers, Professor Ellis Krauss, suggested that China's integration in the Asia-Pacific region is of much more serious concern to the US than bilateral relations.[43]

Second, China has since started to send its graduate students for training in the US, many of whom have remained, creating a substantial brain pool. A large number of them have decided to remain in the US but maintain close family ties and often have deepening professional contacts in China as the country is seen as an emerging scientific and technological superpower. Thus, China is able to use this brain pool to advise and assist in major government programmes, and not a few will be tempted to return to take up key positions in universities, research institutions and high-technology companies.

Third, China has gradually started to acquire foreign companies in high-technology sectors during the past few years. To many it was a stunning announcement in mid-December 2004 when IBM and Lenovo, formerly Legend, agreed to join forces.[44] Actually, it was Lenovo that had acquired the personal computer division of IBM, thereby creating China's fifth-largest company with sales of US$12.5 billion in 2003. Discussions had been going on for some time and were restarted in earnest in March 2004, as IBM was shifting towards services, software and specialized software technology. The Government in China will remain the largest equity holder in the new Lenovo, in which IBM has taken an 18.9 per cent equity stake. At the same time IBM agreed to place 10 000 of its employees, its brand name for five years and some of its prestige in the new company, which will also gain full access to PC-related R&D.[45] More than 70 per cent of its sales will come from outside China, while 70 per cent of its workers will be inside China.[46]

Japan and the EU view the emerging technological power in China from a commercial perspective which naturally has strong nationalistic overtones. However, China's potential to become a technological superpower is, as in the US, a double-edged sword as it has not only commercial implications but equally strong military consequences. This arising concern has become evident in a number of congress hearings and from reports that originated in think tanks such as RAND, or research units more directly connected to the Pentagon. A RAND report[47] of 2001 stated in a summary after having examined commercial technologies in eight industries that have the most potential for supporting military technology development that:

even though China's military will not be the U.S. military's technological equal by 2020, the U.S. still must prepare for a China whose capabilities will steadily advance in the next 10 to 20 years, perhaps developing capabilities in certain 'niches' that will present difficulties for the U.S. military in some potential-conflict scenarios.

The sectors that were examined included microelectronics, computers, telecommunications equipment, nuclear power, biotechnology, chemicals, aviation and space. The report noted that China, at the time, had significant production capabilities in seven sectors with the exception of biotechnology. However, there existed significant limitations to capabilities in all eight, although China's open economic policies have given China access to the most advanced dual-use technologies. China was able to assemble many high-technology products but not yet capture required core technologies. The report noted that while Chinese companies are able to assemble microcomputers comparable to those produced by IBM, they are still dependent on imported microchips.

In the early 2000s R&D globalization has become an important feature of an integrated interdependent market economy and China has recently become an increasingly important player in a global R&D market because of the size of its market for almost all products and the size of its talent pool. The Stimson Center, with its core mission to examine national and international security issues by exploring the important linkages between trade and security, published a profound analysis of foreign high-tech R&D in China in 2003.[48]

This underlines that China's Open Door Policy and its related reforms have given it impressive results in its S&T modernization. It still has a long way to progress before reaching the levels of advanced industrialized countries, but China was able to encourage foreign R&D investment and today the world's leading companies in computers, telecommunications and pharmaceuticals have established their R&D presence in China, many of them having made major R&D investments. The executive summary concludes with the following paragraph.

> Although the United States benefits from a continued net inflow of R&D investment from around the world, US government funding for basic research and education should be increased in order to maintain the US lead in critical high-tech industries and innovation. This is crucial to ensuring the United States remains economically, technologically, and militarily competitive. Additionally, as foreign nationals working in US labs, universities, and high-tech companies become able to find similar work in their own economies due to globalization, the US government must invest more in grade school and secondary education, particularly in basic sciences, mathematics, and engineering, or risk falling behind.

In addition, the growing inflow of foreign R&D investments is to be observed. The author notes that foreign investments in R&D in China play a significant role in the country's long-term development of science and technology. At the same time they are generally focused on applied research and technology development, which has the potential to help China to grasp modern innovative concepts. A document prepared by Beijing City in 2004 lists a total of 167 foreign-invested R&D units in Beijing alone. Almost half of them – 70 – are US units, with another 33 under Japanese control. Taiwan has established ten units and another nine originate in Korea. Furthermore China is in an extraordinarily favourable position to exploit these possibilities as the country has the advantage of a sound and highly-skilled scientific base for which the rapid expansion of the universities will substantially add to the already existing talent pool.

FROM TECHNOLOGICAL SUPERPOWER TO SCIENTIFIC SUPERPOWER

China is still a developing country and is also in transition from a command economy to a market economy; therefore its national innovation system has to be transformed. In the process the responsibility for technological learning has been shifting from the central research and design institutes to enterprises. Thus, industrial experience and technological capacity are nowadays accumulating at the enterprise level, although still slowly.

China is showing considerable prowess in exporting high-technology products, although still with a strong origin in imported technology and components, and primarily based in foreign direct investment. Traditional measures indicate substantial and rapid changes when measuring S&T inputs, although still weakly reflected in technological output. This suggests that the situation and experience of China should be compared with that of Japan in the early 1970s, after which it took the country another 20 years to reach the status of a technological superpower. Thus it might be more relevant to look for technological upgrading in the mushrooming of industrial clusters, particularly in China's coastal areas. A number of such clusters are likely to evolve into strong innovative capability and future centres of excellence. Chinese companies are still learning to turn out innovations that can best serve the domestic market.

The importance in learning lies not only in acquiring technologies, but more in promoting technological progress by releasing market forces. The gaining and accumulating of market experience – institutional settings, effective incentive mechanisms, developing marketable products, selling new product in the market and so on – is more important than just acquiring a particular technology as such. However, this intensive learning process has primarily been concentrated in China's coastal areas – where institutional innovations and expansion of formal training and higher education may be the most significant contributing factors to China's continued progress.

China has already laid the foundation for becoming a formidable power in most fields of industrial technologies by 2020. This will take place through a symbiotic interaction between national R&D programmes and research within the corporate sector, in which the rapid infusion of foreign-controlled activities may not diminish. In accepting a full partnership in a globalizing economy, China has been able to accomplish a major geographical shift of not only industrial production but also industrial R&D into China. Aside from rapidly expanding trade relations with all major industrialized countries, ongoing changes also indicate a shift of R&D gravity towards East Asia.

While changes are underway towards a technological superpower status for China, the country is simultaneously laying the framework for a future aspiring scientific superpower status. This is being done through continued investment in China's human resources. The country could have trained more than 100 million in its colleges and universities by 2020, with a substantial number of students having received masters and PhD degrees. A continued expansion of higher education could make China into one of the most highly educated nations in the world by 2050.

In the foreseeable future the major task of China's scientists will be to provide basic support for continued industrial technological development and to train their successors. They would be able to take on challenging research tasks in subsequent decades and in all likelihood advance the frontiers of knowledge in a large number of existing fields, as well as progressing into new knowledge areas. This will not happen in isolation from the rest of the world but will be carried out substantially in knowledge regions within the national boundaries of China. Beijing and Shanghai are already such budding knowledge metropolises. They will be joined by a number of other knowledge regions in China within the first half of the 21st century.

This will not happen unless China in the medium term formulates and implements far-reaching policies that fully draw the present population majority in rural areas into the modernization process that started in coastal areas some twenty years ago. This would require political insight and foresight as it would result in urbanization and relocation of people on a major scale, bearing a striking resemblance to the industrialization that took place in Britain and Germany during the 19th century – but in a much shorter period of time.

NOTES

1. Pommeranz (2000).
2. Elman (1999).
3. Chen Ellen M. The Dynamics of Tradition, Chapter XIII (n.d.).
4. See Graham (1973), pp. 45–70.
5. Elman (1999).
6. Ibid.
7. Ibid.
8. Mervis (2004), p. 1285 (quotation is by Professor Richard Smalley of Rice University, Houston, TX, USA).
9. 'Chinese to travel in space in 20 years: official', *Xinhua*, 4 November 2004.
10. Ibid.
11. 'Nuclear power generation capacity to triple', *Xinhua*, 5 November 2004.
12. 'China to have 140 million cars by 2020', *China Daily*, 4 September 2004.
13. The Shanghai municipal government still imposes much higher charges on private car buyers than in the other regions in China by auctioning car plates in an effort to control car sales and prevent traffic jams in the city. The average charge for a car plate in Shanghai was more than RMB34 000 in May 2004 (Gong Zhengzheng, 2004b).
14. Gong Zhengzheng (2004a).
15. 'Although the government now requires foreign automakers to have stakes of at most 50 per cent in joint ventures to protect Chinese players, product portfolios and key technology are still tightly controlled by foreigners' (Gong Zhengzheng, 2004a).
16. Fu Jing (2004).
17. Ibid.
18. Gong Zhengzheng (2004b).
19. Gong Zhengzheng (2004a).
20. Gong Zhengzheng (2004b).
21. Total investment in building new auto-making capacity in China will amount to RMB216 600 million by 2007, according to figures released by the National Development and Reform Commission in early 2004 (Gong Zhengzheng, 2004b).
22. Gong Zhengzheng (2004a).
23. Chery in East China's Anhui Province, will start to produce its own brand cars next month in a plant built by its local partner in Iran with an annual capacity of 50 000 units. Zhongxing Automobile, the Chinese mainland-Taiwan joint venture based in North China's Hebei Province, expects to build four to five plants in North Africa and South America. An affiliate of Chang'an has struck a deal with a Vietnamese partner to jointly build a plant, which is expected to start production during the first half of 2005 with a planned capacity of 5000 units within the next two to three years. Geely, the privately-owned car maker based in East China's Zhejiang Province, is also considering building plants overseas (Gong Zhengzheng, 2004a).
24. 'Blueprint for economic growth', *China Daily*, 15 July 2004.
25. Bloom (1990).
26. Ibid.
27. European Commission (2004b).
28. European Commission (2004a).
29. European Commission (2004a).
30. European Commission (2004b), Table 5.1, p. 310.
31. Inter-American Development Bank (2004).
32. Ibid., p. 58.
33. European Commission (2004b), p. 333.
34. Ibid., p. 334.
35. Ibid., p. 354.
36. Ibid., p. 353.
37. A senior official of the Ministry of Economy, Trade and Industry stated that for Japanese

firms to survive they need an annual investment of 200 billion yen in carefully chosen businesses, 70 per cent of sales overseas and a global share of 10 per cent. (Source: 'Firms look to protect sensitive technology at home', *The Japan Times*, 4 October 2004).
38. 'Firms look to protect sensitive technology at home', *The Japan Times*, 4 October 2004.
39. Ibid.
40. 'Sony to invest ¥60bn in '05–06 to boost chip output', *Nikkei Interactive*, 21 December 2004.
41. Bradsher (2004), p. 7.
42. US–China Economic and Security Review Commission (2004).
43. Ibid., p. 176.
44. 'For IBM, deal with Lenovo is "China card"', *International Herald Tribune*, 14 December 2004.
45. 'IBM set up business in China 20 years ago, and now has 4200 employees there. In 1995, IBM opened a research laboratory which now employs 150 Chinese scientists. Five years ago, it established a Chinese software development laboratory, which today has 500 engineers working on Linux projects alone. IBM is the leading corporate supporter of Linux, a free operating system that is an alternative to Microsoft Windows.' (Source: ibid.)
46. Ramstad (2004).
47. Cliff (2001).
48. Walsh (2003).

Appendix: The 2020 Plan on Science and Technology[1]

In 2003 Chinese science and technology agencies, researchers and scientists started to prepare the country's long-term science and technology strategy plan for the period up to 2020. The details of the plan are being released during 2005. The Ministry of Science and Technology (MOST) provided the following outline in connection with the China–EU High-level Forum on S&T Strategy that was held in Beijing on 12–13 May 2005.

PREPARATION OF CHINA'S NATIONAL MEDIUM- AND LONG-TERM S&T DEVELOPMENT PLAN AND ITS PROGRESS

In June 2003 China launched the preparation of its national medium- and long-term S&T development plan. The preparation process has gone smoothly, and the plan is expected to be published soon.

1. Background

The planning efforts launched in June 2003 for preparing a national S&T development plan will produce China's first medium- and long-term S&T development plan in the new century. The new plan also marks China's first S&T development guidance document since the establishment of its socialist market-economy system, and since its accession to the WTO. It will become the first S&T plan to be rolled off and implemented under the new administration. With a clearly defined historical mission to reflect its times, the preparation of the new S&T plan therefore constitutes a major event in China's new-century modernization campaign.

1. *Heading for the goal to build a well-to-do society.* China has defined a development target of creating a well-to-do society by 2020, allowing its population of more than one billion to enjoy a higher living standard. By 2020 China's GDP is supposed to be triple that of 2000. The Chinese government is fully confident of achieving this goal, although it is quite

aware of the tough nature of the task. Numerous elements constraining economic growth, including population, product structures, energy supply, ecological environment and natural resources, have made China choose an industrialization approach different from the traditional industrialization process taken by the developed nations in the past. In this context, the Chinese government has worked out a scientific development line of thinking, placing human resources at the root of everything, and advocating an all-round, coordinated and sustainable development. In the future China's socioeconomic development has to follow a new development pattern, characterized by resources efficiency and environmental protection. China is working on a new industrialization process through which industrialization is propelled by information technology, while information technology is advanced in turn by industrialization. In addition, China is pursuing a coordinated development of both urban and rural areas in the course of agricultural modernization and urbanization. In all areas, it is a comprehensive, coordinated and sustainable socioeconomic development, guided, mobilized and supported by S&T innovations. All these goals have placed urgent demands on Chinese people to accelerate S&T development within the next 15–20 years. Chinese people are asked to use S&T strength creatively in overcoming all the difficulties and constraining elements, promoting coordinated socioeconomic development and realizing future development goals.

2. *Preliminary establishment of the socialist market economy.* Since making the establishment of the socialist market economy one of its major reform goals in 1992, China has achieved substantive progress in economic system reform, both theoretically and in practice. Under the socialist market economy China's S&T activities have witnessed fundamental changes in both structures and modes as compared with the planned economy. China is now enjoying a prosperous S&T development. S&T activities are no longer confined to state-owned research institutes, universities and industrial enterprises. Private S&T activities have become increasingly popular for expanded scales. More and more private businesses are working on technical innovations. As a result, industries have gradually become a major force in R&D activities. A market-oriented S&T resources distribution pattern is shaping up. China's S&T system in the new century, therefore, will be in tune with the development of the market economy.

3. *New opportunities and challenges brought by fast S&T development.* Today's world is spurred by unprecedented fast S&T development. Science and technology has become a major force in stimulating economic development, promoting social advancement, and safeguarding national

security. In the coming decades, the world will have the chance to see new S&T breakthroughs in the fields of information technology, life science, material science, brain and cognition science, earth and environmental science, mathematics and system science. The world may not be surprised to see the birth of new cutting-edge sciences hybriding natural and social sciences. These new developments will trigger new revolutions in human society, and will impose profound impacts on world economic and social development. S&T breakthroughs will create more perspectives for the development of human society. S&T development, however, will also add more uncertainties and increase difficulty for technical vision. How to grasp new opportunities brought by S&T development has become a challenge to every nation. As a developing nation, China has an urgent need to grasp new opportunities brought about by future S&T developments.

4. *Economic and S&T globalization.* Along with deepened economic globalization, R&D activities have also become more and more globalized. As a result, S&T innovation resources, human resources and information have travelled fast worldwide. S&T competition or cooperation have found expression among nations, among regions and among industries. None of the nations in the world can address all its S&T problems in isolation. A nation, when planning for future S&T development, has to consider globalization trends and position itself in a globalization environment according to its capacity, strength and needs. After its accession to the WTO, China has prepared to open up further, making itself fully integrated into international communities. China cannot afford to develop its S&T behind closed doors. It calls for more extensive participation in international S&T cooperation and exchanges. China will seek its own innovations and breakthroughs based on absorbing and learning from the developed nations' advanced technology. In the course of rendering its contributions to world S&T development China will address its own development problems.

The time elements cited above have underlined the necessity and importance for China to prepare a new S&T plan in the new century. Meanwhile, these elements have become timeline demands and major strategic orientations for planning. Since the 1950s, China has, in line with the specific needs of each development phase, published and implemented seven medium- and long-term S&T development plans. These plans, with important promoting roles, have shaped China's S&T position in the world today. The new long-term S&T plan will further advance China's S&T development, laying a solid foundation for building a well-to-do society and accelerating the realization of China's modernization dream.

2. Major Lines of Thinking

The planning will cover a period running from 2006 to 2020. It is an extremely tough and complicated task to work out a medium- and long-term S&T development plan, as it has to meet with China's current state and future socioeconomic development needs, while remaining in line with globalization trends and the natural rules of S&T development.

The Chinese government has paid great attention to the planning. A steering panel personally headed by Chinese Premier Wen Jiabao has been established to provide guidance for and new medium- and long-term S&T development planning. Chen Zhili, state councilor, is the panel's vice-chairperson in charge of planning activities. Premier Wen has made important remarks or instructions about the planning process, showing his concerns and weighted attention. He has also delivered explicit comments on guiding principles, major research orientations, objectives and scientific and democratic processes for planning activities. Furthermore, he has laid down major working principles for the planning. The planning mainly focuses on the following aspects:

1. *Defining national goals.* As a part of government functions, formulating and implementing a national S&T development plan will have clearly defined national goals. In this context, providing a powerful S&T support for materializing the ambitious target of building a well-to-do society by 2020 constitutes a major goal under the current planning. Elements including building a well-to-do society, new industrialization approaches, and 'rejuvenating the nation with science and education' strategy and sustainable development strategy make up main features of the planning. The planning will, in the national interest, address general priorities, distributions and involving systems for China's S&T development, playing a guiding and regulating role in establishing an S&T system tailored to China's situation.

2. *Sorting out strategic priorities.* As a developing nation, China has limited resources available for S&T investment. Such limitation will last for quite a long period of time in the future. Orientations and fields chosen for S&T development therefore have to be highly selective. While taking into full account China's situation and current strength, the planning will focus on promoting national socioeconomic development, industrial restructuring and S&T self-development. Under the principle of 'doing something at the expense of not doing something else', the planning will stick to the medium- and long-term S&T strategic priorities, addressing key issues involving socioeconomic development and

people's, avoiding 'taking care of all', being selective but accurate, highlighting priorities, pooling forces and striving for major breakthroughs.

3. *Encouraging interactions between S&T and socioeconomic development.* In today's world, S&T development is spurred by socioeconomic needs as well as by S&T development itself. In this context, S&T development planning is not only meant to emphasize S&T responsibility for socioeconomic development, but is also meant to address technical problems encountered in socioeconomic development, emphasizing closer ties between science and technology, and emphasizing interactions between S&T and socioeconomic development. The planning will view interactions of science, technology, economy and society, their co-existences and mutual restrictions under the principles of the system theory. General trends and orientations of S&T development are important markers to reflect S&T pioneering roles in socioeconomic development. These elements are meant to forge a close tie between S&T and socioeconomic development, whether in the short, medium or long term, in an attempt to enhance the nation's strategic knowledge and technology reserves, and its sustainable innovation capacity.

4. *Intensifying system innovations.* China's S&T system has to make sure that its system innovations are in line with the demands of the market economy. While continuing to take advantage of the merits of the socialist system, the planning will design a new S&T system and establish a well-functioning national innovation system under the framework of the market economy. Appropriate handling of the relationship between the market mechanism and the state will make itself a focus. This means the government will play a guiding, planning and incentive role in formulating S&T policies and a macro S&T management system, while the market economy does its infrastructure job in S&T resources distribution, and in mobilizing and encouraging all walks of life to contribute to S&T development. Special attention will be given to creating a sound policy and legal environment, enhancing high-tech innovation efficiency and benefits in market-oriented competition and cooperation. S&T-related industrialization and industrial technical innovation capacity-building also constitute a focus.

5. *Regional features.* China is a vast country with large differences region by region. The planning will address uneven regional S&T development, encouraging the search for solutions right for local needs, strengthening the innovation systems characterizing the locality, turning S&T elements into real internal variables in regional economic development, and promoting coordinated regional development with S&T means.

6. *Opening wider.* To be parallel with the economic and S&T globalization process, the planning is asked to have a global vision, with an awareness of globalization, paying more attention to opening up and international S&T cooperation and exchanges. China's S&T development will find its due position in global S&T competition and cooperation. The planning process is open to the international communities.

7. *Public participation.* Along with the improvement of people's lives, the Chinese public have shown their increasing concern over S&T impacts on their work and lives. In addition, they are making more and more diversified demands on S&T development. Promoting public S&T awareness and S&T innovations constitute two basic elements in S&T development, as inseparable from each other as the two wheels on either side of a car and the left and right wings of a bird. To truly mirror the fundamental interests of the majority of the Chinese population, the entire planning process is open to the public. At the same time, the Chinese public are encouraged to be part of the process in an attempt to make the planning a process through which consensus is reached, comments and views are collected, and a democratic and scientific decision-making process is promoted.

3. Priority Strategic Issues

Formulating a medium- and long-term S&T development plan makes itself a major decision-making process for working out China's S&T development strategies. In this context, the strategic study becomes important in providing evidence and grounds for the planning.

Since August 2003 the Chinese Government has sponsored 20 topic-oriented strategic studies involving major strategic and visionary issues concerning China's socioeconomic development and S&T development, with the participation of more than 2000 researchers from Chinese S&T, economic and management communities. The 20 topics are: S&T development strategies in general, S&T system reform and national innovation systems, S&T issues in the manufacturing industry, agricultural S&T issues, S&T issues in energy, resources and marine development, traffic and transportation, modern service industries, population and health, public security, ecological construction, environmental protection and cyclical economy, urban development and urbanization, national defence, strategic high techs and associated industrialization, basic scientific study, S&T conditions and infrastructures, S&T personnel, S&T investment and associated management, S&T laws and policies, innovation culture and popular science, and regional S&T development.

The strategic study, having run for almost two years, has achieved important results. The 20 topic-based studies will not only help to work out the orientations and priorities for China's S&T development in the next 15 years, but will also facilitate the implementation of scientific development concepts and macroeconomic readjustment, providing major evidence for preparing the 11th five-year plan for China's socioeconomic development.

NOTE

1. Official announcement May 2005, distributed by the EU Delegation in Beijing. Simultaneously released was Rodrigues and Yuan (2005).

Bibliography

BCG (2004), 'Capturing global advantage: how leading industrial companies are transforming their industries by sourcing and selling in China, India and other low-cost countries', BCG Report, April.

Bloom, David E., Canning, David and Sevilla, Jaypee (2003), *The Demographic Dividend: A New Perspective of the Economic Consequences of Population Change*, Santa Monica: RAND.

Bloom, Justin L. (1990), *Japan as a Scientific and Technological Superpower*, PB90-234923, Potomac, MA: National Technical Information Service, US Department of Commerce.

Bradsher, K. (2004), 'China casts long shadow over Taiwan's economy', *International Herald Tribune*, 14 December.

Breidne, Magnus (2005), 'Information and Communications Technology in China: a general overview of the current Chinese initiatives and trends in the area of ICT', Vinnova report VR 2005:05, Stockholm: Swedish Institute for Growth Policy Studies.

Broad, W.J. (2003), 'China ready for human spaceflight', *New York Times*, 14 October.

Cao, Cong (2002), *Can Chinese Scientists Win the Noble Prize?*, EAI Background Brief No. 135, Singapore: East Asian Institute.

Cao, Cong (2003a), *China's High-tech Parks in Transition*, EAI Background Brief No. 153, Singapore: East Asian Institute.

Cao, Cong (2003b), *The Emerging Research Institutions of Life Science in China*, EAI Background Brief No. 186, Singapore: East Asian Institute.

Cao, Cong (2004a), *The Brain Drain Problem in China*, EAI Background Brief No. 215, Singapore: East Asian Institute.

Cao, Cong (2004b), *China's Efforts at Turning 'Brain Drain' into 'Brain Gain'*, EAI Background Brief No. 216, Singapore: East Asian Institute.

Cao, Cong (2004c), *China's Scientific Elite*, London: RoutledgeCurzon.

Chang, P.L. and Shih, H.Y. (2003), 'The innovation systems of Taiwan and mainland China: a comparative analysis', *Technovation*, **24**(7), 529–39.

Chen, Qide (2005), 'Intel research centre opens its doors', *China Daily*, 13 May.

Chen, Ellen M. (n.d.), *The Dynamics of Tradition*, Chapter XIII, http://www.crvp.org/book/Series03/III-7a/chapter_xiii.htm (accessed 25 April 2005).

Chen, Hong (2004), 'Striking off non-performing labs is right for research', *China Daily*, 27 December.
Chen, Qide (2004), 'Intel, Shanda to jointly develop digital home technology', *China Daily*, 9 June.
Chen, Shin-Horng (2004), 'Taiwanese IT firms' offshore R&D in China and the connection with the global innovation network', *Research Policy*, **33**, 227–349.
Chen, Xikang, Cheng, Leonard K., Fung, K.C. and Lau, Lawrence J. (2001), 'The Estimation of Domestic Value-added and Employment Induced by Exports: An Application to Chinese Exports to the United States', mimeo, University of California, Santa Cruz and Stanford University.
Chen, Zhiming (2003), 'Home-grown 3G standard encouraged', *China Daily*, 29 August, http://www.chinadaily.com.cn/en/doc/2003-08/29/content_259290.htm (accessed 26 April 2005).
Chen, Zhiming (2004a), 'SMS bonus for Xiaolingtong subscribers', *China Daily*, 24 January.
Chen, Zhiming (2004b), 'TD–CDMA alliance to enlist foreign firm', *China Daily*, 23 October.
Cheung, K.Y. and Lin, P. (2003), 'The spillover effect of FDI on innovation in China: evidence from the provincial data', *China Economic Review*, **15**, 25–44.
China Statistical Yearbook 2003 (2003), Beijing: China Statistics Press, September.
China Universities and Colleges Guidebook (2004), Beijing: Foreign Languages Press.
Clendenin, M. (2003), 'Chinese start-up readies 64-bit processor', http://www.my-esm.com/showArticle.jhtml?articleID=7400203 (accessed 25 April 2005).
Cliff, Roger (2001), *The Military Potential of China's Commercial Technologies*, Santa Monica, CA: Rand Corporation.
Cui, Ning (2003), 'High-tech industries become new money spinners', *China Daily*, 17 September.
Cui, Ning (2004), 'Scientific projects propel nation's growth', *China Daily*, 5 October.
Cui, Ning (2005), 'Foreigners encouraged to seek S&T partners – nation promises easier access for overseas research institutes', *China Daily*, 13 May.
Dahlman, Carl J. (1994), 'Technology strategy in East Asian developing countries', *Journal of Asian Economics*, **5**, 541–72.
Dahlman, Carl J. and Aubert, Jean-Eric (2001), *China and the Knowledge Economy: Seizing the 21st Century*, Washington, DC: The World Bank.

Deng, Xiaoping (1993), *Selections from Deng Xiaoping*, vol. III, Beijing: People's Press.
Dolven, Ben and Neuman, Scott (2003), 'China's aviation dream', *Far Eastern Economic Review*, 25 December, 88–91.
Eberstadt, Nicholas (2004), *Strategic Asia 2003–2004*, National Bureau of Asian Research, http://www.policyreview.org/feb04/eberstadt.html (accessed 27 April 2005).
Elman, Benjamin A. (1999), '"Chinese sciences" and the triumph of "modern science" in China', paper for the conference 'Rethinking Science and Civilization: the Ideologies, Disciplines, and Rhetorics of World History', Stanford University Department of Asian Languages and the Program in History and Philosophy of Science, 21–23 May.
Enright, Michael J. (2000), 'Globalization, regionalization, and the knowledge-based economy in Hong Kong', in John H. Dunning (ed.), *Regions, Globalization, and the Knowledge-based Economy*, Oxford: Oxford University Press, pp. 381–406.
Enright, Michael J., Chang, K.M., Scott, E. and Zhu, W.H. (2003), *Hong Kong and the Pearl River Delta: The Economic Interaction*, Hong Kong: 2022 Foundation, pp. 6–11, 48–79 and 106–30.
Enright, Michael J., Scott, E. and Dodwell, D. (1997), *The Hong Kong Advantage*, Hong Kong: Oxford University Press.
Enright, Michael J., Scott, E. and Chang Ka-mun (2005), *Regional Powerhouse: The Greater Pearl River Delta and the Rise of China*, Chichester: John Wiley & Sons.
European Commission (2002), 'China: European Commission approves Country Strategy Paper 2002–2006', Press Release IP/02/349, 1 March, Brussels: European Commission.
European Commission (2004a), 'Commission report highlights changing competition from China', Press Release IP/04/1400, 25 November, Brussels: European Commission.
European Commission (2004b), 'European competitiveness report 2004', Commission Staff Working Document, SEC (2004) 1397, 8 November, Brussels: European Commission.
Feigenbaum, Evan (2003), *Chinese Techno-warriors: National Security and Strategic Competition from the Nuclear Age to the Information Age*, Stanford, CA: Stanford University Press.
Fisher, Richard D. (2004), 'Foreign military acquisitions and PLA modernization', Testimony before the US–China Economic and Security Review Commission, 6 February 2004.
Freeman, C. (1987), 'National systems of innovation: the case of Japan', in C. Freeman, *Technology Policy and Economic Performance: Lessons from Japan*, London: Pinter.

Fu, Jin (2004), 'Shangai tops for development capacity', *China Daily*, 2 March, http://www.chinadaily.com.cn/english/doc/2004-03/02/content_310718.htm (accessed 27 April 2005).

Fu, Jing (2005a), 'Scientific innovation essential', *China Daily* (Weekend), 2–3 April.

Fu, Jing (2005b), 'China–EU conference reaches high-tech consensus', *China Daily* (Weekend), 14–15 May.

Fujimoto, Takahiro (2002), *Thinking of the Chinese Manufacturing Sector in Architecture Context*, Research Institute of Economy, Trade and Industry, Japan.

Futron Corporation (2003), *China and the Second Space Age*, Futron.

Gida, Aravid (2004), 'Hanging by a thread: textile factories throughout Asia face extinction as a long-standing global trade pact is to expire', *Time*, 1 November.

Gilboy, George J. (2004)), 'The myth behind China's miracle', *Foreign Affairs*, July/August, 38–48.

Gong, Zhengzheng (2004a), 'Fast-growing auto industry draws concern', *China Daily*, 6 October.

Gong, Zhengzheng (2004b), 'New auto industry rules state JV guidelines', *China Daily*, 2 June.

Gong, Zhengzheng (2004c), 'SAIC's Ssangyong bid looks successful', *China Daily*, 24 July.

Gong, Zhengzheng (2005a), 'Carmaker to cut costs by using local components – Volkswagen's joint venture under pressure from the strong euro', *China Daily*, 18 April.

Gong, Zhengzheng (2005b), 'Self-development key for local car makers', *China Daily*, 21 April.

Graham, A.C. (1973), 'China, Europe, and the origins of modern science: Needham's *The Grand Titration*', in Shigeru Nakayama and Nathan Sivin (eds), *Chinese Science: Explorations of an Ancient Tradition*, Cambridge, MA: MIT Press, pp. 45–70.

Gu, Shulin (1999), *China's Industrial Technology*, UNU/INTECH Studies in New Technology and Development, London: Routledge.

Guo, Kesha (2004), *China's Industrial Development Strategy: Policy Options in the New Era*, Social Sciences in China.

Guo, Nei and Chen, Qide (2004), 'Registered trademarks top 2 million', *China Daily*, 12 December.

Hao, Zhidong (2003), *Intellectuals at a Crossroads: The Changing Politics of China's Knowledge Workers*, Albany, NY: SUNY Press.

Hobday, M. (1995), *Innovation in East Asia: The Challenge to Japan*, Aldershot, UK: Edward Elgar Publishing.

Hou, Mingjuan (2001), 'Patented chip industry grows', *China Daily*, 26 April.
Hout, T. and Lebreton, J. (2003), 'The real contest between America and China', *Asian Wall Street Journal*, 16 September.
Howell, Thomas R., Bartlett, Brent L., Noellert, William A. and Howe, Rachel (2003), *China's Emerging Semiconductor Industry: The Impact of China's Preferential Value-added Tax on Current Investment Trends*, prepared by Dewey Ballantine LLP for the Semiconductor Industry Association, Washington, DC.
Hu, Albert (2001), 'Ownership, government R&D, private R&D, and productivity in Chinese industry', *Journal of Comparative Economics*, **29**, 136–57.
Inter-American Development Bank (2004), *The Emergence of China: Opportunities and Challenges for Latin America and the Caribbean*, draft version, Integration and Regional Programs Department, IADB.
International Development Research Centre (Canada) and State Science and Technology Commission (PRC) (1997), *A Decade of Reform: Science and Technology Policy in China*, IDRC.
International Federation of Pharmaceutical Manufacturers Associations (2003), *Accelerating Innovative Pharmaceutical Research and Development in China: A Case Study*, Geneva: IFPMA.
Iritani, E. (2002), 'China's next challenge: mastering the microchip 3002', *LA Times*, 22 October, http://justin.deepdrift.com/smic.htm (accessed 26 April 2005).
Jia, Hepeng (2003), 'Central China throws down gauntlet to coast', *China Business Weekly*, 29 July, http://www.chinadaily.com.cn/en/doc/2003-07/29/content_251063.htm (accessed 27 April 2005).
Kan, Shirley A. (2001), *China: Possible Missile Technology Tranfers from U.S. Satellite Export Policy: Axioms and Chronology*, Washington, DC: Congressional Research Service.
Khazbiyev, Aleksey (2004a), 'China asking Russia to transfer latest military technologies', *CEP Weekly Business Magazine*, 24 May.
Khazbiyev, Aleksey (2004b), 'China takes a swing at the sacred', *CEP Weekly Business Magazine*, 4 May (FBIS translation: Khazbiyev, 2004a).
Kim, Joon-Kyung, Kim, Yangseon and Lee, Chung H. (2004), 'Trade and investment between China and South Korea: towards a long-term partnership', *Journal of the Korean Economy*, **5**(1), 97–124.
Kogan, Eugene (2004), 'Russia–China aerospace industries: from cooperation to disengagement', *China Brief*, **4**(19), 30 September.
Krugman, Paul (1994), 'The myth of Asia's miracle', *Foreign Affairs*, **73**(6), November/December, 62–78.
Lall, S. (1996), *Learning from the Asian Tigers: Studies in Technology and Industrial Policy*, New York: St Martin's Press.

Lardy, Nicholas R. (2002), *Integrating China into the Global Economy*, Washington, DC: The Brookings Institution.

Lee, Zhong (2004), 'Statement on China as an emerging regional and technology power: implications for US economic and security interests', Hearing before the US–China Economic and Security Review Commission, 12–13 February 2004, p. 98.

Leng, Tse-Kang (2002), 'Economic globalization and IT talent flows across the Taiwan Strait: The Taipei/Shanghai/Silicon Valley Triangle', *Asian Survey*, **42**(2), March/April, 230–50.

Li, Weitao (2003a), 'CEC sets up shareholding company', *Business Weekly*, 5 August.

Li, Weitao (2003b), 'Big opportunities from "Little Smart"', *China Business Weekly*, 18 November, http://www.chinadaily.com.cn/en/doc/2003-11/18/content_284299.htm (accessed 25 April 2005).

Li, Weitao (2004a), 'Can TD-SCDMA make a big splash?', *China Business Weekly*, 29 June.

Li, Weitao (2004b), 'Infineon plotting China expansion', *China Business Weekly*, 27 September, http://www.chinadaily.com.cn/english/doc/2004-09/27/content_378183.htm (accessed 26 April 2005).

Li, Weitao (2005), 'Huawei partners with Marconi', *China Daily*, 1 February.

Li, Wenfang (2004), 'Telecom-system firm plans listing 2 units', *China Daily*, http://www.chinadaily.com.cn/english/doc/2004-04/09/content_321824.htm (accessed 25 April 2005).

Li, Zhenyuan (1997), 'The discussion on operational mechanism of aerospace industry', *Aerospace Industry Management*, **7**, 26–9 (in Chinese). Referenced in Weiwei Wu and Bo Yu (2003), *The Study on Organizational Modes of Management of Technology for Aerospace Industry*, School of Management, Harbin Institute of Technology.

Liang, Guoyong (2004), 'New competition: foreign direct investment and industrial development in China', dissertation, Erasmus University, Rotterdam.

Liang, Yu (2004a), 'Embraer wants bigger share of jet market', *China Daily*, 29 March, http://www.chinadaily.com.cn/english/doc/2004-03/29/content_318829.htm (accessed 26 April 2005).

Liang, Yu (2004b), 'Yangtze delta offers an attractive lure for entrepreneurs', *China Daily*, 26 March, http://www.chinadaily.com.cn/english/doc/2004-03/26/content_318091.htm (accessed 27 April 2005).

Lin, Jingtong, Liang, Xiongjian and Wan, Yan (2001), *Telecommunications in China: Development and Prospects*, Beijing: Nova Publishers.

Liu, Baijia (2003), 'In high-tech industry, Beijing focuses on six of the best', *China Daily (HK)*, 9 July, http://www.chinadaily.com.cn/en/doc/2003-07/09/content_244068.htm (accessed 25 April 2005).

Liu, He, Qin, Hai and Yu, Shiyang (2000), 'Urbanization in China: history, retrospection and policy choosing', paper presented at the 'Reforming China's Economy – Data Analysis and a Quantitative Assessment' Conference, Beijing, 9–10 May, International Statistical Information Centre of NBS China and the Stockholm School of Economics.

Liu, Sunray (2000), 'Beijing plans huge microelectronics base', *EE Times*, 2 November.

Liu, X. and White, S. (2001), 'Comparing innovation systems: a framework and application to China's transitional context', *Research Policy*, **30**, 1091–114.

Lu, Ding and Wong, Chee Kong (2003), *China's Telecommunications Market Entering a New Competitive Age*, Cheltenham, UK: Edward Elgar Publishing.

Lu, Qiwen, (2000), *China's Leap into the Information Age: Innovation and Organization in the Computer Industry*, Oxford: Oxford University Press.

Lundvall, B. (1992), *National Systems of Innovation: Towards a Theory of Innovation and Interactive Learning*, London: Pinter.

Luo, Man and Wan, Alexander (2005), 'Auto sector growth sustainable', *China Daily*, 29 April.

Ma, Songde (2000), 'China's international S&T cooperation: importance and promise', *China Science and Technology Newsletter*, **228**, 10 July.

Markov, J. (2004), 'Have supercomputer – will travel', *New York Times*, 1 November.

McKendrick, David (1998), *Dispersed Concentration: Industry Location and Globalization in Hard Disk Drives*, Report 98-03, The Information Storage Industry Center, University of California, San Diego.

McMillion, Charles W. (2005), 'China's high technology development', MBG Information Services, prepared for the US–China Economic and Security Review Commission, 21 April.

Medeiros, Evan, S. (2004), 'Analyzing China's industries and the implications for Chinese military modernization', Testimony to the US–China Economic and Security Review Commission, 6 February 2004.

Mervis, Jeffrey (2004), 'Perceptions and realities of the workplace', *Science*, 28 May.

Meyer, Marshal W. and Lu Xiaohu (2004), 'Managing indefinite boundaries: the strategy and structure of a Chinese business firm', *Management and Organization Review*, **1**(1), 1–30.

Ministry of Science and Technology (1999), *Suggestions on Accelerating the Growth of State-level Development Regions of High-tech Industries*, MOST.

Ministry of Science and Technology (2003), *China Science and Technology Indicators 2002: The Yellow Book on Science and Technology*, vol. 6, Beijing: Ministry of Science and Technology.

MOFCOM (2005), 'Implementation of the strategy to rejuvenate trade through science and technology', MOFCOM Report, 17 May.
Moran, Theodore H. (1998), *Foreign Direct Investment and Development: The New Policy Agenda for Developing Countries and Economies in Transition*, Washington, DC: Institute for International Economics.
Murayama, Hiroshi (2004), 'Taiwan atop East Asia trade triangle' (Nikkei Interactive), 15 March.
Murphy, David (2004), 'When the big stars venture out of China', *Far Eastern Economic Review*, 5 February.
National Bureau of Statistics (2004), *China Statistical Yearbook 2003*, Beijing: NBS.
National Bureau of Statistics and Ministry of Science and Technology (2003), *China Statistical Yearbook on Science and Technology*, Beijing: China Statistics Press.
National Development and Reform Commission (2003), *China Statistics Yearbook on High Technology Industry*, Beijing: China Statistics Press.
National Science Foundation (2004), *Science and Engineering Indicators*, NSF.
Naughton, Barry (1995), *Growing Out of the Plan: Chinese Economic Reform, 1978–1993*, Cambridge: Cambridge University Press.
Naughton, Barry (1999), *Growing Out of the Plan: Chinese Economic Reform, 1978–1993*, Cambridge: Cambridge University Press, paperback edition.
Naughton, Barry (2004), 'The information technology industry and economic interactions between China and Taiwan', in Françoise Mengin (ed.), *China in the Age of Information*, Palgrave Macmillan.
Nelson, R.R. (ed.) (1993), *National Innovation Systems: A Comparative Analysis*. New York: Oxford University Press.
Ng, Loretta (2004), 'China to stockpile oil in states', *International Herald Tribune*, 14 December.
Nolan, Peter (2004), *Transforming China: Globalization Transition and Development*, London: Anthem Press.
OECD (2004a), 'An emerging knowledge-based economy in China? Indicators from OECD databases', STI Working Paper 2004/4, Paris: OECD.
OECD (2004b), *Science, Technology and Industry Outlook 2004*, Paris: OECD.
Orleans, Leo A. (1960), *Professional Manpower and Education in Communist China*, US Government.
Perrins, Robert John (2003), *China: Facts and Figures Annual Handbook*, vol. 28, Gulf Breeze, FL: Academic International Press.
Pillsbury, Michael (2005), 'China's progress in technological competitiveness: the need for a new assessment', report prepared for the US China Economic and Security Review Commission, 21 April.

Pollard, Sydney (1981), *Peaceful Conquest: The Industrialization of Europe 1760–1970*, Oxford: Oxford University Press.

Pommeranz, Kenneth (2000), *The Great Divergence: China, Europe, and the Making of the Modern World Economy*, Princeton: Princeton University Press.

Qi, Jingmei (2004), 'Generating more consumption', *China Daily*, 4 June, http://www.chinadaily.com.cn/english/doc/2004-06/04/content_336519.htm (accessed 27 April 2005).

Qin, Chuan (2004), 'Illegal land development zones cut', *China Daily*, 26 July.

Qin, Jize (2005), 'Nigerian satellite for 2007 launch from Xichang', *China Daily*, 15 April.

Ramstad, E. (2004), 'The technological rise of China was speed – and just the beginning', *Wall Street Journal*, 20 December.

Richmond, R. (2003), 'China: The Next Microprocessor Giant?', http://www.techimo.com/articles/index.pl?photo=16 (accessed 25 April 2005).

Rodrigues, Maria João and Yuan, Zhou (2005), 'Vision paper: a China–EU strategic partnership on knowledge for growth and development', prepared for China–EU High-level Forum on S&T Strategy, Beijing, 12–13 May (outline 2005.05.09).

Roelandt, Theo J.A. and den Hertog, Pim (eds) (1998), *Cluster Analysis and Cluster-based Policy in OECD Countries*, The Hague/Utrecht: OECD-TIP Group.

Romero, Simon (2004), 'Thirsty China eyes Canadian oilfields', *The Business Times (Singapore)*, 24 December.

Rowen, Henry (2004), 'Report summary on Zhongguancun', mimeo.

Science and Technology Statistics Databook 2004 (2005), Beijing: Ministry of Science and Technology Department of Development Planning and Centre for S&T Statistic Analysis.

Semiconductor Industry Association (2002), *An Investigation Report of China's Semiconductor Industry*, Beijing: SIA.

Semiconductor Industry Association and China Centre of Information Industry Development (2004), *A Report on Development Status of Semiconductor Industry in China*, Beijing: CSIA and CCID.

Shanghai Science and Technology Committee (1999), 'Formation of Shanghai knowledge strategies', translated in C.J. Dahlman and J.-E. Aubert (2001), *China and the Knowledge Economy: Seizing the 21st Century*, Washington, DC: World Bank Institute.

Shen, Xaiobai (1999), *The Chinese Road to High Technology: A Study of Telecommunications Switching Technology in the Economic Transition*, Basingstoke, UK: Palgrave Macmillan.

Sigurdson, Jon (2000), 'Knowledge creation and innovation in geographically dispersed organisations', *Asia Pacific Journal of Management*, **17**(2), 297–330.

Sigurdson, Jon (2002), 'A new technological landscape in China', *China Perspectives (HK)*, **42**, August, 37–53.

Sigurdson, Jon (2004a), 'Industry and policy perspectives: technological superpower China?', *R&D Management*, **34**(4), September, 345–7.

Sigurdson, Jon (2004b), *The Innovative China*, EAI Background Brief No. 205, Singapore: East Asian Institute.

Sigurdson, Jon (ed.) (2004c), *Knowledge Systems and their Global Interaction: Summary of Papers*, Lund University, Centre for East and Southeast Asian Studies.

Sigurdson, Jon and Cheng, A. (eds) (2001), 'A new technological landscape in Asia Pacific', *International Journal of Technology Management*, Special Issue, **22**(5/6).

Sigurdson, Jon and Long, V. (2003), *Internationalisation of Chinese IT Industry and Possibilities for Sweden*, Project 2003/114-1040, Invest in Sweden Agency.

Sigurdson, Jon and Palonka, Krystyna (2005), 'Technological governance in ASEAN – failings in technology transfer and domestic research', in Fredrik Sjöholm and Jose Tongzon (eds), *Institutional Change in Southeast Asia*, London: Routledge.

Sigurdson, Jon and Persson, Olle (1998), 'The new technological landscape in Pacific Asia: an inquiry into the dramatic changes in patenting and scientific publishing and its underlying forces and effects', *Research Evaluation*, **7**(1), 31–8.

Simon, Denis (ed.) (1996), *Techno-security in an Age of Globalization: Perspectives from the Pacific Rim*, Armonk, NY and London: M.E. Sharpe.

Simon, Denis (2001), 'The microelectronics industry crosses a critical threshold', *The China Business Review*, **28**(6), 8–20.

Simon, Denis F. (2005), 'Presentation for Hearing on China's High Technology Development', US–China Economic and Security Review Commission, 21 April.

Simon, Denis and Goldman, Merle (1989), *Science and Technology in Post-Mao China*, Cambridge, MA: Harvard University Press.

Simon, Denis and Rehn, D. (1987), 'Innovation in China's semiconductor components industry: the case of Shanghai', *Research Policy*, **16**, 259–77.

Song, Wenwei (2003), 'Forum eyes private education', *China Daily*, 23 July.

Stickel, Victor G. (2003), 'Recent HPC activities in China', ATIP – (First) Chinese HPC Workshop, 16 November, Phoenix, Arizona (PPT presentation).

Suttmeier, Richard P. (1980), *Science, Technology, and China's Drive for Modernization*, Stanford, CA: Hoover Institution Press.

Suttmeier, Richard P. and Yao Xiangkui (2004), *China's Post-WTO Technology Policy: Standards, Software, and the Changing Nature of Technonationalism*, The National Bureau of Asian Research.

Tamada, Tatsuru (2000), 'The challenge of creating "cluster industries": in search of new sources of added value', *NRI Quarterly*, Winter, 2–17.

Tang, Tong B. (1984), *Science and Technology in China*, London: Longman.

Thurow, Lester C. (1997), *The Future of Capitalism: How Today's Economic Forces Shape Tomorrow's World*, New York: W. Morrow.

UNCTAD (2004), *World Investment Report 2004*, United Nations.

UNCTAD (2005), *World Investment Report 2005*, Geneva: UNCTAD.

United Nations Population Division (2002), *World Population Ageing, 1950–2050*, UN.

US–China Economic and Security Review Commission (2004), 'China as an emerging regional and technological power: implications for US economic and security interests', Hearing before the US–China Economic and Security Review Commission, 12–13 February 2004, Washington, DC, http://www.uscc.gov (accessed 25 April 2005).

Vatikiotis, M. (2004), 'Outward bound', *Far Eastern Economic Review*, 5 February.

Vogel, Ezra (1990), *One Step Ahead in China: Guangdong Under Reform*, Cambridge, MA: Harvard University Press.

Walcott, Susan M. (2003), *Chinese Science and Technology Industrial Parks*, Aldershot, UK: Ashgate Publishing.

Walsh, A. (2003), 'Hearing on China's industrial, investment and exchange rate policies: impact on the US', Senate Commission on US–China Economic and Security Review, 25 September 2003.

Walsh, Kathleen (2003), *Foreign High-tech R&D in China: Risks, Rewards and Implications for US–China Relations*, Washington, DC: The Henry L. Stimson Center.

Walsh, Kathleen A. (2003), 'Testimony in the hearing on China's industrial, investment and exchange rate policies: impact on the US, Panel III – China's investment strategies', Senate Commission on US–China Economic and Security Review, 25 September 2003.

Wang, Chunfa and Wang Changlin (n.d.), 'WTO boundary and policy adjustment of Chinese high technology industry', *Chinese Science and Technology Industry*, **152**.

White, S. and Liu, X. (1998), 'Organizational processes to meet new performance criteria: Chinese pharmaceutical firms in transition', *Research Policy*, **27**, 369–83.

Williamson, Jeffrey G. and Higgins, Matthew (2001), 'The accumulation and demographic connection in East Asia', in Andrew Mason (ed.), *Population*

Change and Economic Development in East Asia, Stanford, CA: Stanford University Press, pp. 123–54.

Wong, John and Chan, Sarah (2003), 'China's outward direct investment: expanding worldwide', *China*, **1**(2), September, 273–301.

World Bank Institute (2000), *China's Development Strategy: The Knowledge and Innovation Perspective*, Washington, DC: World Bank.

Wu, Yixue (2003), 'The way forward: IT, urbanization', *China Daily (HK)*, 3 November.

Xiao, Cao (2003), 'Plugging into high-tech', *China Daily*, 20 September.

Xiao, Huo (2004), 'Industrial alliance boosts IT sector', *China Daily*, 4 October.

Xie, Wei and White, S. (2004), 'Sequential learning in a Chinese spin-off: the case of Lenovo Group Limited', *R&D Management*, **34**(4), 407–22.

Xin, Dingding and Tian, Huaiyu (2003), 'Preparations made for business assault', *China Daily*, 2 September.

Xiong, Yuxiang (2004), 'Military spending for defence purpose', *China Daily*, 29 January.

Xu, Dashan (2004a), 'New aviation firm takes off in Harbin', *China Daily*, 20 May, http://www.chinadaily.com.cn/english/doc/2004-05/20/content_332340.htm (accessed 26 April 2005).

Xu, Dashan (2004b), 'Plane parts industry takes off', *China Daily*, 15 January, http://www.chinadaily.com.cn/en/doc/2004-01/15/content_299000.htm (accessed 26 April 2005).

Xu, Guanhua (2003), 'Speech on technology strategy', Closing Ceremony of the Ministerial Forum on Industrial Policies of China, Beijing International S&T Industries Fairs, 15 September.

Xu, Yan and Gong, Yan (2003), 'National innovation system and its implications for 3G development in China', *Communications and Strategies*, **52**(4th quarter), 155–74.

Yan, Dai (2001), 'Wanxiang nabs helm of US firm', *Business Weekly (China)*, 25 September.

Yu, Q.Y. (1999), *The Implementation of China's Science and Technology Policy*, Westport, CT: Quorum Books.

Zhang, Ed (2003), 'Time to turn rural towns into mega-cities', *China Daily*, 10 January, http://www.chinadaily.com.cn/en/doc/2003-01/10/content_150913.htm (accessed 27 April 2005).

Zhang, Fan and Zheng Jingping (1998), 'The impact of multinational enterprises on Economic Structure and Efficiency in China', paper submitted to The Washington Center for China Studies (WCCS), August.

Zhang, N. (1998), 'On the technological development system of China's automobile industry', *Science and Technology Management*, **3**, 49–54.

Zhang, Yong (2003a), 'Jet production set to start', *China Daily*, 7 May, http://

www.chinadaily.com.cn/en/doc/2003-05/07/content_164336.htm (accessed 26 April 2005).

Zhang, Yong (2003b), 'DuPont opens R&D workshop', *China Daily (HK)*, 5 November.

Zhao, Huanxin (2004), 'Nation plans satellite "constellation"', *China Daily*, 6 January.

Zheng Minzheng (2002), 'China's IC industry under fast development', in China Semiconductor Industry Association (CSIA) (ed.), *An Investigation Report of China's Semiconductor Industry*, Beijing: China Semiconductor Industry Association.

Zheng, Yongnian, (2004), *Globalization and State Transformation in China*, Cambridge: Cambridge University Press.

Zhu, Boru (2004), 'US expert drums up support for supercomputing', *China Daily*, 26 October, http://www.chinadaily.com.cn/english/doc/2004-10/26/content_386388.htm (accessed 25 April 2005).

Index

Abott 177
academic research
  commercialization 13–14
  Japan 294
Academy of Mathematics and System Science 165
access to technology, overseas investment 82
advance manufacturing technology
  GRINM 138
  national research programme 40
  overseas investment 82
  R&D institute reform 102
advance materials, S&T programmes 43, 50
Advanced Micro Devices 164
aerospace industry 192–5, 284
  R&D funds 201
  Russia–China cooperation 201–3
  technological capability 198
  see also space
Aerospatiale 169
ageing society 4
agglomeration, regional 226
Agreement on Basic Telecommunications 86–7
Agreement on Information Technology Products 86
Agreement on Textiles and Clothing (1995) 158
agriculture
  collectivization 3
  graduates by field of study (1996–2001) 56
  S&T programmes 43, 50
  student enrolments, higher education (1996–2001) 56
Aiben, H. 250
Airbus 168–9, 201
Airbus A300/A310 169
Airbus France 169

aircraft industry 167–9
  global integration 169–71
  technological collaboration 200–201
  see also military aircraft
Alcatel 14–15, 109, 116
Amersham 269
Analysis of Strategies and Technologies Centre (AST) 203
Ancai Group 15
Anhui 20
application-specific integrated circuits (ASICs) 115, 144, 184, 292
Arca Technology 45, 139
ARJ21 regional jet aircraft 167, 168, 170, 171
Asia, supercomputer systems 164
astronomy 280
Auto China 188
automation, 863 S&T Programmes 43
automobile industry 77–8, 83, 84–5, 187–9, 231, 285–6, 296
automotive semiconductor market 162
Aviation Industry Development Research Centre 167
AVIC-1 Commercial Aircraft Co. (ACAC) 201
AviChina Industry and Technology Co. Ltd 169, 170

ballistic missiles 198
Bangalore research centre 114, 115
Baoshan Steel Project 7
BASF 270
Bayer 76, 270
Beidou navigation system 207–8
Beijing
  economic statistics 238
  education 233, 234
  Geely University 63–4
  IC industry 142–3
  migrant workers 20

328

Index

patent applications 235
point-of-sales (POS) systems 45
projects and expenditures, 863
    Programmes 44
R&D centres 32, 88, 90, 91
semiconductor industry 130–31, 133, 144–5
Beijing Economic-Technical Development Area 164
Beijing Electronics 199–200
Beijing Genomics Institute (BGI) 165
Beijing Jiaotong University 166
Beijing Oriental Enterprise (BOE) 16, 83, 84, 119, 199
Beijing Putian Founder Communications, Inc. 110
Beijing Semiconductor Manufacturing Corporation (BJSMC) 45, 163
Beijing University 12, 31, 45, 58, 92, 114, 131, 178, 225–6
Beijing University Founder Group Corporation 14
Beijing University Medical Hospital 225
Beijing University of Post and Telecommunications (BUPT) 13, 114
Beijing University Science Park 14
Bell Laboratories Advanced Technology Research Institute 90
Belling Co. (Shanghai) 140
BHA Aero Composites Co. 168
Bio-Engineering Gene Incubator 224
biochemistry, 863 S&T programmes 43
Biomass Energy 204
biomedicine, Shanghai 260
biotechnology 171–9
    aim of development 172–3
    an assessment 178–9
    collaborative agreements 172
    government policies and initiatives 174–5
    integrated companies 171–2
    intellectual property rights 173
    pharmaceutical industry 176–8
    R&D personnel 173
    resource allocation 284
    Shanghai 175–6, 268–9
    substantial support for 32
    traditional Chinese medicines 178
Bird 15, 287
blade supercomputing 165, 166

BLX IC Design 45–6
BOE Hydis Technology Co 84
Boehringer Ingelheim 177, 269
Boeing 168, 201
BoHai Rim (BHR) 19, 215, 218, 241
Bombardier 169
Boston Consulting Group (BCG) 77, 78
Botelho, M. 167
Boyuan, F. 44
brain drain 182
brain gain, future 64–7
brainpower industries 4
branding 119–20
Brazil, technological cooperation with 169, 208
British Petroleum 270
budget, defence 199
bureaucratic model, science and technology 34
Bush, G.W. 191, 299

Cao, Cong 65
Caohejing National Industrial Park 262–3
capital-intensive industries 186
'capturing global advantage' 78
cars, predicted number (2020) 285
Carter, J. 204
cathode ray tubes (CRT) 15, 80
CDMA 1X system 116
Celsar 269
central processor units 45, 50, 139
Changhong Electronic Group 85, 120, 122
ChemExplorer Co. Ltd 268
Chen, E. 279
Chen, S. 67, 165, 166
Chen, Zhili 187, 309
Cheng, Jianpei 52
Chengdu 170
Chengdu Jian10 (J-10) 211
Chengdu University 92
Cheung Kong Scholars Programme 65
China
    (2020) 284–8
    as engine of growth 157–8
China Academy of Launch Technology (CALT) 195
China Academy of Space Technology (CAST) 195, 284

Index

China Academy of Telecommunications Research (MII) 160
China Aerospace Corporation (CASC) 193, 195
China Aerospace Machinery and Electronics Corporation (CAMEC) 195
China Aerospace Science and Industry Corporation (CASIC) 193
China Aerospace Science and Technology Corporation (CASTC) 195
China Aviation Industry Corporation I (CAIC I) 167, 168, 170, 171
China Aviation Industry Corporation II (CAIC II) 167, 170
China Basic Research Institution of Bell Laboratories 90
China Centre of Information Industry Development (CCID) 127, 139–40, 141, 144, 162
China Corporate Research and Development Centre 269
China Development Bank 151
China Eastern Airlines 169
China Electronics Corporation 147, 261
China Great Wall Industry Corporation (CGWIC) 195
China International Marine Container Group (CIMC) 16
China Kejian 120
China Mobile 115, 147, 159, 160–61
China National Aero-Technical Import and Export Corporation 169
China National Bluestar 82–3
China National Chemicals Import and Export Corporation (Sinochem) 83
China National Space Administration 191–2, 193–5
China National Space Agency (CNSA) 192
China Netcom 160
China Posts and Telecommunications Industry Corporation 109
China Precision Machinery Import and Export Corporation (CPMEIC) 195
China Putian Group 109–10, 117, 122, 147, 148, 160
China Putian Institute of Technology (CPIT) 110
China Railcom 150

China Semiconductor Industry Association Report (2002) 134–5
China Telecom 147, 160
China Torch Internet Innovation Centre 273
China Unicom 115, 146, 147, 159–60
China University of Geosciences 166
China University of Mining and Technology (CUMT) 60–61
China Urban Labour Employment and Labour Flow 237
China Warship Design Institute 198
China-Brazil Earth Resources Satellite (CBERS) 208
China–EU High-level Forum on S&T Strategy 187
China-Europe Global Navigation Satellite System Technical Training and Cooperation Centre (CENC) 206
China-Singapore Suzhou Industrial Park 161
Chinese Academy 282–3
Chinese Academy of Engineering (CAE) 5, 10
Chinese Academy of Sciences (CAS) 5, 10, 29, 31, 33, 34, 45, 92, 175
Chinese Academy of Social Sciences (CASS) 29–31, 33, 238
Chinese companies
  design sector, semiconductor industry 129, 139
  as global players 14–17
  IPR position 284
  overseas investment 82–5
  patent portfolio, in US 17
Chinese diasporas 64–5
Chinese Economic Area (CEA) 179, 180–81
Chinese Heavy Industry Ministry 135
Chinese language, Linux version 293–4
Chinese National Human Genome Centre 175
chip fabrication 128, 130, 139
chip laboratories 115
Chongqing 133, 238
Chongqing Chongyou Information Technology Group 148
Chongqing University 13
cities, competitive 238

330   Index

city-building programmes 242
civil aircraft 167, 168
civilian production, aerospace industry 193
civilian–military integration (CMI) 199–200
civilization, China 279–94
Climbing Programme 102
coastal provinces 215
　advances 287–8
　industrial structures 79
　industrialization 219, 222–6
　intensive learning process 303
　mega-cities 239
collaborative research agreements 90, 92, 172
combat vessels 203
Commercial Aircraft Co., Ltd (ACAC) 171
Commission on Science and Technology and Industry for Defence (COSTINF) 197
COMMIT 110, 148, 149
comparative advantage, knowledge-intensive products 123
competence blocks, regional 22–3
competition
　and cooperation 23–6
　and standardization 17–18
competitive advantage, EU views on China's 295–6
Competitiveness Report (EU) 295
Computer Associates 112
computer sector 127; see also high-performance computers; personal computers; supercomputers
consumer electronics 80, 117, 122, 274, 299
consumption
　semiconductor equipment (2000–03) 134
　urban–rural divide 239
cooperation, and competition 23–6
Corning 109
corporate R&D, increase in 101
corporatization, research institutes 102, 103
Cortech Corporation 139
COSMOBIC 115
Country Strategy Paper (EU) 205

country-specific knowledge, lack of 120
Cross-Century Talent Programme 65
crude oil imports 5
cruise missiles 198
cultural networks 231
Cultural Revolution 3, 29, 34, 54, 58, 65, 70, 283

Daewoo Motors 83, 286
Dalian 221
Dalian Qigong University 114
Datang 110, 129, 139, 146, 147, 148, 149, 150
Dawning Information 165, 166
Decision on Accelerating S&T Progress 102
Decision on Profound S&T Reform 102
Decision on the S&T System 101–2
defence industry, see military industry
Dell 111
demographic transition, savings and investment 294
Deng, Xiaoping 8, 42–3, 54, 64, 102, 204
Department of Computer Science and Technology 165
Department of Microelectronics, Fudan University 253–5
design centres, FDI investment 74
design sector, IC industry 128, 129, 139, 144
destroyers 198, 203–4
developing countries, inferior position, S&T 87
Developing the Western Region 131
development zones
　creation of serious problems 241
　as planning instrument 246
　role of 220–22
Digital Audio Broadcasting 204
digital signal processing 50, 139
domestic market, growth of 157
domestic oil, consumption and production 5, 6
Dong Bei 177
Dongfanghong 209
Dongguan 21–2, 72, 109
Dongji University 249
Dongnan University 114
Dragon microprocessor 45

Dupont 76, 88, 269
DVD players 17, 24, 80, 82, 107

Earth observation system 192
East China Normal University 176
East China University of Science and
    Technology 176
East Wind rockets 210
EC120 helicopters 169
economic development 24
    car industry 286
    National 211 Project 13
    Shanghai 248
economic globalization 308
economic growth 26
    China as engine of 157–8
    Shanghai 258
economic integration, with neighbouring
    countries 179–86
economic performance, transformed
    R&D institutes 104
economics
    graduates by field of study
        (1996–2001) 56
    student enrolments, higher education
        (1996–2001) 56
education
    need for upgrading 282
    regional innovation systems 233–6
    see also higher education; universities
Eight-Year Plan for Science and
    Technology (1978–85) 36
electronics industry
    changing trade patterns 159–61
    development 186, 297
    employment effects 80
    export categories 80
    foremost supplier of goods 24
    globalization 182
    migration of production to China 291
    Taiwanese support 184
    technological capability 212
    see also semiconductor industry
Elman, B. 279–80, 281
Elpida 266
EMB170 jet aircraft 167, 170–71
Embraer 167, 169, 170, 171, 201
employment
    defence industry 197
    growth in exports 80
    IT sector 126, 144
    rural 236–7
    semiconductor industry 133
    Zhangjiang Hi-tech Park 264
    ZTE Corporation 267
    see also manpower
energy
    S&T programmes 43, 50
    see also oil
engineering 186
    graduates by field of study
        (1996–2001) 56
    student enrolments, higher education
        (1996–2001) 56
engineers
    in higher learning institutions 58
    in LMEs (1995–2001) 57
    in R&D institutions (1996–2001) 58
entertainment, digital 274
entrepreneurial business-oriented
    Chinese community 64
equality 72
Ericsson 21, 76, 90, 109, 116, 151
EUREKA 8, 42
Eurocopter 169
Europe
    foreign R&D concerns 94
    semiconductor equipment from 131
    supercomputer systems 164
    technological collaboration with
        204–8
European Aeronautic Defence and Space
    Company (EADS) 168–9
European Union 2, 180, 181, 191
    Country Strategy Paper 205
    Fifth Framework Programme for
        Science and Technology 205
    S&T relationships with 209
    views on China's competitive
        advantage 295–6
Explorer I 193
export processing zones 271
export-oriented development 244
exports
    from Shenzhen 20
    from Taiwan to China 182–3
    high-tech 26, 48, 79, 80, 185, 186
    IT sector 119, 120, 126
    Putian Group 110
    semiconductor industry 134

telecommunications 185–6
  to US 81, 183
  value-added 80–81
family planning programme 4
Fang, Yi 36
farmers
  migration into non-agricultural areas 238
  privately-owned workshops 242
Fengpu Industrial Park 270–71
Fengxian Modern Agriculture Park 274
Fengxian Technology Innovation Centre 273
Fifth Framework Programme for Science and Technology (EU) 205
fighter aircraft 200, 203, 211
financial resources, research and development 284
Finland, salary comparison with 293
First Automotive Works (FAW) 188
Five year plans 31
flat screen panels 16
floating population 20, 239
floppy disk drives 107
foreign affiliates
  attracting overseas brain talent 94–5
  low turnover 26–7
foreign companies
  automobile industry 285
  expansion of 97
  in high-tech industry 300
  IT sector 127
  relations with R&D institutions 74
  Shenzhen 21
foreign direct investment (FDI)
  attraction of industrial 70
  inward, *see* inward investment
  negative results of 97
  outward, *see* outward investment
  research and development 74, 88–92, 301–2
  shaping of economic regions 219
  technology transfer 74
foreign markets, exploiting 119–20
foreign research, in China 93–5
foreign technology, exploiting 73–88
Formation of Shanghai Knowledge Economy Strategies 220
Founder 102

Framework Programmes for Research and Technological Development 209
France
  cooperative research programme with 204
  foreign R&D investment in China 76, 90
Fraunhofer Gesellschaft 116
Fudan University 31, 92, 176, 249, 251–3, 265
  Department of Microelectronics 253–5
  Graduate School 255
  Medical School 274
  science park 256–8
Fujian 20
Fujitsu 266
funding
  research and development 32, 198–9, 201, 283
  science and technology activities 105
  Shanghai incubators 272
Fuqiu Vocational Studies College 250
FutureWei 116, 151
Fuzhou 223

Galactic Computing 165–6
Galileo Project 27, 206–8
Geely 287
Geely University 63–4
General Accounting Office, 2002 report 162
General Administration of Civil Aviation of Shanghai 169
General Armaments Department (GAD) 197
General Electric 88, 201
General Motors 88, 286
Germany
  foreign R&D investment in China 76, 90
  salary comparison with 293
GlaxoSmithKline 269
Global Digital Telecommunication 204
global innovation system (GIS) 27, 290–94
global integration 169–71, 211–12
global players 14–17
global politics 208–11

Index

global positioning system (GPS) 107, 206–7
global production networks (GPN) 181–2, 227
global rivalry 23–6
Global Software Group 92
global sourcing 120–21
globalization, science and technology 308, 311
Glonass system 206, 207
Godson 45
government agencies, role in attracting FDI 244
government control, foreign investment 82
government policies, biotechnology 174–5
Gow, I. 63
Grace Semiconductor Manufacturing Corporation (GSMC) 140, 163
Graduate School, Fudan University 255
graduates
  industrial parks for overseas Chinese 94
  life sciences 178
  living in the United States 65
  need to reduce expectations 54–5
  salaries 287–9
  see also postgraduates; undergraduates
Great Dragon 120, 146
Great Leap Forward 3, 29, 34, 69
  China's second 71–3
Greenvalley 269
Grikin Advanced Materials Co. Ltd 137
GRINM Semiconductor Materials Co. Ltd (GRiTek) 134–8
gross domestic product (GDP)
  auto industry as driver of 285
  by 2020 306–7
  by region 238, 239
gross expenditure, research and development 283
gross production, by region 238
Guangdong 20, 44, 133, 223, 235
Guangzhou 95, 238

Haier Group 81, 85, 120, 121, 122
Hainan 221
Hangzhou 95, 115, 131, 244, 276

Harbin Aircraft Industry Group Corporation 169, 170
Harbin Aviation Industry (Group) Co. Ltd 170
Harbin University of Engineering 225, 226
hard disk drives, manufacture 226–7
He Jian Technology Corporation 163
health, S&T programmes 50
Hebei 133, 186, 218
Henan 20
high definition TV (HDTV) 40
high-performance computers 31, 40
high-performance computing 164
High-Performance Computing Centres 165
high-tech companies 31
high-tech industry
  Chinese threat 296–8
  clusters 95, 217, 220
  development, Shanghai 276
  development zones 220–21
  Dongguan 22
  exports 26, 48, 79, 80, 185, 186
  foreign companies in 300
  inward investment 77, 79–80, 180–81
  IT networks 181–2
  output value, enterprises (1991–2002) 48
  outward investment 83
  regional diversity 219–20
  salaries 293
  treaties relating to 86
  see also semiconductor industry
high-tech parks, Shanghai 262–9
high-tech pioneering undertakings 46
High-Tech Research Development Programme (863) 8–9, 40, 42–6, 129, 145
higher education
  institutions 234
  S&T personnel in 58
  Ningbo 287–9
  Shenzhen 225–6
  student enrolments (1978–2002) 55
  see also universities
highway construction 285
history
  graduates by field of study (1996–2001) 56

student enrolments, higher education (1996–2001) 56
Hitachi 88
HJD-04 146
Honda 88
Honeywell 201
Hong Kong
  entrepreneurs, Shenzhen 222–33
  foreign R&D investment in China 90
  as source of FDI 76
household appliances 24
household family contract responsibilities 40
HT-7U Superconducting Tokamak Fusion Experimental Equipment 52–3
Hua Bei 177
Huada 129, 139
Huahong-NEC Electronics 115, 129
Huali 147
Huang, Jigong 150
Huaqiang Group 107–9, 122
Huaqiang Netcom Co. 108
Huawei Technologies 1, 18, 21, 113–16, 120, 123, 140, 146, 147, 149, 151, 160
Hubei 20, 44
Hughes 210
Hui, Mr 268
human resources
  investment 303
  research and development 284
  see also manpower
Hunan 20, 44
Hydis 120, 200
Hynix 16, 83, 84, 119, 174, 292

IBM 1, 31, 76, 88, 111–13, 116, 271, 300
IC Design Technology Innovation Centre 273
Ideabank 139
IMEC 266
Imperial Peking University 281
imports
  crude oil 5, 6
  high-tech products 79–80
  Putian Group 110
  Russian share of Chinese arms 202
  skill-intensive 295
  US restrictions on textiles 159

Inchon Oil Refinery 83–4
income, urban–rural divide 239
incubator systems 14, 21, 46, 249, 271–4
Indonesia, overseas investment 85
industrial groups 70–71
industrial parks 248, 270–74
industrial policy, evolution of 69–73
industrial revolution, Europe 23
industrial standards 17
industrial structure
  coastal provinces 79
  transformation of 70
industrial workers, salaries 293
industrialization
  approach 307
  coastal 219, 222–6
  Communist China 3
  current 23, 73
  FDI-based 79–80
  need for less energy-intensive 7
industry
  reliance on foreign technology 282
  Shanghai 258–62
industry-specific knowledge, lack of 120
Infineon Technologies 139, 161–2, 266
information, projects, S&T programmes 43, 50
information and communication technology
  branding 119–20
  China's future 151–2
  consumer electronics as driver of 117
  entrepreneurialism 117–18
  foreign capital 117
  foreign markets
    exploiting 119–20
    surfing on R&D manpower 121–3
  foreign R&D investment in China 88–90
  global R&D landscape 152–3
  growth rate 116–17
  internationalization 118–19
    hurdles to 120–21
  market expansion 106–16
  as pillar industry 32
  Shanghai 259
  standards 18
  technology transfer 74
  see also telecommunications

Information Engineering College 145–6
Information Science and Technology (IST) programme 205
information service industry, Shanghai 259
information technology, *see* information and communication technology
infrastructure development 236–7
innovation
 technological capability, R&D institutes 104
 see also global innovation system; national innovation system; regional innovation systems
Institute of Acoustics (CAS) 50
Institute of Biochemistry and Cell Biology 176
Institute of Computer Technology (CAS) 31, 111, 112, 165
institutional networks 231–3
integrated circuit (IC) industry
 ASICs 115, 144, 184, 292
 Beijing 142–3
 demand 127–8
 design sector 128, 144
 foreign direct investment 162
 GRiTek 134–8
 large-scale integration 144–5
 policy and progress 139–40
 regional characteristics 129–34
 role of 126–9
 Shanghai 260
 VAT rebate 140–42
integrated product development (IDP) 116
integrated sourcing strategy 77–8
integrated supply management (ISM) 116
integration, *see* civilian–military integration; economic integration; global integration
Intel 141, 267, 271, 274, 292
intellectual culture 283
intellectual property rights
 biopharmaceuticals 173
 globalization of S&T 96
 Huawei–CISCO conflict 115–16
 Shanghai 274–5
 weak position, Chinese companies 17, 284

intelligence-intensive regions 220
Inter-American Development Bank report 296–7
International Business Incubator (IBI) 273
investment
 in education 250
 human resources 303
 see also foreign direct investment; savings and investment
inward investment 75–7
 encouragement of 216
 high-tech industry 77, 79–80, 180–81
 local government agencies role in attracting 244
 production facilities 186
 research and development 25, 74, 76, 88–92, 301–2
 semiconductor industry 162–3
Israel–China relationship, military technology 211
ITER nuclear fusion project 27, 53, 284, 290

Jamestown Foundation 201
Japan 2
 Chinese research as threat to 298–9
 experience as technological superpower 294–5
 influence on China's regional development 185–6
 outward investment in 84–5
 as source of FDI 76
 supercomputer systems 164
Japan as a Scientific and Technological Superpower 294
Japanese companies, declining presence in China 100
Jiang, Lintao 160
Jiangnan Shipyard 198
Jiangsu
 education 55, 233, 234
 employment, semiconductor industry 133
 high-tech exports 219–20
 migrant workers 20
 patent applications 235
 projects and expenditure, 863 Programmes 44
Jiangxi 20

Jiaotong University 59, 139, 166, 176, 249, 265, 274
Jiaozuo School of Railroad and Mines 60–61
Jingfeng Investment Co. 108
Jinpeng Group 146
Jinqiao Export Processing Zone (JEPZ) 271
Johnson and Johnson 177
joint ventures
  aircraft industry 168–9, 201
  auto sector 83, 188
  CRT glass tubes 15
  pharmaceutical industry 177
  semiconductor industry 161–2, 265–6
  technology transfer 74, 106, 122
    Huaqiang Group 108–9
    Huawei Technologies 114, 115–16
    IBM and Lenovo 111–13
    Putian Group 109, 110, 117
  telecommunications 14, 15, 149
Julong Telecom 146

Kehui Technology Innovation Centre 273
key institutions, national innovation system 30
key laboratories 52, 176, 252–3, 254
Key Technologies Research and Development Programme 8, 38–40
Kirin 269
knowledge, lack of industry- and country-specific 120
knowledge centre, transformation of Taiwan as 182
knowledge creation, Shanghai 274–5
knowledge economy 26–8
knowledge environment 95–8
knowledge sources, industrial technical development 157–8
knowledge-intensive products, comparative advantage 123
kombinat character, state companies 70
Konka Electronic Group 15, 85, 119, 140
Korea
  Chinese investment in 83–4
  economic partnership with 218
  foreign R&D investment in China 76, 90

  influence on China's regional development 185–6
Korean War 76
Krauss, E. 300

labour costs, capital-intensive industries 186
labour-intensive industries 4, 7, 27, 72
land, indiscriminate use of 241
Langchao 165
large and medium enterprises (LMEs)
  employment in 57
  new technology enterprises 102
  R&D units 104–5
large-scale integration (LSI), design and development 144–5
laser jet printers 80
law
  graduates by field of study (1996–2001) 56
  student enrolments, higher education (1996–2001) 56
Law on Private Education Promotion (2002) 250
learning, importance in 302–3
LED semiconductors 139
Legend Group 102, 111–13, 147, 165
Lenovo 1, 15, 31, 111–13, 122–3, 147, 165, 300
Lewis, Sir Alfred 157
LG-Philips 15, 141
LHWT 144–5
Lian, Sheng 141, 142
Liaoning 44, 133, 186, 218
licence fees, DVD manufacturers 17
licences, 3g system 148, 150
life sciences, see biotechnology
light manufacturing, labour-intensive 7
Liking University 114
Lilly 268
Linux version, Chinese language 18, 293–4
liquid crystal displays 80
listening posts 75
literati 280–81
literature
  graduates by field of study (1996–2001) 56
  student enrolments, higher education (1996–2001) 56

Little Smart 146–7, 150
Liu, Gaozhu 168
Liu, Guanqiu 85
living standards, urban residents 239
local governments, economic and technological development 219, 220
Lockheed Martin 210
Logistics Centre, Huawei 116
Long March rockets 191, 210
Long and Medium-term Programme on S&T Development 37
Loral 210
Lucent Technologies 90, 109
Lucent-Bell Laboratories 90
Luen Thai 159
Luhu-class missile destroyers 198
Lujiazui Finance and Trade Zone (LFTZ) 271

Makiyenko, Mr 203
management culture, China 255
management skills 72
manned space flights 26, 191, 192
manned spacecraft 284
manpower
  mobilization 71–2, 73
  technological and scientific 53–64
  see also employment; human resources
manufacturing
  863 S&T programmes 43
  entrepreneurial fervour in ICT 117–18
  S&T Plan (2020) 4–5
  technological prowess 100
Mao, Zedong 4, 71, 72–3
Marconi 116
marine shipping containers 16
markets, integration global and civilian 211–12
Masta Engineering and Design Inc. 82
mathematics 280, 281
Matsushita (Panasonic) 82, 100, 109, 115, 117, 119
Maxwell 144
Medical Instrument Research Institute 225
Medical School, Fudan University 274
medicine
  graduates by field of study (1996–2001) 56

student enrolments, higher education (1996–2001) 56
Medtronic 269
mega-cities, coastal areas 239
Mega-Projects of Science Research (MPSR) 52–3
mergers and acquisitions 74–5, 123
Micro-Nano Electronic Platform 254
microelectronics industry 162
Microsoft 88
migrant workers 20, 237
military aircraft 200, 202, 203, 211
military doctrine 208
military industry 196–9
  electronics production 291
  modernization 210–11
  production system 193
  S&T plan (2020) 5
  technological collaboration 200, 201–4
  US concern at progress of 25
  within political system 194
  see also civilian–military integration
military security, supercomputing as threat to 165
Minhang District 274
Ministry of Communication 285
Ministry of Information Industry (MII) 15, 127, 129, 141, 142, 146, 148, 152, 284, 292
Ministry of Land and Resources 105, 241
Ministry of Metallurgy Industry 135
Ministry of Post and Telecommunications 12–13, 110, 117
Ministry of Science and Technology (MOST) 8, 9, 10, 31–2, 33, 41, 100, 141, 143, 167, 240
Ministry of Water Resources 105
MIPS Technologies 45
missile programme 35, 198
mobile applications 160–61
mobile multimedia technology alliance (MMTA) 159–60
mobile phone industry 14, 15, 18, 24, 78, 159, 297
mobile radio technology 151
mobilization, manpower 71–2, 73
mobilization model, science and technology 34

modernization
  goals 3–4
  human resources 64
  military 210–11
  regional disparities and social tensions 23
molecular biology 172, 176
Motorola 21, 77, 78, 88, 90, 91–2, 107, 109, 111, 131, 146, 147, 210, 267
Motorola Advanced Technology Centre (MATC-Asia) 92
Motorola China Research Centre 92
Motorola Tianjin Integrated Semiconductor Manufacturing Complex 163
multimedia 139, 159–60
multinationals
  attraction of low cost labour and brainpower 24
  attraction of R&D functions 32–3
  challenges posed by 96–7
  investment in China 25, 76, 219
  low turnover of foreign affiliates 26–7
  semiconductor foundries 163

Nanjing 95, 131
Nanjing University 92, 114
Nankai University 59–60, 166, 225
nanotechnology 10, 284
National 211 Project 12, 13, 58–64, 289–90
National Centre for Drug Screening 175
National Development and Reform Commission 141, 152, 165, 286
National Engineering Research Centre for Semiconductor Materials 138
National Engineering Research Centres (NERCs) 51
National Engineering and Technology Research Centre for Nonferrous Metal Matrix Composites 138
national goals, S&T plan 309
national innovation system 33–4
  evolution 245
  FDI streams 75
  key institutions 30
  structure 229
National Key Basic Research Programme (973) 9–10, 32, 47–50
National Key Laboratory on Application-Specific Integrated Circuits 254

National Natural Science Foundation 9, 32, 33
National New Products Programme 52
National Programme of S&T Development (1978–85) 37
National Research Centre for Science and Technology for Development (NRCSTD) 5
national science conference 36
national security, S&T as threat to US 96, 191
National Semiconductors 112
natural resources, overseas investment 82
Naughton, Barry 6, 184–5
naval technology 198, 203–4, 281
navigation, satellite-aided 198, 206–8
NEC 16, 21, 115, 140, 143
Needham, J. 280
networking 123
New Culture Movement 281
New and High Technology Industry Development Zones (NHTIDZ) 220–21
new product lines, benefits of outsourcing 78
999 Company 177
Ningbo 54, 276, 287–9
Ninth Five Year Plan and Long-term Programme on S&T Development (–2010) 37
NOAK 203, 204
Nokia 22, 78, 107, 109, 116, 147, 151
nonferrous metals, research and development 135
Nonferrous Metals Department 135
Nonferrous Metals Institute 135
Norson Telecom Consulting 113–14, 151
Nortel 88, 109, 116, 147, 148, 149, 150
North American Free Trade Area (NAFTA) 180, 181
North China Microelectronic Industrial Base 142
North China Pharmaceutical Group 177
notebook computers 18, 24, 182
Nottingham-Ningbo University 62–3
Novozymes 89–90
NTCH Inc. 116
NTT DoCoMo 144

nuclear power 27, 53, 204, 284–5, 290

OECD
  definition of clusters 227
  foreign R&D concerns 94–5
oil 5–7
Oki 144
old product lines, outsourcing 78
open coastal zones 221
Open Door Policy 4, 64, 158, 200, 301
openware software 18, 293–4
operational clusters 226–8
Outline of Chinese Science and Technology Development (1956–67) 135
outsourcing
  development stages in 77–9
  impact of 94–5
outward investment 81–5, 180
Overseas Chinese High-Tech Venture Park 224

Panasonic (Matsushita) 82, 100, 109, 115, 117, 119
PAS Xiaolingtong 147
passenger aircraft, technological collaboration 201
patents
  applications by region 235
  pharmaceutical areas 178
  portfolio, Chinese companies in US 17
  Shanghai 274–5
Pearl River Delta (PRD) region 19–20, 215
Peking University 131
People's Liberation Army (PLA) 36, 195, 197, 199, 202
People's Republic of China 73, 75
personal computers 31, 111–12, 113, 297; *see also* notebook computers
personal digital assistants (PDAs) 18
Perspective Programme of S&T Development (1956–67) 37
pharmaceutical industry 171, 176–8, 268–9
PhD students, Fudan University (2000–04) 252
philosophy
  graduates by field of study (1996–2001) 56

student enrolments, higher education (1996–2001) 56
physical networks 231
Pilot Project of a Knowledge Innovation Programme (PPKIP) 34
Pioneer 269
PLA IEC 146
planning process (2006–20) 309–11
point-of-sales (POS) systems 45
Policies for Encouraging the Development of Shenzhen Software Industry based on the National Programme 225
Policies for the Promotion of Science and Technology 42
policy makers, foreign R&D concerns 93
politics, global 208–10
Pommeranz, Kenneth 279
population
  BHR region 19
  Dongguan 21
  floating 20, 239
  median age (2003 and 2045) 4
  official statistics 237
  PRD region 19
  projects, 973 Programme (1998–2002) 50
  Shenzhen 20
  YRD region 19
postgraduates
  by field of study (1996–2001) 56
  dynamic expansion 12
  enrolments, by field of study (1996–2001) 56
priority strategic issues, science and technology 311–12
private enterprises
  import of high-tech products 80
  Shenzhen 21, 241
private universities/colleges 55, 249, 250
processing and assembly (P&A), exports 80, 81
production lines, outsourcing 78
professional model, science and technology 35
Programme of S&T Development (1963–72) 37
provincial governments, economic and technological development 219, 220

public interest-related institutes (PIRI), transformation of 103, 105–6
public participation, science and technology 311
PVO (Russian missile defence) system 203

Qian, Xuesen 35, 66–7
Qigong University, Dalian 114
Qingdao 221
Qinhuangdao 221
Quantum Design Institute (QDI) 112

Rand report (2001) 300–301
Rand report (2004) 196, 197, 198
raw materials, overseas investment 82
Reform Movement (1898) 281
reforms 5, 6
  universities 12, 31, 54
regional development, Korean and Japanese influence 185–6
regional innovation systems 19–23
  diversity of 215–28
  future of 228–36
regional jet aircraft 167, 168, 169–71
regional pillar industries 41, 42
regional S&T development 310
rent responsibility system 102
Report on R&D Plan on High Technology, The 8–9
research and development
  advancing enterprises and search for technologies 106–16
  agglomerations 226
  attracting multinational activities 32–3
  biotechnology 179
  centres
    Huawei Technologies 114, 115
    IC industry 132–3
    serious issues posed by foreign 93
    Shanghai and Beijing 32, 88, 90, 91, 260, 266–8, 269
    see also National Engineering Research Centres
  expenditure (2003) 33
  financial and human resources 284
  funding 32, 283
    aerospace industry 201
    military industry 198–9

globalization 301
information and communication technology 152–3
institutes
  aerospace industry 193
  Beijing 143
  China–France cooperative 204
  distribution of human resources (1996–2001) 58
  foreign 76, 187
  out of date 73–4
  reform
    assessing 103–5
    GRITEK case study 134–8
    since 1978 101–3
  relations with foreign companies 74
  Shanghai 175–6
inward FDI 25, 74, 76, 88–92, 301–2
Japan 294
mandate from Taiwan-based IT firms to Chinese subsidiaries 184
military industry 208–9
networks, molecular biology 172
nonferrous metals 135
personnel
  biotechnology 173
  in higher learning institutions 58
  in LMEs 57
  promotion system, GRINM 136
  salaries 104
  transfer to corporate sector 283
programmes, *see* science and technology, programmes
reform 31
renovating structures of 100–106
semiconductor industry 128–9, 139, 143
Shanghai 248, 249
Shenzhen 21
telecom sector 145–6
  as threat to Japan 298–9
Research and Development Centre of Northern Telecom (Nortel) 76
research universities 249
residence control system 233
resources and environment, S&T programmes 43, 50
returned students 65, 66
revealed comparative advantage (1986–2001) 185

Rhone-Poulenc 76
Roche 177, 269
rockets 191, 210
Rockwell Collins 201
rural economy, development of 40–41
rural employment 236–7
rural masses, mass mobilization 72, 73
rural urbanization 239–43
Russia 2
   technological cooperation with 201–4, 284

salaries
   comparison, traditional and high-tech industries 292–3
   graduates 54, 287–9
   R&D institutes (2001) 104
Samsung 139, 292
Sanjiu 269
Sankyo 269
Sanyo 100, 109, 117
satellite launch vehicles 192, 193, 210
satellite launches 26, 191, 192, 195, 209–10, 284
satellite projects 27, 169, 198, 206–8, 208
savings and investment 294
Schneider 15, 119, 120
science
   China 279–94
   graduates by field of study (1996–2001) 56
   student enrolments, higher education (1996–2001) 56
science parks 14, 249, 256–8
science and technology
   activities in enterprises 104, 105
   globalization 308
   impact of globalization on IPR institutions 96
   inferior position of developing countries 87
   institutes, reform 102
   national strategy 8–11
   new opportunities and challenges 307–8
   personnel 53–64
      in higher learning institutions 58
      in LMEs (1995–2001) 57
      in R&D institutions (1996–2001) 58
   policy formulation, stages in 33–6
   policy instruments to develop 187
   programmes
      phases 37–8
      Key Technologies Research and Development Programme 8, 38–40
      Spark Programme 8, 40–42
      High-Tech Research Development Programme (863) 8–9, 40, 42–6, 129, 145
      Torch Programme 9, 46–7
      National Key Basic Research Programme (973) 9–10, 32, 47–50
      other national 51–3
   strategy 291
   threat to national security 96, 191
   winning and losing sectors 187–8
Science and Technology Commission, Shanghai 175, 262, 272
Science, Technology and Industry Outlook (OECD, 2004) 94–5
Science and Technology Plan (2020) 4–5, 10, 38
   background 306–8
   major lines of thinking 309–11
   priority strategic issues 311–12
Science and Technology Progress Law 102
Science and Technology Statistics Yearbook (2003) 101
Science and Technology University (Hefei) 112, 114
scientific superpower 302–3
scientists
   in higher learning institutions 58
   in LMEs (1995–2001) 57
   R&D institutions (1996–2001) 58
secondary schools
   specialized 234
   student enrolments (1978–2002) 55
SEMATEC project 133
Semiconductor Illumination Engineering Group 139
semiconductor industry
   development and shortcomings 127
   equipment consumption 134
   exports (2003) 134
   research and development 10, 139

research organization involvement 128–9
sales 128
special support for 32
Taiwan–China relations 183
technological base and trade pattern 161–4
see also integrated circuit (IC) industry
Semiconductor Industry Association (US) 163, 164
Semiconductor Manufacturing International Corporation (SMIC) 44, 45, 92, 130, 131, 140, 141, 149, 163, 265–6
SH Huahong 139
Shanda Networking 274
Shandong 44, 55, 133, 186, 218
Shanghai
  biotechnology 175–6
  creating a knowledge metropolis 248–9
  economic statistics 238
  education 233, 234
  high technology parks 262–9
  high-tech industry 219, 276
  IC manufacturers 140
  industrial engineering and production 143
  industrial parks 270–74
  industry and technology 258–62
  intellectual property rights 274–5
  migrant workers 20
  patent applications 235
  projects and expenditure, 863 Programmes 44
  R&D centres 32, 88
  semiconductor industry 133, 164
  Taiwanese entrepreneurs 218
  universities 249–58
Shanghai Automotive Industry Corporation (SAIC) 83, 286
Shanghai Aviation Industrial (Group) Corporation 170
Shanghai Bankcard Industry Park 268
Shanghai Biological Chip Engineering Research Centre 175
Shanghai Bristol-Myers Squibb 177
Shanghai Centre for Bioinformation Technology 175

Shanghai Chemical Industrial Park (SCIP) 270
Shanghai College of Traditional Chinese Medicine 269
Shanghai Comprehensive Industrial Development Zone (SCIDZ) 270
Shanghai Fengpu Export Processing Zone 271
Shanghai Hutchinson Pharmaceuticals 177
Shanghai IC R&D Centre 260
Shanghai Institute of Biochemistry 176
Shanghai Institute for Biological Sciences 175
Shanghai Institute of Cell Biology (SICB) 176
Shanghai Intellectual Property Administration (SIPA) 275
Shanghai International IPR Forum 275
Shanghai Jiaotong University 59, 139, 176, 249, 265, 274
Shanghai Lei Yun Shang 177
Shanghai Ligong (S&T) University 112
Shanghai Medical Instrument Co., Ltd 177
Shanghai Medical Instrument Group Company 177
Shanghai Medical University 255
Shanghai New Drug Research and Development Centre 175
Shanghai People's Government Industrial Group 270
Shanghai Pharmaceutical Co., Ltd 177–8
Shanghai Pharmaceutical Group Corporation (SPGC) 177
Shanghai Pudong Software Park (SPSP) 261, 267–8
Shanghai Research Centre for Biomodel Organism 175
Shanghai Second Medical University 176
Shanghai Shuichan University 112
Shanghai Silicon Intellectual Property Rights Transaction Centre 260
Shanghai Supercomputer Centre 164, 165
Shanghai TCM Compound Purification Centre 177
Shanghai Technology Innovation Centre (STIC) 272

Shanghai University of Traditional
    Chinese Medicine 176
Shantou 221
Shanxi 44, 133
Shaxing Silicon 139
Shell Electric Manufacturing Holdings
    165, 166
Shenyang 170
Shenzhen 20–21
    attraction for IC manufacture 140
    coastal industrialization 222–6
    dominance of private sector 241
    high-tech exports 219
    import of advance manufacturing
        technology 102
    industrial engineering and production
        143
Shenzhen High-Tech Industrial Belt
    (SHIB) 225
Shenzhen High-Tech Industrial Park
    (SHIP) 224, 225
Shenzhen Huaqiang Communication Co.
    108
Shenzhen Institute of International
    Technology Innovation 225
Shenzhen Software Park (SZSP) 224–5
Shenzhen Special Economic Zone 224
Shenzhen University 225
Shenzhen Virtual University (SVU)
    225
Shenzhen Yunwang Intelligent System
    Co. 108
Shenzhen-Hong Kong Institute of
    Industry-Education Research 225
Shenzhou V 192
Shenzhou VI 284
Shougang Iron and Steel Corporation
    82, 84
Sichuan 20, 44, 133
Siemens 21, 78, 88, 116, 147, 148, 149,
    177
Sigma-Jinghua 139
Signal Research Group (US) 149
Silan 129, 139
Silicon Valley 120
Singapore, as source of FDI 76
Sino-Japanese War (1894–95) 281
Sino-US Agreement on S&T 204
Sivin, N. 280
skills shortage 54

Skyworth Group 119–20
Slowing Down of the Engine of Growth,
    The 157
social networks 231
socialist market economy, establishment
    of 307
socialist system, ability to surpass
    industrialized world 35
socioeconomic development, interaction
    between S&T and 310
software design houses 144
software development, outsourced to
    China 184
software industry
    Shanghai 260–62
    Zhongxing Technologies R&D Centre
        266–8
Software Institute (CAS) 90
Software Park, Shenzhen 224–5, 261
Songjiang 140
Songjiang Export Processing Zone
    (SJEPZ) 271
Songjiang Industrial Zone (SJIZ) 271
Songjiang Science and Technology park
    263
Songshan Lake Science and Technical
    Industrial Park 22
Sony 82, 100, 117, 119, 144
Soutec 147
Sovremennyy class destroyers 203
space
    ambitions in 191–2
    efforts 26
    global integration 169
    see also aerospace industry
Spark Programme 8, 40–42
special economic zones (SEZ) 70, 102,
    224, 258
specialized incubators, Shanghai 273
Spreadrum Communication 148
SPS 143
Ssangyong Motors 83, 286, 287
SSMEC 139
standardization 17–18, 160, 293–4
state agencies, control of foreign
    investment 81
State Commission of Science, Technol-
    ogy and Industry for National
    Defence (COSTIND) 32, 167, 195,
    197, 204

State Council Circular (Number 18) 130, 140–42
State Defence Science and Technology Commission 5
State Development and Reform Commission (SDRC) 83
State Economic and Trade Commission (SETC) 174–5
State Forestry Administration 105
State Intellectual Property Office 275
State Key Laboratories 52, 176, 252–3, 254
State Meteorology Administration 105
State-owned Assets Management Commission 109
State-owned Assets Supervision and Administration Commission (SASAC) 71, 285
state-owned enterprises
 failure of technological capability 285
 imports of high-tech products 79–80
 military industry 196
 need for competitiveness 87
 overseas investment 83–4
state-owned institutions, rent responsibility system 102
steel industry 7
STMicroelectronics 147
stored-programme-computer (SPC) switching 40, 146
strategic alliances 87–8
strategic competitor, China as 299–302
Strategic Defence Initiative (SDI) 8, 42
strategic priorities, S&T plan 309–10
students
 attraction of US universities 182
 enrolments
  higher education 12, 27, 55
  secondary education 55
 in higher education, Shanghai 249–50
 returnees (1978–2002) 66
 studying abroad (1978–2002) 66
 see also graduates; postgraduates; undergraduates
Su-27 fighters 203
Su-30 fighters 203
submarines 203
Suggestions on Tracing the Development of World Strategic High Technology 8, 42

SUN 76
supercomputers, usage and production 164–6
Suzhou 131, 219, 276
 export-oriented development 244
 GDP growth (2003) 239
 semiconductor industry 140, 161
Suzhou Industrial Park 164
Suzhou Matsushita Semiconductor 163
Suzhou Technology Centre 92
Suzhou University 92
SVA 16
synthesis and intersection, 973 S&T Programme (1998–2002) 50
system innovations, science and technology 310
system-on-chip (SOC) projects 10, 143

T3G 148
Taiji 269
Taiwan, relations with China
 conditions determining 180–81
 foreign R&D investment 76, 90
 technological integration 181–5
Taiwan Semiconductor Manufacturing Corporation (TSMC) 140, 141, 163
talent recruitment 250, 287
talent-oriented plans 65
Tasly 269
tax (VAT) rebate scheme 140–42
TCL Communication 14, 15, 119, 120, 140
TD-SCDMA Forum 147, 149
TD-SCDMA Industry Alliance 110, 147, 149
TD-SDCMA technology 147–51
technological aspirations 3–5
technological capability
 aerospace industry 198
 assessing changes in 104, 105
 auto industry 188, 285, 286
 civilian electronics industry 212
 IC industry 163–4
 innovation, R&D institutes 104
technological clusters 95, 217, 220, 226–8
technological collaboration
 Brazil 208
 Europe 204–8
 Israel 211

Russia 201–4
technological integration, China and Taiwan 181–5
technological policy, evolution of 29–53
Technologies Aerospace 169
technology
 exploiting foreign 73–88
 Shanghai 258–62
 see also information and communication technology; science and technology
technology exploitation institutes, corporatization 102, 103
technology parks, Shanghai 248
technology transfer
 aerospace industry 202
 decision to promote 74–5
 fee for 101–2
 joint ventures, *see* joint ventures
 military industry 203, 211
 need for change 35
telecommunications 145–51
 exports 185–6
 Shenzhen 20–21
 see also mobile phone industry
television sector 14, 15, 18, 24, 40, 82, 85, 107, 111, 122
Telfort BV 116, 151
Ten Year Programme of S&T Development (1991–2000) 37, 51
Tenth Five Year Plan on S&T Development 37
tertiary industry system, aerospace industry 193
Texas Instruments 111–12
textiles
 changing trade patterns 158–9
 employment effects 80–81
Thailand
 competition from clusters 228
 overseas investment 85
Thelander, M. 149
Thomson 15, 82, 119
3Com 115
3G system, *see* TD-SCDMA technology
3GPP 147
TI 109, 267
Tianjin 218, 221
 economic statistics 238
 education 233, 234

employment, semiconductor industry 133
high technology clusters 95
manufacturing facilities 92
Tianjin University 13
Tongji University 265
topic-oriented strategic studies 311–12
Toppan 266
Torch Programme 9, 46–7
Toshiba 88, 109, 266
towns, rural 242–3
township-and-villages industries 40
Toyota 78
trade patterns, changing 158–64
Trade Related Aspects of Intellectual Property Rights (TRIPS) 86, 173, 275
Trade Related Investment Measures and the Agreement on Subsidies and Countervailing Measures 86
traditional Chinese medicines (TCM) 177, 178
traditional clusters 226
Tsinghua Tongfang 139
Tsinghua University 12, 31, 45, 60, 90, 92, 112, 114, 131, 166, 178, 225–6
Tsinghua Xuetang 60
Tsumura and Co. 177, 269
Tu, Hailing 136, 137

Ultimate Semiconductor 163
undergraduate teaching 31
undergraduates
 by field of study (1996–2001) 56
 enrolments, by field of study (1996–2001) 56
 Fudan University (2000–04) 252
 see also graduates
Unilever 76
United Kingdom, foreign R&D investment in China 76, 90
United Manufacturing Corporation 140
United States 2
 –China Economic and Security Review Commission 300
 arms embargo against China 166
 buying semiconductor equipment from 131
 China as strategic competitor 299–302

Chinese graduates living in 65
Chinese S&T as threat to national security 191, 210
Chinese supercomputing as threat to military security 165
exports from China to 81, 183, 185–6
government investment in blade computing 166
patent portfolio of Chinese companies in 17
R&D investment in China 76, 90
relations with China 25
relief from foreign competition 121
restrictions
  access to semiconductor technology 162
  on textile imports 159
supercomputer systems 164
threat of shift of R&D to Asia 93–4
universities, attraction of 182
Universal Automotive Industries (UAI) 84
universities
  affected by 211 project 58–64
  attraction of US 182
  in Beijing 142
  expansion 54
  leap forward 289–90
  reforms 12, 31, 54
  role of 11–14
  science parks and research centres 65–6
  Shanghai 249–58
  see also graduates; higher education; *individual universities*; private universities/colleges
University Teaching Quality and Reform Project 94
University Town 21
urban–rural divide 236–9
urbanization
  (1949–98) 232–3
  rate 237–9
  rural 239–43
UTStarcom 147, 150

value-added, in exports 80–81
VAT rebate, IC manufacturing 140–42
Verheughen, G. 295
VeriSilicon Holding 149

Vimicro Company 139, 160
Virgin Islands, as source of FDI 76
Virtual University Incubator 224
Volkswagen 188

W-CDMA 114–15, 148
wafer production 130–31, 260
Waigaoqiao Free Trade Zone (WFTZ) 271
Wang, Dianzuo 136
Wang, Jian 236, 242
Wang, Xiaodong 178
Wanxiang Group 84–5
WAPI protocol 18
Wassenaar agreement 162
weapons technology, foreign 197
well-to-do society, development target 306–7
Wen, Jiabao 309
wireless local loop (WLL) system 146–7
World Trade Organization (WTO), impact on foreign technology absorption 85–8
Wuxi 219, 244, 276
Wuxi CSMC-Huajing 163
Wuxi Huazhi Semiconductor Co. 163
Wuxi Semico 139
Wyse Technology 45

Xiamen 221, 223
Xian 95, 140, 170
Xian Aircraft Company 169
Xian Dianzi (Electronics) Jishu University 112
Xiaolingtong 146–7, 150
Xike 129
Xu, Guanhua 187
Xu, Qin 44
Xu, Xincai 250
Xuihui District 274
Xuihui Software Park 273

Yandong 143
Yang, Fujia 62–3
Yangpu Technology Innovation Centre 272
Yangtze River Delta (YRD) region 19, 184, 215, 239
Yantai 221
Yue Xiu Enterprises 146

Zhang, Xinsheng 148, 149
Zhangjiang Biotech and Pharmaceutical Base Development Co. 268, 269
Zhangjiang Hi-tech Park 164, 248, 249, 262, 263–5, 269
Zhangjiang Hi-tech Park Development Corporation 261
Zhangjiang Professor Programme 250–51, 262
Zhangjiang Scholar Award 254
Zhangjiang Semiconductor Industry Base (ZSIB) 164
Zhaoqiang Blue 107
Zhejiang 20, 55, 133, 235
Zhejiang University 60, 92
Zhejiang Wanli University (ZWU) 61–3
Zhongguancum 111, 144, 221–2
Zhongxing Technologies Corporation (ZTE) 18, 21, 81, 115, 123, 140, 146, 147, 148, 151, 160, 264, 266–7
Zhongzhi 45
Zhou, D. 254, 255
Zhuhai 221
Ziyuan-2 satellite 192
Zizhu Science-Based Industrial Park 263